T0135490

Photomodulation of Supramolecular Systems Containing Bioactive Small Molecules and Biopolymers

Zur Erlangung des akademischen Grades eines

DOKTORS DER NATURWISSENSCHAFTEN

(Dr. rer. nat.)

bei der Fakultät für Chemie und Biowissenschaften

des Karlsruher Instituts für Technologie (KIT) – Universitätsbereich

vorgelegte

DISSERTATION

von

M.Sc. Johannes W. Karcher

aus Oberachern

KIT-Dekan: Prof. Dr. Reinhard Fischer

Referent: Dr. Zbigniew L. Pianowski

Korreferent: Prof. Dr. Hans-Achim Wagenknecht

Tag der mündlichen Prüfung: 23.07.2019

Bibliographic information published by the Deutsche Nationalbibliothek

The Deutsche Nationalbibliothek lists this publication in the Deutsche Nationalbibliografie; detailed bibliographic data are available on the Internet at http://dnb.d-nb.de .

ISBN 978-3-8325-5099-8

Logos Verlag Berlin GmbH
Comeniushof, Gubener Str. 47,
10243 Berlin
Tel.: +49 (0)30 42 85 10 90
Fax: +49 (0)30 42 85 10 92
INTERNET: https://www.logos-verlag.de

Every revolutionary idea seems to evoke three stages of reaction. They may be summed up by the phrases:
1) It's completely impossible.
2) It's possible, but it's not worth doing.
3) I said it was a good idea all along.
Clarke 1985[1]

The present work was carried out at the Karlsruhe Institute of Technology, Faculty of Chemistry and Biosciences, Institute of Organic Chemistry (IOC) in the period from January 1st 2016 to June 12th 2019 under supervision of Dr. Zbigniew L. Pianowski.

Die vorliegende Arbeit wurde in der Zeit vom 01.01.2016 bis 12.06.2019 am Institut für Organische Chemie (IOC) der Fakultät für Chemie und Biowissenschaften am Karlsruher Institut für Technologie (KIT) unter der Leitung von Dr. Zbigniew L. Pianowski durchgeführt.

I hereby declare truthfully that I have prepared this thesis autonomously except for the help explicitly stated in the treatise itself; that I have exhaustively and accurately indicated all auxiliary means; and that I have marked all portions taken, verbatim or in altered form, from the work of others or my own publications.

Hiermit versichere ich wahrheitsgemäß, die vorliegende Dissertation selbstständig verfasst und ohne unerlaubte Hilfsmittel angefertigt, andere als die angegebenen Quellen und Hilfsmittel nicht benutzt und die den verwendeten Quellen wörtlich oder inhaltlich entnommenen Stellen als solche kenntlich gemacht habe. Die Arbeit wurde bisher in gleicher oder ähnlicher Form noch keiner anderen Prüfungsbehörde vorgelegt und auch nicht veröffentlicht. Ich habe die Regeln zur Sicherung guter wissenschaftlicher Praxis am Karlsruher Institut für Technologie (KIT) in der gültigen Fassung beachtet.

Karlsruhe, 12.06.2019

Johannes Karcher

Table of contents

Abstract

Molecular photoswitches can change their conformation upon irradiation with light and convert this molecular movement into macroscopic changes of sophisticated materials. By combining state-of-the-art azobenzenes with supramolecular low-molecular-weight gelators, based on cyclic dipeptides, photomodulation of hydrogels was achieved. Photochromic amino acids with structures related to the natural *L*-phenylalanine were an essential element. A significant red-shift from the UV-irradiation regime to triggering with green or red light was achieved by modifying the initially used unsubstituted azobenzene structure into tetra-*ortho* substituted fluoro- and chloro-azobenzenes.

These photochromic building blocks were synthesized in gram quantities and their use is not restricted to this single purpose. Rather, a broad variety of photomodulation of biomolecules, in particular peptides, can be accessed with the two functional groups of amine and carboxylic acid and their formation of amid bounds. Particularly, the modular synthesis of tetra-*ortho*-fluoroazobenzenes with various linkers could be fitted to three other applications for the artificial DNA analogue PNA, metal-organic framework and catenanes as asymmetric catalysts. There, the crucial advantages of tetra-*ortho*-fluoroazobenzenes, as a robust photoswitch with reversible isomerization between the photoisomers with green and violet light, respectively, are the high photostationary state of 90-95%, the lifetime of two years at room temperature and the NMR active ^{19}F atoms.[2]

The supramolecular hydrogels are multi-responsive towards pH, ions, temperature, and, in particular, light. Hereby the gelator F_2-PAP-DKP-Lys can be turned into a sol upon irradiation with green light and within this transition a drug, for example an antibiotic, was released. As the gelator does not decrease cell viability in cell cultures of *E. coli*, below concentrations of 5 mM, the hydrogel with an encapsulated antibiotic (ciprofloxacin) could have been used for light control of the bacterial growth. The further improved supergelator Cl_4-PAP-DKP-Lys_2 is reversibly switchable with red and violet light in the range of 0.2-0.5% in an isotonic Ringer's solution. At a concentration of 1%, the hydrogel consisting of Cl_4-PAP-DKP-Lys_2 remains stable in a boiling water bath. Thus, we envision that the enhanced stability of the supramolecular assembly of Cl_4-PAP-DKP-Lys_2 and orthogonal biocompatibility of F_2-PAP-DKP-Lys, could enable further applications, spanning from regenerative medicine to electronic devices,[3] besides our future vision towards photopharmacology.

Kurzzusammenfassung

Molekulare Photoschalter können ihre Konformation durch Bestrahlung mit Licht ändern und diese molekulare Bewegung in einen makroskopischen Wandel hochentwickelter Materialien übersetzen. Durch die Kombination modernster Azobenzole mit supramolekularen, niedermolekularen Geliermitteln, basierend auf zyklischen Dipeptiden, wurde die Photomodulation der Hydrogele erzielt. Photochrome Aminosäuren, mit Strukturen ähnlich des natürlichen *L*-Phenylalanins, sind ein essenzielles Element. Eine signifikante rot Verschiebung vom UV Bereich zu grünem und rotem Licht wurde durch die tetra-*ortho* Substitution mit Fluor und Chlor des anfangs unsubstituierten Azobenzols ermöglicht.

Diese photochromen Funktionsbausteine wurden im Gramm Maßstab synthetisiert und ihre Verwendung ist nicht nur auf diesen einzelnen Nutzen beschränkt. Stattdessen können umfassende Variationen photochromer Biomoleküle, insbesondere Peptide, mithilfe der zwei funktionellen Gruppen Amin und Carbonsäure und deren Knüpfung von Amidbindungen, ermöglicht werden. Insbesondere die modulare Synthese der tetra-*ortho*-Fluoroazobenzole mit verschiedenen funktionellen Gruppen kann für drei weitere Anwendungen angepasst werden; der künstlichen DNS Analoga PNS, der Metallorganischen Gerüste und Catenane zur asymmetrischen Katalyse. Der entscheidende Vorteil der tetra-*ortho*-Fluoroazobenzole, als robuster Photoschalter mit reversibler Isomerisierung mit grünem und violettem Licht, liegt im hohen photostationären Zustand mit 90-95 %, der Halbwertszeit mit zwei Jahren bei Raumtemperatur und der NMR-aktiven ^{19}F Atome.[2]

Die supramolekularen Hydrogele reagieren sowohl auf pH als auch auf Ionen, Temperatur und insbesondere Licht. Hierdurch kann das Hydrogel F_2-PAP-DKP-Lys durch Bestrahlung mit grünem Licht in ein flüssiges Sol überführt werden und innerhalb dieser Transformation kann ein Medikament, zum Beispiel Antibiotikum, freigesetzt werden. Da das Geliermittel die Zellviabilität von *E. coli* unter einer Konzentration von 5 mM nicht reduziert, konnte das Hydrogel mit dem verkapseltem Antibiotikum Ciprofloxacin zur lichtgesteuerten Kontrolle des Wachstums einer Bakterienkultur verwendet werden. Das weiter verbesserte Supergeliermittel Cl_4-PAP-DKP-Lys_2 wurde reversibel mit rotem und violettem Licht im Bereich von 0.2–0.5 % in isotonischer Ringer Lösung vom Gel zum Sol geschaltet. Bei einer Konzentration von 1 % bleibt das Hydrogel Cl_4-PAP-DKP-Lys_2 selbst im kochenden Wasserbad stabil. Durch die erhöhte Stabilität der supramolekularen Struktur von Cl_4-PAP-DKP-Lys_2 und der orthogonalen Biokompatibilität von F_2-PAP-DKP-Lys werden weitere Anwendungen von regenerativer Medizin bis zu elektronischen Geräten[3], zusätzlich zu unserer Vision in Richtung Photopharmakologie, ermöglicht.

1 Introduction

1.1 Supramolecular modulation towards dynamic materials

In nature the recognition of enzymes or the folding of biomolecules is based primarily on the supramolecular interactions in an aqueous environment between them.[4] Even though the hybridization of DNA, information processing of cascades or the activity of enzymes are well-orchestrated natural processes, their modulation often remains a challenge and a remarkable achievement.[5] Like the gene editing with CRISPR/Cas[6], that leads to the total synthesis of *E. coli* with a recoded genome.[7] The modulation of supramolecular systems in water can mimic the dynamic nature of life, which requires a dynamic chemistry out of equilibrium. A closely related medicinal target is the formation of β-amyloid polypeptides. The abnormal aggregation leads to Alzheimer's disease and despite multiple attempts no drug has been developed to cure the disease yet.[8] In regard to Alzheimer's β-amyloidosis the light-induced suppression of $A\beta_{42}$ self-assembly with methylene blue prevents this equilibrium and leads to the disintegration of their aggregates.[9]

In contrast to the destabilisation of supramolecular interaction, the majority of recent reports describe the bottom-up design of supramolecular systems, e.g. biomaterials.[10] The necessity of compartments for life is not only fundamental, but also relevant for intracellular delivery. The self-assembly of amphipathic molecules into colloidal particles is a broad field[11] – a low-molecular-weight molecule based on a ααβ-tripeptide is a recent example of this class. This cationic nanoparticle has a diameter of 675 ± 30 nm in aqueous media and a zeta potential of $+34.7 \pm 0.4$ mV due to the protonated guanidinium groups derived from the arginine side chain.[12] Additionally, from a synthetic point of view, the modular synthesis of this tripeptide in over five steps and with 65% yield has an advantage[12] in comparison with other supramolecular systems in water.[13]

The assembly of scalable porous protein cages derived from the capsid-forming enzyme lumazine synthase is at the upper limit of complexity (Figure 1). This versatile encapsulation system assembles into icosahedral symmetric particles from a bacterium and even the wild-type AaLS is remarkably thermostable with $T_m = 120$ °C with a similar structure in relation to virus capsules.[14]

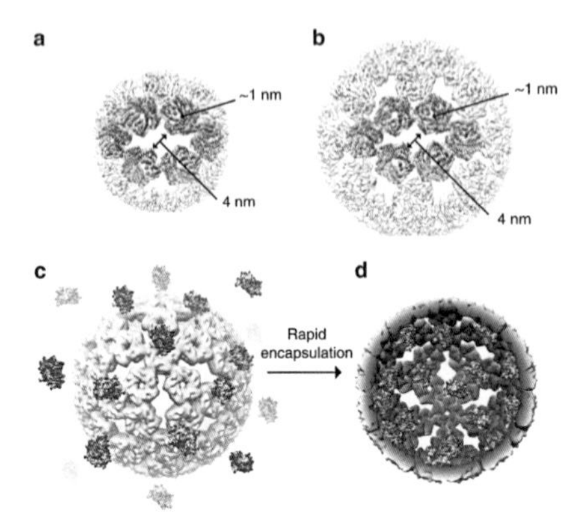

Figure 1. a, b) Keyhole-shaped opening of scalable porous protein cages derived from lumazine synthase from *Aquifex aeolicus*; b) each slot surrounded by 16 monomers, four shown in blue lack the inter-pentamer interactions; c, d) spontaneous encapsulation of supercharged GFP d) association of the guest with the cage interior;[15] licensed under CC BY 4.0.

By increasing the negative charges the protein shell was increased and the symmetric structures were constructed from pentameric units that enable the encapsulation of complementary charged guests.[15] Related bacterial protein compartments can be used for selective diffusion and even for the development towards synthetic organelles.[16] In a capsid of the bacteriophage P22 the oxygen-tolerant [NiFe]-hydrogenase was encapsulated as a biomolecular catalyst for hydrogen production. The hydrogenase cargo is protected in the capsid and enables a 100-fold increase in the activity.[17] For the modification of proteins, supercharging has proven possible and even quaternary structures and higher order architectures may be possible hereby.[18]

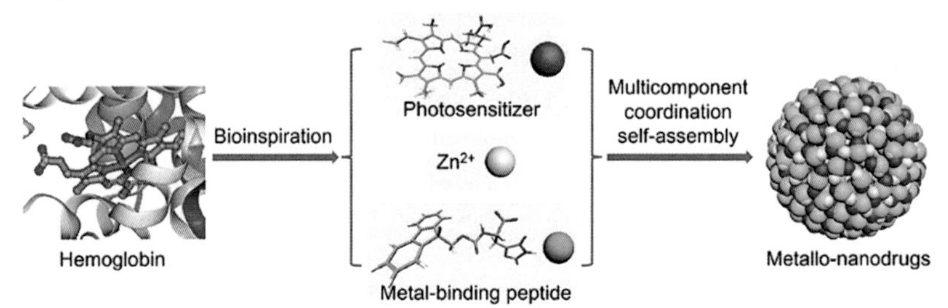

Figure 2. Supramolecular metallo-nanodrugs self-assemble by cooperative coordination of small peptides and photosensitizers in the presence of zinc ions; reprinted with permission.[19] Copyright 2018 American Chemical Society.

To come back to nanospheres based on peptide-based supramolecular assembly, a photodynamic metallo-nanodrug can be formed with a photosensitizer instead of a photoswitch.

The metallo-nanodrug from the group of YAN consists of the cooperative coordination of the heme group of human haemoglobin to histidine through the metal ion Zn^{2+}. The histidine group is part of a dipeptide Z-HF or an amphiphilic histidine derivative Fmoc-H. Both complexes between histidine and Zn^{2+} are responsive to pH and vanish below a pH of 5.0 or above 8.5, and are in addition responsive to 3 mM glutathione (GSH). Both nanoparticles have a narrow distribution size of 75 ± 16 nm with Fmoc-H and 78 ± 21 nm with Z-HF and a negative zeta potential with −20 ± 1.3 mV and −20 ± 2.5 mV, respectively. The robust colloidal stability under physiological conditions and the burst release in tumour microenvironments are key properties. The negligible toxicity and prolonged lifetime in blood are further promising advantages of this system. In combination with the porphine chlorin e6, a photosensitizer with antitumor activity, the metallo-nanodrug inhibits cell viability in combination with irradiation with a 635 nm laser in comparison with the unencapsulated Ce6.[19]

These supramolecular assemblies are in general dynamic essentially due to their non-covalent interactions and can be modulated by pH or temperature. This kind of modulation was reported for a supramolecular helix that can reversibly transform the chirality upon changing the pH. The inversion of the chirality can also be inverted with a temperature from 20 to 80 °C with an inversion point at ca. 70 °C. The molecules themselves consist of a symmetric naphthalenediimide with two *L*-glutamates attached with amide bounds (Figure 3).[20]

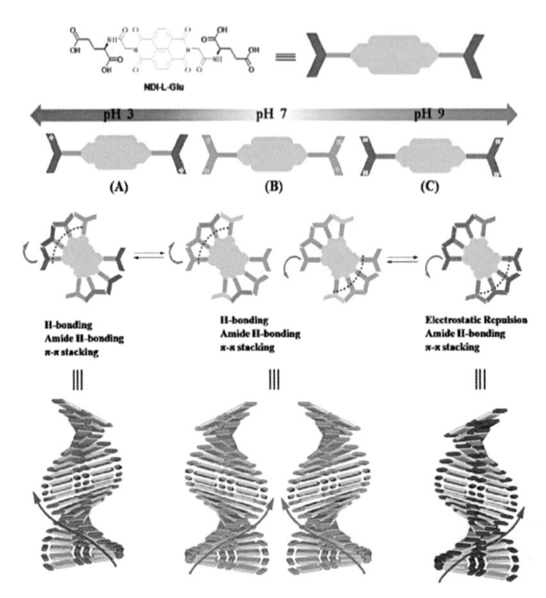

Figure 3. Illustration of NDI-*L*-Glu aggregation pathways and their stacking of dimers; A) acidic pH results in left-handed helix; B) self-assembly under physiological pH both left- and right-handed superstructure are present; C) basic pH resulting in right-handed helix. The self-assemblies depend on the hydrogen bonding and changes the dihedral angular twist direction of both helical superstructures;[20] licensed under CC BY 4.0.

1.2 Molecular photoswitches and motors

The photomodulation of biological systems with photoswitches or molecular motors on the other hand is in general orthogonal with most *in vitro* and biological models like cell cultures, zebra fish and mice. The high spatial and temporal control of light is a major benefit and advantageous in comparison with other triggers like pH, temperature, electrochemical or magnetic fields. In general, photoswitches can be used as a key ingredient to drive systems out of the thermodynamic equilibrium.[21]

The energy of light can be transformed reversibly into flexibility changes, molecular motion and changes in polarity. By means of its wavelength, amplitude modulation and intensity, light can deliver information. It is a mild trigger that does not intrinsically lead to contamination of the sample. The variation of reported molecular adapters increases almost from day to day. Classical scaffolds, like azobenzenes, spiropyrans and diarylethenes are widely used, but improved versions of these and emerging photoswitches like indigoids slowly change the research area.[22] Molecular motors, in turn, are photochromic compounds with a locked synchronous rotor motion.[23] Depending on the molecular architecture, the integration of such a motor can lead to molecular motion and other functions, thus being in a way functionally related to regular molecular photoswitches. For example, DNA hybridization can be switched with light using 4,4′-bis(hydroxymethyl)-azobenzene[24] or a molecular motor as a powerful multistate switch in a biological environment (Figure 4).[25]

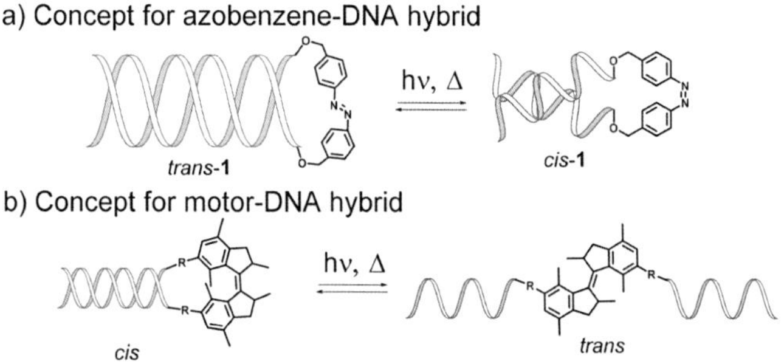

Figure 4. Overview of the concept of photochromic DNA hairpins; A) design including a azobenzene by Sugimoto *et al.*;[23] B) rotary molecular motor based on overcrowded alkene;[25] licensed under CC BY 4.0.

In a related manner, the dynamic change of chirality can be induced by self-assembly of double stranded helices with light. The unidirectional rotary motor may also be connected with oligobipyridyl ligands, which assemble into metal helicates. This motor initiates the inversion of the chirality of dynamic self-assembled helices.[26]

For all the aforementioned classes of molecular switches, the underlying principle is photochromism, defined as a reversible phototransformation of chemical species of two forms, both with different absorption spectra. This photoisomerization usually changes physicochemical properties of respective isomers as well, like geometrical structure, the dielectric constant, the refractive index or oxidation/reduction potential.[27] The non-substituted azobenzene with a N=N double bond, connecting two phenyl rings, isomerizes with UV light from the stable *E*-isomer to the less thermodynamically stable *Z*-isomer. This isomerization can be reversed by irradiation with blue light or by thermal equilibration.[28] In the UV range from 320 – 360 nm the *E*-isomer exhibits a strong absorption, which is associated with the $\pi \rightarrow \pi^*$ transition. The second maximum at ca. 450 nm has a lower absorption and is associated with the $n \rightarrow \pi^*$ transition. Unsubstituted azobenzenes reach a photostationary state ($PSS_{365\,nm}$) of ca. 80% of the *Z*-isomer. Irradiation with blue light (460 nm) can isomerize this mixture reversibly back to ca. 80 – 90% *E*-isomer. As an alternative, the relaxation of the *Z*-isomer occurs thermally within a few days to hours at ambient temperature.[29] The introduction of substituents in azobenzene can significantly affect the lifetime of the *E*-isomer depending on the temperature and solvent. It can range from microseconds up to years. Electron-withdrawing substituents increase the lifetime in general and electron-donating substituents lead to its decrease. A combination of electron withdrawing and electron donating substituents is a push-pull system that has characteristically short *E*-isomer lifetimes and the absorption is redshifted by a bathochromic shift.[2, 30]

The lifetime of the photoswitch is a fundamental factor in their application and azobenzenes with a long lifetime are beneficial to control processes by light in material science[31] and biology.[28, 32] But photoswitches with a short lifetime can be applied in microsecond processes, for example, the photoactivation of ion channels.[33] In optoelectronics, multicolour displays and holography the compounds can be activated by light and the back-switching occurs thermally.[29, 34] Within the last decade, numerous modifications were introduced which enabled the photoisomerization with visible light in both directions. In particular, multiple substitutions in *ortho* positions of aromatic rings (relative to the azo bond) with methoxy,[35] chloro[30c] and fluoro groups[30a] offer higher stability to photobleaching, hydrolysis and prolonged lifetime of the *Z*-isomer. These modifications enhanced perspectives for applications *in vivo* particularly in photopharmacology.[30b]

However, the synthesis of these tetra-*ortho*-azobenzenes with standard azo coupling became inefficient due to either steric hindrance or electron-poor aromatic substituents. Thus, fluoroazobenzenes were synthesized using a non-standard condition modification of the Mills reaction[2]. Chloro substituents can be either introduced by late-stage functionalization of non-substituted azobenzenes,[36] or synthesized by the *ortho*-lithiation of chlorinated aromatic substrates followed by the coupling reaction with aryldiazonium salts.[37] Despite these advances, the majority of recent structures of photopharmacology *in vivo* still use the non-substituted azobenzenes.[38] By the azologization of drugs of, e.g. styrenes, (hetero)aryl amides, *N*-phenyl benzamides and related moieties, photoswitchable drugs are accessible,[39] but the addition of the aforementioned *ortho* substitution might compromise their pharmaceutical properties on top of the increased synthetic effort. Nevertheless, the optical control of the GIRK channel was achieved with tetra-*ortho*-fluoroazobenzenes[40] and the glutamate receptor ion channel with tetra-*ortho*-chloroazobenzenes, both without the necessity for triggering systems with UV light.[41]

1.3 Spiropyrans for chiroptical switches

Another class of photoswitches are spiropyrans, which, in comparison to azobenzenes, are less common in regard to photopharmacology.[38] One of the few demonstrations is a photochromic version of the antibiotic ciprofloxacin – spirofloxacin, which sadly suffers from photodegradation.[42] This fatigue might be due to instability of the merocyanine form in aqueous solution. After visible light irradiation of the merocyanine, a retro-aldol reaction leads to a degradation and subsequently forms a 4-nitro-salicylaldehyde and a Fischer's base.[43] The unique character of spiropyrans and their wide application in dynamic materials lie in the vastly different physiochemical properties of the spiropyran SP and the merocyanine MP form.[44] The dipole moments of SP (~4 – 6 D) and MC (~14 – 18 D) are relevant as well as their significant structural difference from the smaller SP and to more basic and polar MP form. These differences are linked to the multiple responsiveness of spiropyrans[44] and include temperature, polarity of the solvent, redox potential,[45] pH,[46] and mechanical force.[47] For example, spiropyrans show a negative photochromism in a hydrophobic environment in the cavity of a cage.

Hereby the encapsulated spiropyrans can be converted upon irradiation with blue light, which corresponds to the minimum absorption, to the colourless form.[48] Spiropyrans can be switched through a two-photon process, in which two beams have to overlap in the time and space. This was applied e.g. in a 3D optical storage memory.[49] In a recent publication an amphiphilic spiropyran with a C_{18}-chain is described, which co-assembles with the gelator PULG into a photochromic supramolecular gel.[50] Scheme 1 shows the molecular structure of PULG with its cationic amphiphile bearing pyridinium and the fatty acid moiety with a glutamide core. With a trace of water a chiral self-assembled nanostructure is induced with PULG alone and with increasing water content a different pitch length or various helical structures were observed.[51] In comparison to a co-assembly of PULG with an azobenzene[52] the spiropyran system can be used as a chiroptical logic circuit and shows the aforementioned multi-responsiveness with irradiation and the acid-base treatment.[50]

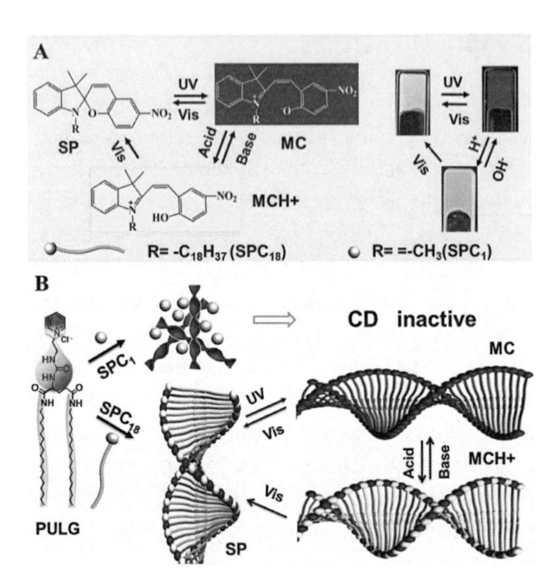

Scheme 1. A) structural change of spiropyran C_{18} by isomerization and acidichromic process. The co-organogel consists of PUL(D)G and spiropyran; B) illustration of the molecular structure of the dual chiroptical switch. With the doped PULG gel the chirality was transferred to the assemblies by entanglement of its hydrophobic chains. Upon photoirradiation and acid-base treatment the CD signal switched depending on the ring state of the spiropyran (closed SP, open MC and protonated MC^+);[50] with permission. Copyright 2016 John Wiley and Sons.

In an analogue approach of the same group of LIU both molecules, the artificial fatty acid derivative and the spiropyran, were combined. This enantiomeric glutamate gelator self-assembles into nanohelices and upon irradiation the circular dichroism, circularly polarized luminescence, fluorescence and UV-Vis spectra change. This quadruple optical and chiroptical switch can be used as a rewritable material that was fabricated from their xerogels on glass.[53]

1.4 Dithienylethenes towards autoamplification

Other chiroptical switches are constructed on dithienylethenes (DAE) without the supramolecular assembly instead of a chiral photoswitch to control cholesteric liquid crystals. Then the challenge lies in a design of this chiral dopant that leads to a control over chiral amplification by irradiation with light. Here and in general the chiral dopants with dithienylethenes are thermally stable in comparison with azobenzene-based chiral dopants.[54] At least with the integration of perfluorocycloalkenes a thermally irreversible photochromic system is available with a high resistance against photodegradation.[55] Upon light-induced cyclisation the changes in geometry from the colourless open form to the closed colourful form are minor.[22] Overall, DTEs have a high fatigue resistance and are bistable without thermal backswitching.[56] By substitution of the aryl groups with thiophene, furan, selenophene, or thiazole rings, both isomers can be stable up to 80 °C.[57] DAEs can be tuned with extended conjugated substituents and can be reversibly switchable with visible light with promising functionality for optical memory systems.[58] Recently, a sterically hindered DAE with high rotational barrier was introduced due to the benzobis(thiadiazole) ethene bridge that enables a full separation of five isomers.

In case of these BBTEs the parallel ring-opening is photochemically inert and according to the Woodward-Hoffmann rules the photochemical conrotatory reaction of 4n + 2 π-electron system allows the anti-parallel ring-opening and reversible transformation to the ring-closed enantiomers. The BBTEs are bistable chiroptical switches and a non-destructive readout is possible with polarised light because their absorption can be outside their electronic absorption band. This eliminates any photo-excitation that would result in an induced structural change and is suitable for photomemory material; in other words, it prevents the intrinsic racemization in common diarylethenes.[59] The racemization of diarylethenes was used for the autoamplification of molecular chirality induced by supramolecular chirality. For this, diarylethenes were functionalized with amide moieties which allow their assembly through hydrogen bonding and result in a low-molecular-weight gelator.[60]

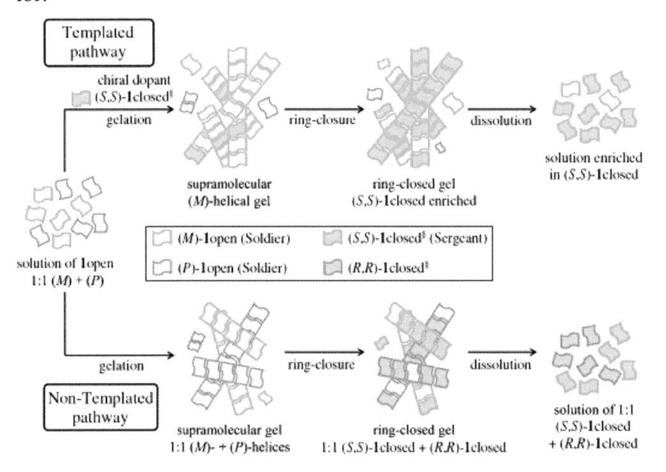

Figure 5. Concept for autoamplification of the molecular chirality by the induction of the supramolecular chirality;[60] with permission. Copyright 2014 John Wiley and Sons.

A minor amount of the chiral compound induces the overall chirality of a mainly achiral material (Figure 5). Here the small amount of the chiral sergeant ($SP_{closed}^{\S}$) was added to the precursor (SP_{open}) without compromising the gel formation. Thus, the formation resulted in either a *P*- or *M*-helical gel. The chiral information is locked at the molecular level as the spiropyran undergoes the ring closure upon irradiation of this gel and the aforementioned enantiomeric excess is formed.[60]

1.5 Emerging photoswitches and molecular motors

Besides the three common classes of azobenzenes, spiropyrans and diarylethenes, there are multiple emerging photoswitches.[22] For example, donor–acceptor Stenhouse adducts (DASAs) have a large change in polarity after their photoisomerization and thermal conrotatory 4π-electrocyclization forming a cyclopentenone.[61] Together with azobenzenes they can be used for orthogonal reversible photoswitching systems.[62] Another novel class are the acylhydrazones with high fatigue resistance, easy preparation and high tunable thermal half-life of the *Z*-isomer.[63] Hereby a photochromic heteroditopic receptor was designed, in which both cation and anion binding sides can be switched on and off reversibly in acetonitrile.[64] The isomerization of indigoids and imines, by analogy with the overcrowded alkenes, inspired the development towards new molecular motors.

The concept of photodriven molecular motors with imines was first proposed[65] and later realized based on the isomerization of C=N double bounds for a synthetic molecular motor. Here the concept is widely applicable and in principle every suitable chiral imine can be used and functionalized as a molecular motor with light, relaxing back to the thermal equilibrium.[66] In a similar way, indigoids are long-known chromophores derived from the indigo dye, offering bistable photoswitching near the bio-optical window.[67] Hemi-indigo photoswitches consist of indigo and a stilbene fragment with very high photostationary states of respectively > 90% *E*-isomer with green light (470 to 530 nm), or 99% *Z*-isomer with red light (590 to 680 nm) can be reached. In addition, the half-life of the less-stable isomeric state is up to 83 years at rt.[68] Indigoid photoswitches inspired the development of a molecular motor based on the hemithioindigo structure,[69] which bears a stereogenic centre at the sulphur and a sterically demanding *tert*-butyl group at the stilbene side. This three-step motor interconverts in a fixed sequence upon visible light irradiation, which results in a monodirectional and unidirectional rotation with up to 98% directionality.[70] The tool boxes of photoswitches and molecular motors are growing and offer a broad variety of advantages, especially for supramolecular photomodulation of selected systems.

1.6 Photomodulation of vesicles

The formation of vesicles in water was established with amphiphilic molecular motors. The design consists of a hydrophilic headgroup with a quaternary amine, hydrophobic bis-dodecyl residue and, in between, the molecular motor, which requires a challenging synthesis. This amphiphilic molecular motor and 1,2-dioleoyl-*sn*-glycero-3-phosphocholine (DOPC) co-assemble and lead to the formation of nanotubular structures with a bilayer wall of 4 nm and a diameter of 15-25 nm in water. The fast, molecular motor with a half-life of 40 ns did not lead to reorganization, but with a slower motor with a half-life of 270 h, vesicles were formed with a diameter of 130 nm after irradiation with UV light. By heating the system to 50 °C for 16 h the half-life of the molecular motor is reduced and the nanotubes are reformed.[71] There are numerous other photoswitchable vesicles reported,[72] because vesicles count as a basic building block of life and can be used for synthetic compartments like microreactors[73] or for light-triggered transport and release.[74]

1.7 Photochromic supramolecular assemblies

Except from peptides,[75] the related supramolecular assembly of spheres has a similar amphiphilic design, e.g. a diarylethene with an unpolar octyloxycarbonyl and *N*-octylcarbamoyl group and bis-ethylene glycol groups. These assemblies undergo macroscopic morphological transformations by division of the microspheres after irradiation. The ether-linked closed ring isomer revealed nanofibres and the amide-like closed-ring isomer showed polymorphism of supramolecular assembly.[76]

By an inversion of this amphiphilic design into a point-symmetric molecule, a related solvent-induced morphological transition from microrods in toluene to spheres in an aqueous DMF solution was reported. Here the molecule consists of three point-symmetric diarylethene molecules that are connected with a benzene-1,3,5-tricarboxamide (BTA) as a modifiable core.[77] The BTA forms hydrogen bonds, being a supramolecular building block that was used for chiral supramolecular polymers in water,[78] amphiphilic organoplatinum(II) metallacycles,[79] for holographic gratings by substitution with three azobenzenes[80] or the induction of macroscopic chirality of supramolecular gels with achiral BTA derivatives.[81] In an analogous way, rod-like assemblies were formed which result in an organogel macroscopically. Upon irradiation with UV light the microrods dissociate and reversibly regenerate upon irradiation with visible light (450 nm).[77] With the same C_3-symmetrical structure of BTA core a light-driven supramolecular nanowire actuator was developed (Figure 6). Instead of the diarylethenes in the previous example, the nanowire consists of unsubstituted azobenzenes that display a photoinduced reversible bending. One nanowire together with a PS nanowire were positioned next to each other and acted as a photochemical tweezer that can grip and detach a microparticle from a silicon substrate.[82] For the formation of these nanowires the supramolecular material was processed with a meniscus-guided solidification method using a micropipette.[83]

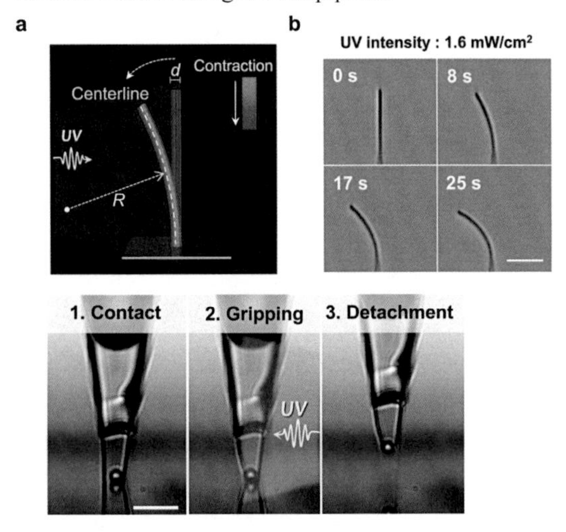

Figure 6. a) Illustration of the photoinduced bending of the nanowire upon irradiation; b) microscope image showing bending (450 nm (*d*)×22.5 µm (*l*)), scale bar 10 µm; 1-3) contact, gripping detachment tweezers grips a PS microparticle, scale bar 20 µm;[82] licensed under CC BY 3.0.

Another common building block for supramolecular architectures are cyclodextrins with drug and gene delivery applications.[84] Several photoswitchable supramolecular hydrogels are based on the concept of the interaction between cyclodextrins and azobenzenes.[85] But overall, cyclodextrin-based supramolecular assemblies have a broad variety of bioactive applications[86] and are multistimuli-responsive to pH, redox, enzyme, irradiation or magnetic field.[87] By using paclitaxel-modified cyclodextrin, the microtubule assembly was mediated (Scheme 2).[88]

Also, the strikingly distinctive binding affinities of arylazopyrazole with the hydrophobic cavities of β-cyclodextrins are the basis of the controlled, photoresponsive, reversible supramolecular architecture. Both cyclodextrin and arylazopyrazole were modified with paclitaxel,[88] an anticancer drug that stabilises microtubules (MT) and reduces their dynamicity, promoting mitotic arrest and cell death.[89] Microtubules assemble and disassemble dynamically from two ends – plus and minus – and consist of two proteins, the α- and β-tubulin. They are essential in biological processes like intracellular transport and cell division and are part of the cytoskeleton of eukaryotes.[90] The intertubular aggregation of MT was regulated with this supramolecular system by light irradiation with different wavelengths. With TEM microscopy the nanoparticle aggregates were visualized with an average size of 286 nm by the cross-linked complex. After irradiation to the *Z*-isomer, nanospheres with 175 nm were observed. In a cell culture with A549 the *E*-isomer led to a higher cell death of 12% in comparison with 7.8% for the *Z*-isomer. The cell death was associated with a shrinkage of the cell, most likely because of the intracellular aggregation of the supramolecular cross-linked MTs.[88]

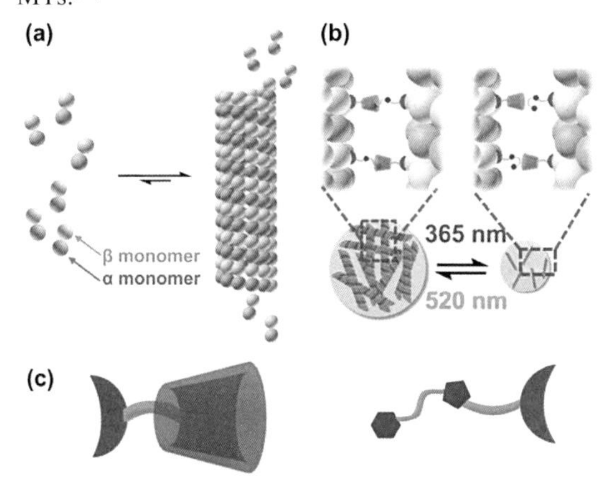

Scheme 2. Illustration of molecular structure and ternary supramolecular assembly; a) dynamic assembly of microtubule; b) *trans*-azobenzene and their interaction with β-cyclodextrin (left), expelled *cis*-azobenzene from the cavity of β-CD; c) paclitaxel-modified β-cyclodextrin (left), paclitaxel-modified photochromic arylazopyrazole (right)[88] with permission. Copyright 2018 John Wiley and Sons.

As a comparison, the photomodulation of the dynamics of microtubules can be regulated with a photochromic analogue of combretastatin A-4 by substitution of the stilbene substructure with an azobenzene. This outstanding pharmacophore in the field of photopharmacology showed an impressive difference of submicromolar toxicity after irradiation with 390 nm, while its photoisomer is > 250 times less toxic in the dark. The photostatins control mitosis *in vivo* with single-cell precision and modulate the MT dynamic with a response time of a few seconds.[91] Another essential protein such as the myosin in muscle tissue inspired the research into synthetic molecular muscles.[92] This long standing goal of preparing artificial muscles was set in an attempt to control motion with organised supramolecular assemblies,[93] rotaxanes[94] and host guest complexes.[95]

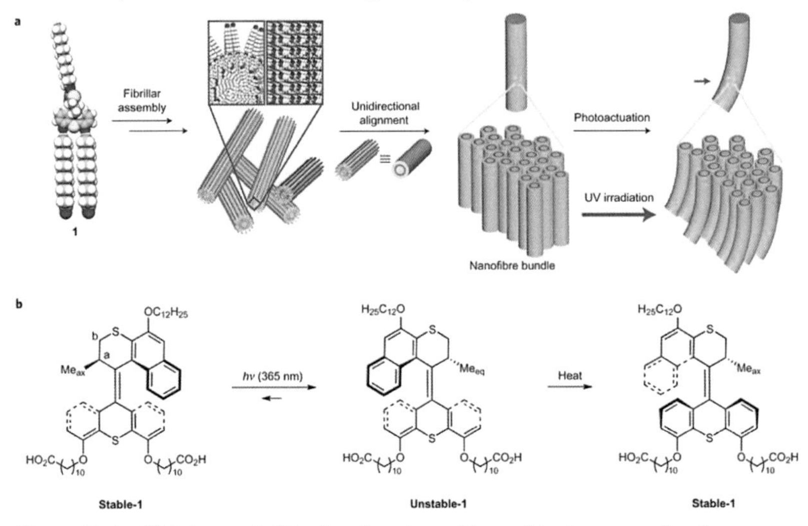

Figure 7. Artificial muscle-like function from hierarchical supramolecular assembly. a) photoresponsive rotary motor and fibrillar assembly into nanofibres. The generated strings are able to bend upon irradiation; b) Inversion steps of motor by irradiation and heat, both isomers Stable-1 are identical and shown from different angles;[96] with permission. Copyright 2018 Springer.

The design of a supramolecular muscle was renewed with an amphiphilic molecular motor as the core unit (Figure 7). The amphiphile is composed of an unpolar dodecyl chain for the upper half and two polar carboxyl groups for enhanced water solubility, connected *via* two alkyl-linkers for the lower half. The amphiphilic molecule undergoes self-assembly in water and is processed to avoid crystallization by drawing from a pipette into aqueous calcium solution generating shear flow.[96] This method converts isotropic solutions of peptide amphiphile solutions to liquid crystals that group the molecules into filaments of bundled nanofibres and form monodomain fibrous gels.[97] The nanofibres have a length of one micrometre and a diameter of ~5-6 nm with 0.5 wt%, and a distinctive difference can be observed after processing nanofibers with 5 wt% to unidirectionally aligned nanofibre bundles with cryo-TEM. The resulting string with its hierarchical supramolecular organization bends from 0° to 90° within 60 s irradiation towards the light. The string returns to the original conformation after incubation at elevated temperatures of 50 °C for 3 h because of the half-life of the molecular motor with $t_{1/2}$ = 2.7 h.[96]

1.8 Photoresponsive supramolecular gels

Low-molecular-weight gelators[98] are intensively studied due to their broad application scope in regenerative medicine, responsive hydrogels, biomineralization or smart materials, like surface modification, optical technologies or electronic devices.[3] They might serve as an inspiration for supramolecular architectures,[99] because of their tunability and diversity.[100] From a comprehensive summary of supramolecular gelators, it is clear that photoresponsive gels with azobenzenes are the most common examples with a diverse spectrum of applications.[101] Even without photoisomerization, only the presence of azobenzenes in the molecular co-assembly can improve the mechanical properties of a hydrogel.[102]

A photoswitch and the gelator can be covalently coupled or co-assembled from two units by electrostatic interaction between cations and anions based on cationic surfactants[103], amphiphilic glutamides[52], Fmoc-Phe-OH,[104] or chiral chaperone gelators[105] and phenylalanine derivative gelators.[106] A spiropyran functionalized with a sulfonate can induce a reversible photomodulation of phase transition of an otherwise non-photoresponsive dipeptide Phe–Phe.[107] Another non-photochromic gel co-assembled with zwitterionic amphiphile C_{16}IPS and Mo_7 and ultimately formed a photochromic supramolecular hydrogel. Herein the UV light induces a blue colour in the transparent hydrogel and a repeated pattern of writing and erasing with air could be used for reversible information storage.[108] As an alternative to salt bridges, the halogen bond-driven co-assembly can also be utilized for gelation between azopyridine and the halogens of TFDIB in acetonitrile.[109]

As mentioned before, the photochromic gels with cyclodextrin and azobenzenes were intensively investigated. Gels with cyclodextrin can consist of co-assemblies,[110] but most gels with cyclodextrin contain polymers[85a, 111] and have properties like self-healing ability,[112] multi-stimuli sensitivity,[113] even molecular recognition,[114] or chiroptical reversibility[115] and red-light responsivity with tetra-*ortho* methoxyazobenzenes.[116]

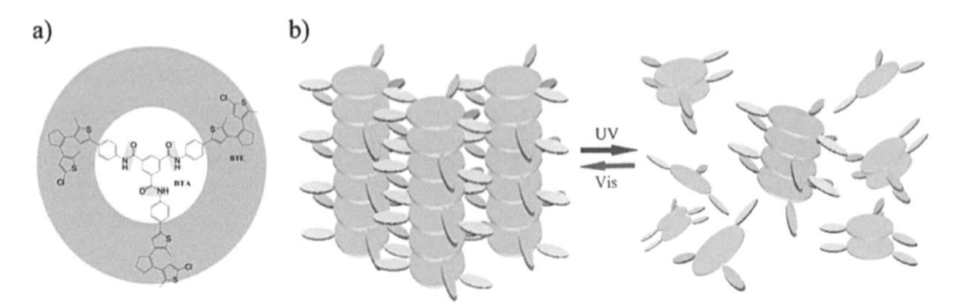

Figure 8. a) Structure of symmetric C_3-BTE; b) Rod-like assemblies of building blocks and morphological transformation with light;[117] with permission. Copyright 2016 John Wiley and Sons.

Besides these examples, numerous covalent combinations were designed with azobenzenes and a gelator, e.g. peptides with protected Asp[118] or dipeptides with Gly-Gly[119] or with Val-Val.[120] Here in particular the design with low-molecular-weight gelators of dichromonyl[121], cysteine[122] and 1,3,5-benzenetricarboxamide[117] peaked our interest with their rather unusual symmetric structure of the amphiphiles in contrast to the more common linear structure with a hydrophobic and hydrophilic headgroup.[123] Based on carbohydrates a macrocycle was synthesized with an unsubstituted azobenzene within. This photoswitchable organogel adopts preferential *P* - helicity for the azobenzene moiety and has a higher thermal stability of 51 d at rt in contrast to acyclic photoswitches. As a chiroptical switch, the chirality can be switched with temperature and by isomerization with light.[124]

The emerging molecular motors find their way into the photoresponsive gels and in this manner a bis-urea-based low molecular weight gelator was combined with an overcrowded alkene. Hereby a hexyl-urea derivative gelates in toluene down to a concentration of 0.4 mg/mL. The gel dissipates upon irradiation with UV light and the resulting *trans*-to-*cis* isomerization initiates a reversible gel-to-sol transition.[125] In another example, a luminescent hydrogel was formed with a fluorescent molecular rotor, a dithienylethene-bridged bispyridinium dye and inorganic Laponite (Scheme 3). Based on this composition of an organic-inorganic hybrid hydrogel the authors envision reversible luminescence switching, transparency and mouldability.[126]

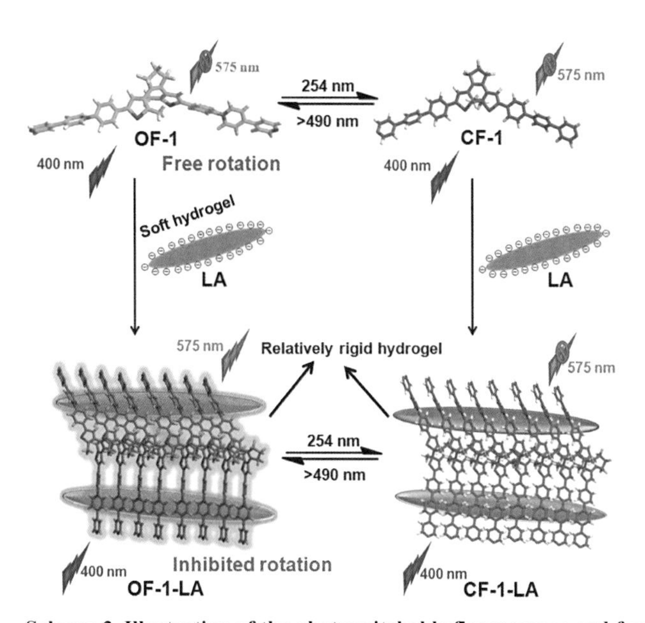

Scheme 3. Illustration of the photoswitchable fluorescence and formation of the hydrogel; structures of the molecular rotors OF-1 and the reversible photoisomerization of the dithienylethene core CF-1 (top), cooperative integration with Laponite and formation of transparent hydrogel OF-1-LA (bottom);[126] with permission. Copyright 2017 John Wiley and Sons.

In summary, photoresponsive supramolecular gels form a class of advanced materials with their typical gel-to-sol transformation by light. Moreover, this triggered macroscopic gelation can serve as an inspiration for other photoswitchable supramolecular systems,[127] biomimetic light-harvesting nanoarchitectonics[128] or biofunctional materials.[129]

1.9 Non-covalent photomodulation of oligonucleotides

In the recent development of photochromic oligonucleotides, the covalent integration of photoswitches into DNA remains a common strategy either into the backbone as nucleosides or nucleotides besides nucleosides surrogates and end-capping. Despite several successful examples of photomodulation with covalently bound photoswitches, the development of photochromic antisense agents or their application in a biological environment remains limited.[130] Nevertheless, artificial nucleic acids based on peptides with guanidine[131] or the photomodulation of PNA/DNA hybridization by light[132] offer a potential for *in vivo* modulation of genes. Below we will shortly focus on the photomodulation of oligonucleotides by non-covalent interactions. In general, a related amphiphilic structure of surfactants with azobenzene can reversibly modulate DNA condensation. This influence was observed upon visible light irradiation with a surfactant-to-DNA base pair ratio of ca. 7, both electrostatic and hydrophobic forces being important in this process.[133] Several other surfactants were reported in the meantime,[134] in the context of either their photocontrol of genomic DNA,[135] or for gene delivery with phototriggered release.[136] For the photoswitchable gene transfection itself an azobenzene-containing polycation based on poly[2-(dimethylamino)ethyl methacrylate] can condense plasmid DNA into nanocomplexes and lead to increased gene expression.[137]

With an increased number of positive charges, the photosensitive polyamines demonstrate high efficiency and the interactions with oligonucleotide chains is directly correlated to their molecular charge. With three charges, the photochromic cation compacts kilobase pair DNA at millimolar concentrations and at neutral pH and is subsequently and completely unfolded upon irradiation with UV light for 3 min. With blue light the refolding of DNA can be induced reversibly into the compact A-form of DNA .[138] With a similar water-soluble 4-(phenylazo)benzoic acid derivative with 3 amines (Azo-3N) the DNA-binding was investigated with a CD and the results suggest an intercalation between the DNA base pairs. In addition, the data points to a conformational change of the B-to-A transition in the *trans*-form and in the *cis*-form the DNA seems to remain in its natural B-form (Figure 9).[139]

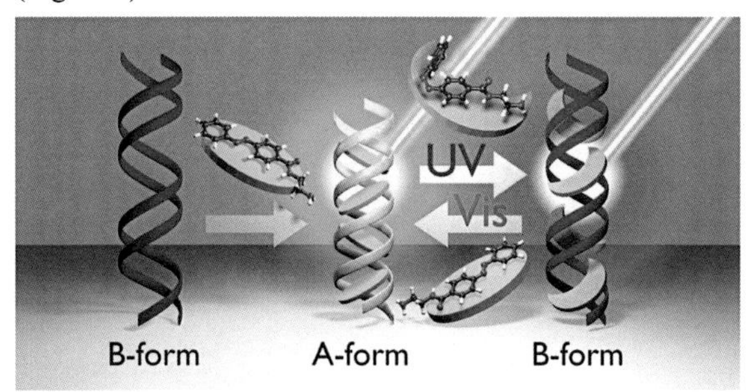

Figure 9. Possible interaction between Azo-3N and the DNA template; the photochromic azobenzene Azo-3N induces the A-form, upon irradiation the DNA is turned into the B-form reversibly;[139] licensed under CC BY 4.0.

A related ether-based surfactant with one guanidine group does not change the melting temperature significantly. But the symmetric equivalent with an azobenzene in between and two guanidine groups at both ends strongly increased the melting temperature by intercalation of the *trans*-isomer in the double helix. After irradiation the bis-cationic azobenzene is ejected from the intercalation and the melting temperature decreases to 18 °C.[140] The identical symmetric azobenzene with quaternary amines instead of the guanidine groups does not lead to an intercalation.[141] A minimum of two positively charged amine groups is also a requirement for the intercalation with spiropyrans. Here, a single quaternary modified spiropyran cannot bind to DNA. Upon irradiation the planar, aromatic and polycyclic merocyanine form with the second charged amine leads to an intercalation into the DNA helix.[142]

In a related spiropyran, the quaternary amine was substituted with guanidine and additionally a guanidine group was introduced in the para position of the hydroxy group of the unpolar merocyanine. Also, the closed spiroform does not bind significantly to DNA, but the merocyanine form intercalates into DNA. But the intercalation depends on the protonation of the merocyanine and the binding of the protonated form is ca. 50 times stronger.[143] As pointed out earlier, for another dually activated spiropyran derivative, this only simultaneous activation with UV irradiation and the presence of high proton concentrations might serve as a biologically logic AND gate.[144]

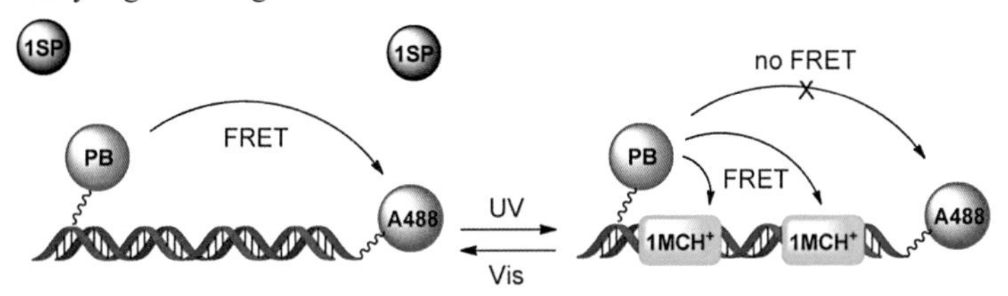

Scheme 4. Reversible switching of FRET from on to off; the spiropyran 1SP does not bind to the DNA strand (left); After irradiation, the merocyanine 1MCH$^+$ binds to DNA through intercalation and FRET of Alexa488 is suppressed (right);[145] licensed under CC BY 4.0.

The same group of ANDRÉASSON used a related guanidinium spiropyran for a reversible energy-transfer switching (Scheme 4). The FRET between pacific blue and alexa488 occurs freely and shuts down by UV-induced isomerization to the merocyanine form by quenching the pacific blue excited state. In this system the FRET efficiency can be tuned continuously by varying the light exposure time, which results in variations in the binding density of MCH$^+$ and is not merely a binary effect.[145]

Also, with diarylethenes supramolecular interactions between DNA were studied, for example as a chiroptical photoswitchable DNA complex. This dithienylethene switch was terminated with two 2-aminoethyl groups, which led to an electrostatic binding of these protonated amines to the DNA double helix. In this case, both the open and the closed forms bind to DNA and DNA chirality is transmitted to this switching unit upon supramolecular complexation. But the other way round, the switch modifies the DNA structure.[146] With a similar diarylethene with terminal quaternary amines of two methylquinolinium, both forms bind to *ct*-DNA through intercalation. In this model system, an enantiomeric enhancement with a direct chirality transfer from DNA was observed.[147] A thiazole orange modified diarylethene does not have photochromism and weak fluorescence in aqueous solution, but upon binding to DNA the fluorescence is enhanced drastically and switching becomes possible.[148]

In contrast to this intercalation, DNA hybridization can be induced in face of a GG mismatch pair in short 11-mer. With a symmetric linking of naphthyridine carbamate dimer with an azobenzene the melting temperature increased by 15.2 °C after irradiation.[149] Recently, the photomodulation of another guanidine-azobenzene-derivative (GuaAzo) was reported that binds in its *trans* form to the minor groove of the DNA.[150] Together with another guanidinium-based DNA ligand (GuaBiPy),[151] the ligands form a heteromolecular supramolecular complex with DNA (Figure 10). This "primitive" supramolecular network[152] can be remotely controlled by one photoswitching compound over the DNA-templated assembly of another compound.[150]

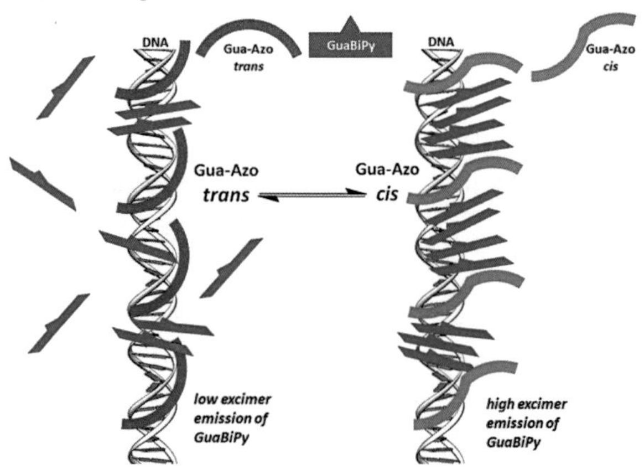

Figure 10. Illustration of the photochromic DNA-templated self-assembly of DNA with GuaBiPy/GuaAzo;[150] with permission. Copyright 2017 John Wiley and Sons.

1.10 Cell penetration and photochromic peptides

The photocontrol of peptides and proteins is diverse and here azobenzenes remain the first choice photoswitch as well,[153] though several emerging photoswitches tend to be an alternative,[22] as pointed out before. One attractive application is the photoswitchable control of cell-penetrating peptides. Several cell-penetrating peptides[154] are derived from the Tat protein of the human immunodeficiency virus (HIV)[155] and are mostly polycationic with a high arginine content or of amphiphilic nature.[154, 156] But there are other cell-penetrating systems even closely related to the following supramolecular hydrogels. An example is that of cell-penetrating peptides, based on 2,5-diketopiperazines, recently reported (Figure 11).[157] These cell-penetrating peptidomimetics also have cationic moieties, a high serum stability and are non-toxic to HeLa cells up to 500 μM.

The transfection efficiency of one derivative Kkd-5 was ca. 50% higher than the positive control with stearyl-R_8.[157] The uptake process leads to fluorescence in the cytoplasm and an endocytosis-driven cellular uptake was indicated by comparison of incubation of the cells with the FITC labelled diketopiperazine derivative at 37 °C and 4 °C.[157-158] However, the endocytosis pathway is not suitable for RNA because of the degradation in the endosome.[159] This led to the development of different strategies for RNA and even the blood-brain barrier was crossed with a 29-amino-acid peptide derived from rabies virus glycoprotein. Hereby the iRNA was delivered to the central nervous system, resulting in efficient gene silencing *in vivo*.[160]

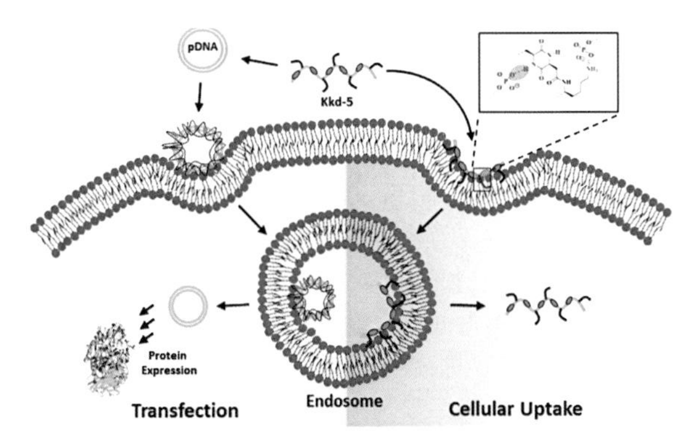

Figure 11. Cyclic dipeptide based cell-penetrating peptidomimetics for effective DNA delivery;[157] with permission from The Royal Society of Chemistry.

For the transfection of mRNA, oligo-serine ester was reported for *in vitro* and *in vivo* application through an unprecedented mechanism. The mRNA is encapsulated by this cationic cargo, which forms complexes readily. After cellular entry of this complex, the charge of the amine groups of serine is altered. This leads to a degradation *via* an O–N acyl shift to 2,5-diketopiperazines and the subsequent release of the mRNA. The transfection efficiency in HeLa cells exceeds 95%, which is significantly higher than the commercial transfection reagent lipofectamine L2000 (55% – 71%).[161] In order to control the uptake, light is one favourable option, for example, for the drug delivery.[162] In regard to photochromic cell penetration, the group of MÖLLER synthesized a hairpin structure of a peptide with an azobenzene at the turning point. One side of the hairpin consists of a positively charged poly-arginine and the other side of a negatively charged poly-glutamate.[163]

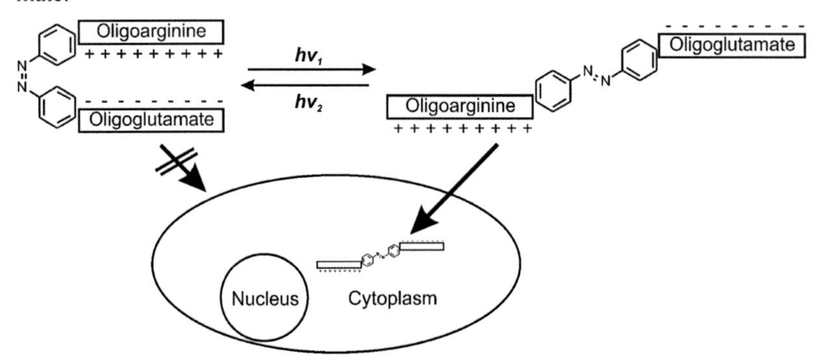

Figure 12. Photoswitchable cell penetrating peptides composed of a hairpin with an oligo-arginine and oligo-glutamate side;[163] with permission. Copyright 2015 John Wiley and Sons.

In its closed form, there are contacts between the NH groups of arginine and the H^{β} and H^{γ} of glutamic acids. Thus, the positive charges are shielded and the hairpin is not active, which means it cannot penetrate the cell membrane of HeLa cells. After irradiation with cyan laser with 488 nm, the *cis*-azobenzene is switched locally to the *trans*-azobenzene and thereby the folded (closed) hairpin is converted to an open hairpin (open chain, or *trans*-hairpin). This *trans*-hairpin with its positively charged arginine side chain can penetrate the cell membrane of HeLa cells. An endosomal uptake was proposed and followed by a fluorescent label with a fluorescent microscope. Unfortunately, this fluorophore rhodamine B led to a significantly reduced *trans*-to-*cis* conversion of ~ 40%.[163]

A similar reduction of PSS was also reported for another peptide containing an azobenzene and labelled with a different fluorophore - fluorescein Fl (Figure 13).

Here, the peptide itself is not cell-penetrating. The construct was injected directly into zebrafish embryos. The azobenzene *in vivo* half-life of 7.5 min was reported, which was lower than the *in vitro* half-life of 10.7 min.[164]

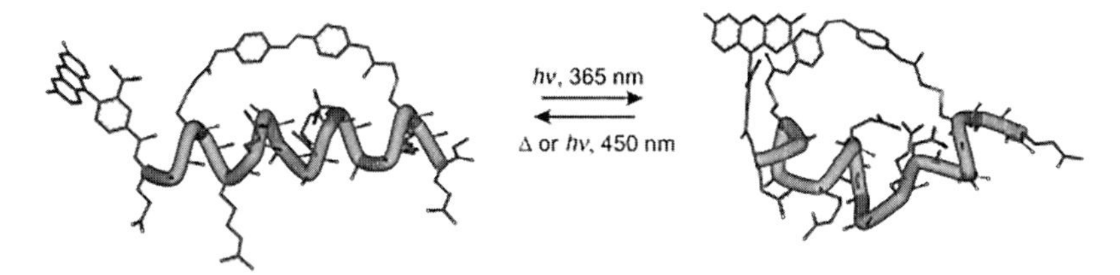

Figure 13. Fluorescent peptide reporter with sequence FI-EACAREAAARE-AACRQ-NH_2; Azo-linker connected at both cysteines; Fluorescein FI;[164] with permission. Copyright 2011 John Wiley and Sons.

This peptide consists of *D*-amino acids with an arginine-rich and lysine-rich sequence, and folds into a helix. The fluorescein was attached at the start of the sequence and the azobenzene-linker was integrated as a loop by covalent coupling with the two cysteines. This construct was stable and photoswitchable over the period of 2 days in the early zebrafish development.[164] The side-chain cross-linked azobenzene installed at two cysteine residues was applied in this strategy.[165] A similar design was applied for the light regulation of protein–protein interaction, which is involved in clathrin-mediated endocytosis.[166] On top, this strategy with photoswitchable click amino acids was combined with genetical encoding by suppression of the amber codon.[165e, 167] With this technique, an azobenzene was incorporated into calmodulin in *E. coli* or in mammalian cells.

The conformation of this calmodulin was changed by light and led to a decrease in the helix content.[165e] In contrast to the photomodulation of proteins, in smaller peptides, the integration of the photoswitch into the backbone is synthetically possible. For a derivative of the cyclic gramicidin S the antimicrobial activity was controlled by an azobenzene, which was integrated into the backbone.[168] An enhanced cell penetration with arginine groups[169] led to an improved antimicrobial activity against *S. aureus*.[168] Besides these covalent integrations of azobenzenes into peptides, the emerging field of photopharmacology offers photomodulation of native proteins with a photochromic drug.[39] Most of these drugs require sophisticated re-design and evaluation of the changed properties,[170] but these photochromic drugs often require less effort for their synthesis, in comparison to the integration of a photoswitch into a protein to reproduce the photochromic effect.[38]

2 Aim of the project

The aim of this project is the development of photochromic materials based on supramolecular structures containing small bioactive molecules and biopolymers. One part of the project focuses on the syntheses of photoswitchable azobenzene derivatives, in particular the recently improved modifications with tetra-*ortho* fluoro and chloro substitutions. These derivatives can be photoisomerized reversibly with visible light, their less stable photoisomers have longer lifetime than non-halogenated analogues, and they can be applied *in vivo* studies.[30a, 30c] Additionally, the syntheses and characterisation of pyridinium azobenzenes will be investigated. These, in turn, exhibit fast thermal back isomerization, improved water solubility and isomerization with green light. Within this thesis, we explored possibilities of incorporation of these photoswitches into a broad range of structures, for example, in MOFs, peptides and catenanes. The applied changes to the azobenzene core are not merely a derivatisation, instead they lead to a substantial improvement of the photophysical properties. The long-term goal is incorporation of these photoswitches into drugs and antisense agents.

The main part of the thesis describes photomodulation of supramolecular hydrogels. A low-molecular-weight hydrogelator consisting of a dipeptide of phenylalanine and lysine, which forms strong supramolecular gels in aqueous solutions, inspired us to produce its photochromic analogues.[171] The properties of the hydrogels were characterised and their application, in particular for drug release, as well as their biocompatibility evaluated. In this process, all hydrogels will be tested and compared with each other. Apart from macroscopic cross-linking of supramolecular fibres resulting in gelation, this versatile peptidic supramolecular structure might be utilised for other applications, as amphiphilic scaffolds and photomodulation of supramolecular assemblies are basic design principles for chiroptical switches, vesicles, auto-amplification of chirality, transfection of oligonucleotides, photochromic peptides or artificial muscles mentioned before.

3 Results and Discussion

In this thesis, I present the first class of photochromic supramolecular hydrogels based on the integration of azobenzenes into a low-molecular-weight gelator (LMWG) based on cyclic dipeptides – 2,5-diketopiperazines (DKP). The four novel photochromic gelators are derived from a previously reported non-photochromic LMWGs Phe-DKP-Lys **5** from NACHTSHEIM (this notation, widely used below, corresponds to a cyclic dimer of *L*-phenylalanine and *L*-lysine)[172] and Tyr-DKP-Lys **6** from the group of FENG (Figure 14).[171] In the diverse field of peptide-based supramolecular gelators, cyclic dipeptides stand out due to a firm hydrogen bonding network and structural rigidity provided by the central ring – in comparison to other LMWGs or other peptide-based gelators.[101] The great variety of diketopiperazines as bioactive natural products is another incentive for the development of their photochromic analogues with potential applications in photopharmacology.[173] As already discussed, the combination of azobenzenes with supramolecular systems by co-assembly with electrostatic interaction[52, 103] or by the covalent binding to supramolecular gelators can yield multi-responsive hydrogels[123] or photoresponsive soft nanotubes.[174]

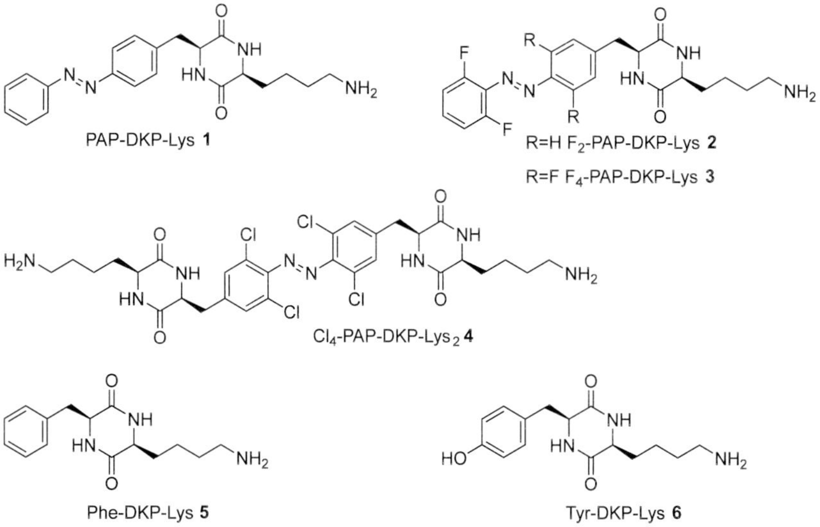

PAP-DKP-Lys **1**

R=H F_2-PAP-DKP-Lys **2**

R=F F_4-PAP-DKP-Lys **3**

Cl_4-PAP-DKP-Lys_2 **4**

Phe-DKP-Lys **5**

Tyr-DKP-Lys **6**

Figure 14. Structures of photochromic supramolecular hydrogels PAP-DKP-Lys 1, F_2-PAP-DKP-Lys 2, F_4-PAP-DKP-Lys 3, Cl_4-PAP-DKP-Lys_2 4, Tyr-DKP-Lys 5[172a] and Phe-DKP-Lys 6.[171]

With natural dipeptides laminated nanoribbons[119] or fibrous systems[120] can self-assemble and undergo reversible reorganisation that leads to a gel-to-sol transition with light. These examples and their potential applications, from regenerative medicine to electronic devices,[3] inspired the following research as well.

3.1 Photoresponsive self-healing supramolecular hydrogel

The design of gelator **1** is based on Phe-DKP-Lys **5**, which forms a transparent hydrogel, whereas derivatives of the gelator comprised glycine, serine, cysteine and histidine instead of lysine form opaque hydrogels, when prepared in an aqueous PBS buffer (phosphate-buffered saline).[171] This information is relevant for designing photochromic analogues because an opaque hydrogel would have a limited penetration depth of light (scattering would occur instead). Dissipation of energy instead of its absorption would decrease efficiency of triggering such a photochromic hydrogel. Moreover, while gelator Phe-DKP-Glu **55** also forms transparent hydrogels, the basic lysine residue was favoured over the glutamic acid residue because the positively charged ε-amino group can form salt bridge interaction with the negatively charged phosphate backbones of oligonucleotides – an important feature for potential delivery systems of therapeutic RNA or DNA fragments. Azobenzenes were selected as the photochromic motif not only due to their structural similarity with the side chain of phenylalanine, present in the original gelator **5**, but also because the change in geometry and polarity upon photoisomerization is expected to strongly destabilize the supramolecular structures building fibres. There we hoped that the irradiation will cause a rapid and reversible gel-to-sol transition. Thus, the rational molecular design of the photochromic amphiphilic LMWG **1** is based on the data from literature,[171-172, 175] rather than on the screening of a library of various gelators with a mere trial and error test to optimize the gel by structure relation, as likewise approached in the literature.

3.1.1 Synthesis and photoisomerization of the supramolecular gelator 1

The synthesis of supramolecular gelator PAP-DKP-Lys **1** starts with a MILLS reaction of nitrosobenzene **7** with commercially available chiral *N*-Boc-4-amino-*L*-phenylalanine Boc-Phe(4-NH_2)-OH **8** to form the asymmetric photochromic amino acid derivative Boc-PAP-OH **9** (where PAP stands for (4-phenylazo)-*L*-phenylalanine residue).[175a] This photochromic amino acid was coupled with H-Lys(Boc)-OMe·HCl using HBTU amide coupling protocol to yield the linear dipeptide Boc-PAP-Lys(Boc)-OMe **10**. After removal of the Boc-protecting group dipeptide **10** was cyclized to form the 2,5-diketopiperazine moiety. This modular synthetic strategy can be adapted to generate various derivatives and this synthesis provides gelator **1** on a gram scale. This enables the screening of different gelation conditions like solvents, additives, or different buffers without the need to recycle or resynthesize gelator **1**. PAP-DKP-Lys **1** is soluble in MeOH, DMF and DMSO, but insoluble in ACN or PhMe.

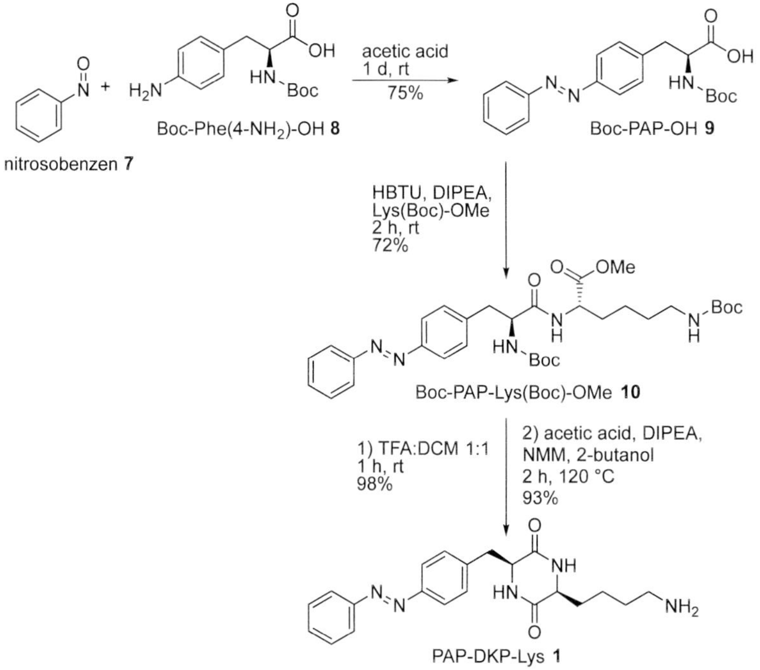

nitrosobenzen **7** + Boc-Phe(4-NH_2)-OH **8**
acetic acid, 1 d, rt, 75%
Boc-PAP-OH **9**
HBTU, DIPEA, Lys(Boc)-OMe, 2 h, rt, 72%
Boc-PAP-Lys(Boc)-OMe **10**
1) TFA:DCM 1:1, 1 h, rt, 98%
2) acetic acid, DIPEA, NMM, 2-butanol, 2 h, 120 °C, 93%
PAP-DKP-Lys **1**

Scheme 5. Synthesis of the supramolecular hydrogelator PAP-DKP-Lys 1.

3.1.2 Properties of supramolecular hydrogel 1

PAP-DKP-Lys **1** dissolves upon short boiling in aqueous solutions with tested concentrations of up to 3% (Table 9). Solutions with concentrations at and above the gelation limit of 1.5% of gelator **1** form hydrogels after cooling down to rt. At lower concentrations (below 1.5% of **1**) hydrogels are fragile upon shaking by hand, but the vial with the hydrogel can be carefully inverted without dissipation of the gel. By addition of 1% TFA or 50 mM NaCl, the stability of the gel can be increased. This stabilizing effect was observed by heating the gel sample until it turned into a sol. The gel-to-sol transition temperature T_{g-s} of a 1.5% PAP-DKP-

Lys in diH_2O is 38 ± 4 °C and in 50 mM NaCl 51 ± 4 °C (see Table 9). The same effect was observed at a higher concentration of 2% PAP-DKP-Lys with an increase from 51 ± 4 °C in diH_2O (gel **A**) to 70 ± 1 °C in 50 mM NaCl (gel **B**). With the addition of 1% TFA, $T_{g\text{-}s}$ can be increased with a 2% sample to 68 ± 1 °C, which is a comparable gain. Hypothetically, the stability of the hydrogel is tuneable according to the Hofmeister effect[176] as long as the pH of the aqueous solution is low enough to keep the lysine residue protonated. This estimation stands in contrast to the characteristics of Phe-DKP-Lys·TFA salt **5**, which did not form a hydrogel in the range between 2-5% in this buffer. A sample containing 5% of this gelator **5** formed a hydrogel in 20 mM NaOH in PBS buffer with a $T_{g\text{-}s}$ of 53 ± 4 °C. The gel-to-sol transition temperature increased to 62 ± 3 °C in 100 mM NaOH in PBS buffer. These results are in agreement with the literature of Phe-DKP-Lys **5**, which forms a hydrogel at 1.5% in 1 mol/L phosphate buffer with a pH of 10-10.5.[171]

In analogy to monoprotic acids like TFA, polyacids like DNA should increase the stability of hydrogels based on PAP-DKP-Lys similarly to the cooperative effect of salt bridges. This hypothesis was investigated with commercially available double stranded DNA oligomers (htDNA, ca. 1300 bp). In comparison with the 2% hydrogel in diH_2O, the addition of 0.2% htDNA increased the $T_{g\text{-}s}$ from 51 ± 4 °C to 64 ± 8 °C (gel **C**). In case of the 1.5% hydrogel this beneficial effect by addition of 0.2% htDNA results in an increase from 38 ± 4 °C to 44 ± 3 °C. The exchange of the buffer to aqueous 50 mM NaCl increased the $T_{g\text{-}s}$ to 60 ± 3 °C (gel **D**), too. However, the combination of both additives did not increase the $T_{g\text{-}s}$ in case of the 2% hydrogel in 50 mM NaCl (70 ± 1 °C) significantly in comparison to the further addition of 0.2% htDNA (71 ± 2 °C). In conclusion, the addition of sodium chloride or TFA increases the gel-to-sol transition temperature, and the hydrogels' properties can be tuned within the stated range to a suitable stability against mechanical force or temperature.

The selected formulation of **1**, hydrogels A to D (Table 1), were characterized further *via* scanning electron microscopy (SEM). The images in Figure 15 show xerogels of gelators A, B, C and D, which were prepared by lyophilization of the respective hydrogels.

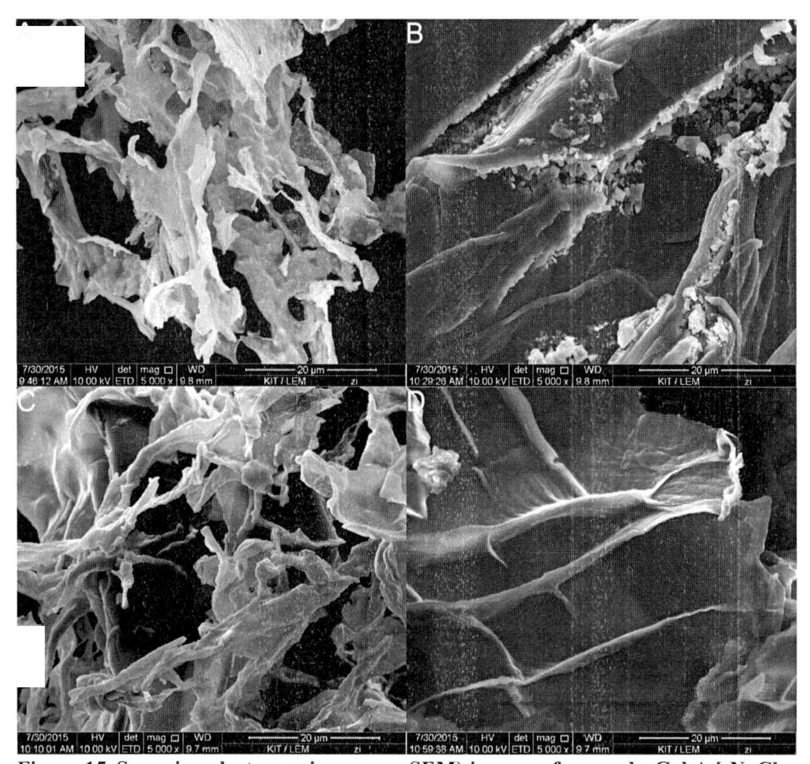

Figure 15. Scanning electron microscopy (SEM) images of xerogels. Gel A [-NaCl, -DNA]: 20 g/L of 1 in diH_2O; B [+NaCl, -DNA]: 20 g/L of 1 in 50 mM aqueous NaCl; C [-NaCl, +DNA]: 2 g/L htDNA and 20 g/L of 1 in diH_2O; and D [+NaCl, +DNA]: 2 g/L htDNA and 15 g/L of 1 in 50 mM aqueous NaCl.

The same hydrogels A to D were measured in the environmental scanning electron microscope ESEM and are displayed in Figure 16. The ESEM microscopy can prevent the complete loss of water with a saturated water atmosphere, but observation is limited to the surface of the hydrogel. In all hydrogel mixtures the ESEM pictures are similar and show round lamellar layers vertical to the surface, which are folded one over another repeatedly. The structures are comparable to other hydrogels with a Boc-PAP-OH **9** moiety.[118] The complex arrangement of xerogels is characteristic for gelator **1**, but there is no correlation between the supramolecular structure of the gelator in aqueous solution because of the absence of water in high vacuum. In principle, the self-assembly of gelator **1** is not restricted to aqueous solutions and might be related to structures exhibited in vacuum by cyclo-diphenylalanine.[177]

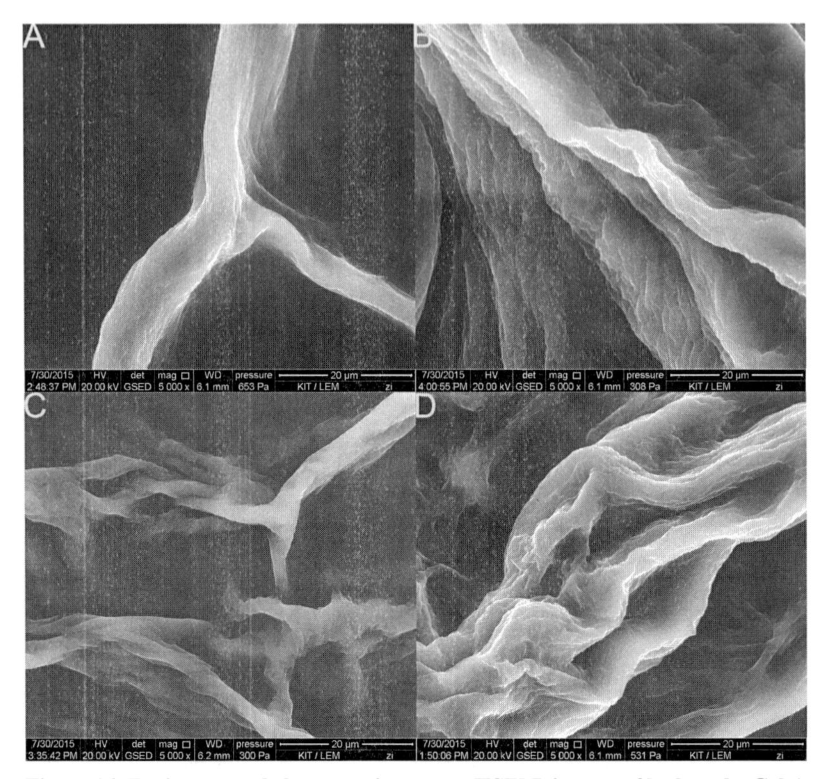

Figure 16. Environmental electron microscopy (ESEM) images of hydrogels; Gel A [-NaCl, -DNA]: 20 g/L of 1 in diH_2O; B [+NaCl, -DNA]: 20 g/L of 1 in 50 mM aqueous NaCl; C [-NaCl, +DNA]: 2 g/L htDNA and 20 g/L of 1 in diH_2O; and D [+NaCl, +DNA]: 2 g/L htDNA and 15 g/L of 1 in 50 mM aqueous NaCl.

The difference between photoisomers of the hydrogels was investigated by irradiating gel D for 30 min with 365 nm and comparing the ESEM image of the sample to the non-irradiated sample (Figure 17). After irradiation the laminar structures are less pronounced than before and the surface is planar with several cracks on the surface.

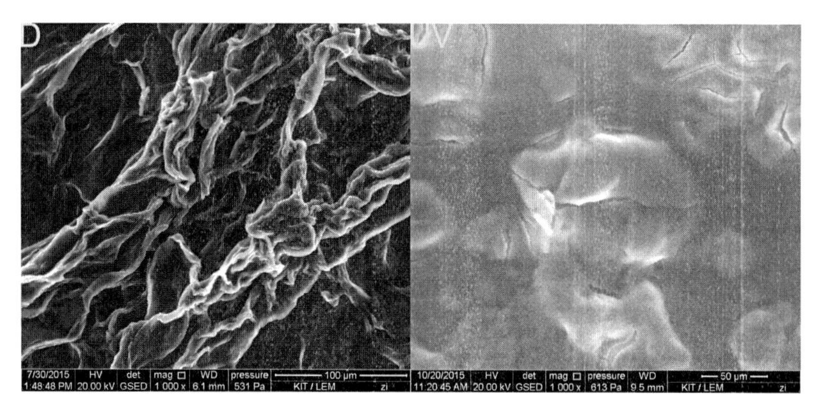

Figure 17. Environmental electron microscopy (ESEM) images of hydrogel D before and after irradiation with 365 nm for 30 min. 2 g/L htDNA and 15 g/L of 1 in 50 mM aqueous NaCl.

In addition to SEM microscopy a diluted sample of 5 g/L PAP-DKP-Lys in 50 mM NaI was prepared by ultrasound on a carbon-coated copper grid and measured with a TEM microscope (Figure 18). Here, a fibrous network was detected, which is similar to the fibrous network of a xerogel of Tyr-DKP-Lys.[172a]

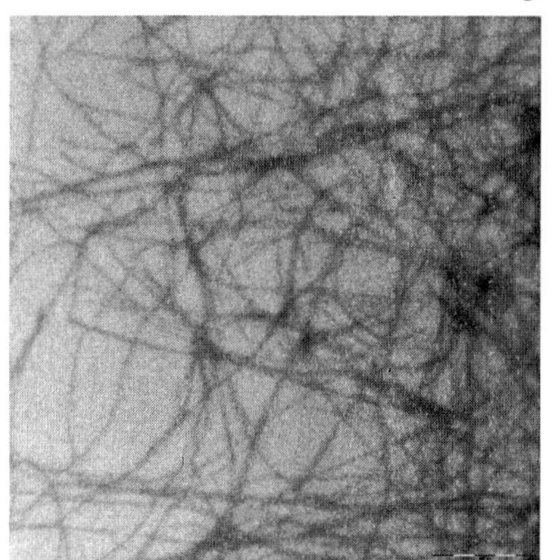

Figure 18. TEM image of the sample 5 g/L PAP-DKP-Lys 1. Staining with 50 mM NaI. Scalebar=200 nm.

The UV-Vis spectrum of the non-irradiated sample PAP-DKP-Lys in diH_2O is similar to other azobenzene derivatives,[118] and so is the isomerization with UV light at 365 nm (Figure 19). Regarding the light source, two 10 W LEDs were placed inside a water-cooled aluminium block and the sample was irradiated from a distance of 3 cm from both sides. The diluted 60 µM sample reached the photostationary state after 10 s. Gelator **1** can be isomerized reversibly 100 times or irradiated with 365 nm (10 W LED) for 24 h without significant degradation (Figure 47).

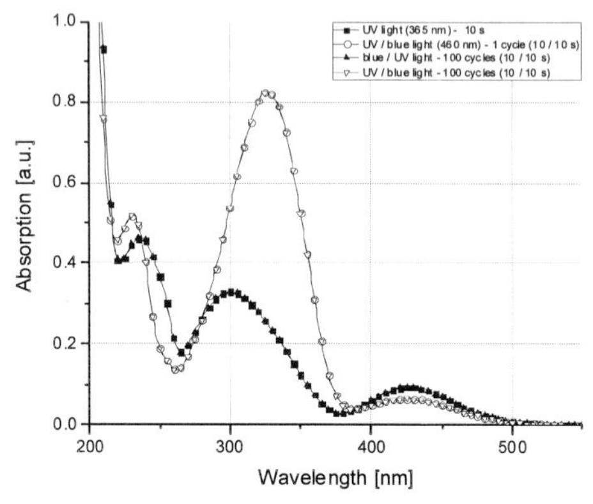

Figure 19. UV-Vis spectra: 60 μM PAP-DKP-Lys 1 in diH_2O; Photostationary states (10 sec irradiation) at 365 nm (10 W, black squares) – and 460 nm (10 W, empty dots); The same sample after 100 full switching cycles (10 sec each) ended with UV-light irradiation (black triangles), or blue light irradiation (empty triangles).

3.1.3 Supramolecular hydrogels dissipating with light

The same device with 2×10 W LEDs (365 nm) was used to investigate the gel-to-sol transition of the hydrogel upon irradiation with 365 nm. Gel D can be dissipated to the sol upon irradiation for 30 min. In Table 1 the influence of the concentration onto photo-induced gel-to-sol transition is demonstrated. In summary, the hydrogels dissipate after irradiation for 30 min, but there is an upper concentration limit for the photomodulation depending on the stability. Both are dependent on the $T_{s\text{-}g}$ and the rheological rigidity of the hydrogel. The lower end of this window is restricted by the gelation limit, which is required to form a hydrogel in the first place. The upper concentration limit is proportional to the gel-to-sol transition temperature and the photomodulation is only possible below a $T_{s\text{-}g}$ of 70 °C, according to the samples in Table 1. This photochromic window depends on the photostationary state of the azobenzene, too. Without any irradiation the azobenzenes are thermally equilibrated to > 95% of the *E*-isomer. After 30 min irradiation with 365 nm 77% of the PAP-DKP-Lys is isomerized to the *Z*-isomer. In course of isomerization the supramolecular interaction between *Z* and *Z* or *Z* and *E* of the molecules is reduced, but supposedly not completely excluded. Within this window, properties like gelation time and mechanical stability can be tuned and modified for a certain application. An advantageous gelation time of 5 min, a sufficient gel-to-sol transition temperature of 60 ± 3 °C and a gel-to-sol irradiation of 30 min were observed for the gel D mixture with 1.5% PAP-DKP-Lys, 0.2% htDNA in 50 mM NaCl.

Table 1. Duration of gelation, rigidity against shaking and melting temperatures (T_{g-s} i.e. gel-to-sol transition temperature) of hydrogels with PAP-DKP-Lys (1) with different dopants.

Sample	Gelation rt, 5 min	Shaking stability	T_{g-s}	365 nm, 30 min
Gel **A** 20 g/L **1**	gel	fluid	51±4	inst. gel (10 min)
30 g/L **1**	gel	gel	69±2	gel
Gel **B** 20 g/L **1**, 50 mM NaCl,	gel	gel	70±1	gel
Gel **C** 2 g/L htDNA, 20 g/L **1**	gel	fluid	64±8	inst. gel (60 min)
2 g/L stDNA, 20 g/L **1**	gel	fluid	68±5	inst. gel (10 min)
15 g/L **1**	fluid	fluid	43±5	fluid (10 min)
15 g/L **1**, 50 mM NaCl	fluid	fluid	51±4	fluid
2 g/L htDNA, 15 g/L **1**	gel	fluid	44±3	fluid (10 min)
Gel **D** 2 g/L htDNA, 15 g/L **1**, 50 mM NaCl	gel	fluid	60±3	fluid
2 g/L stDNA, 15 g/L **1**, 50 mM NaCl	gel	fluid	68±6	inst. gel
2 g/L htDNA, 20 g/L **1**, 50 mM NaCl	gel	gel	71±2	gel
1.5 g/L htDNA, 15 g/L **1**, 50 mM NaCl	gel	fluid	66±4	inst. gel
2 g/L stDNA, 20 g/L **1**, 50 mM NaCl	gel	fluid	73±3	gel
1.5 g/L stDNA, 15 g/L **1**, 50 mM NaCl	gel	fluid	70±5	inst. gel (10 min)
2 g/L DOX, 15 g/L **1**, 50 mM NaCl	fluid	fluid	56±1	fluid

The mechanical stability of the hydrogels is characterised preliminarily by inversion of the vial or shaking and, as mentioned before, by the gel-to-sol transition temperature. Here rheology characteristics for gels A to D were measured for a comparison with other LWMGs. The value determined for the storage modulus G' of the hydrogel D was 1.5×10^4 Pa and the loss modulus G'', 4.6×10^3 Pa (Table 14, gel A-C: Figure 48, Figure 49 and Figure 50). A higher T_{g-s} leads to an increase of the storage modulus (gel A, B and D). The exception of gel C might be due to the high derivation of the T_{g-s}. In general, the values of storage modulus G' are comparable to the 1.5% Phe-DKP-Lys **5** hydrogel in 1 mol/L phosphate buffer. As stated before, the incubation period of the gel, the buffer and concentration of the gelator have a significant influence on the storage modulus.[171] Regeneration of the storage modulus of samples B and D to their initial value occurs within 1 min (Figure 49, Figure 20, bottom). For samples A and C the regeneration of the storage modulus is slower and exceeds 1 min in case of 2% gelator **1** in diH_2O (Figure 48, and Figure 50). In total, an exact correlation of all factors might be too complex for a supramolecular system with several significant variables, and it would require more measurements. But for one variable, e.g. a salt like NaCl, a correlation with regard to the change in mechanical properties can be demonstrated.

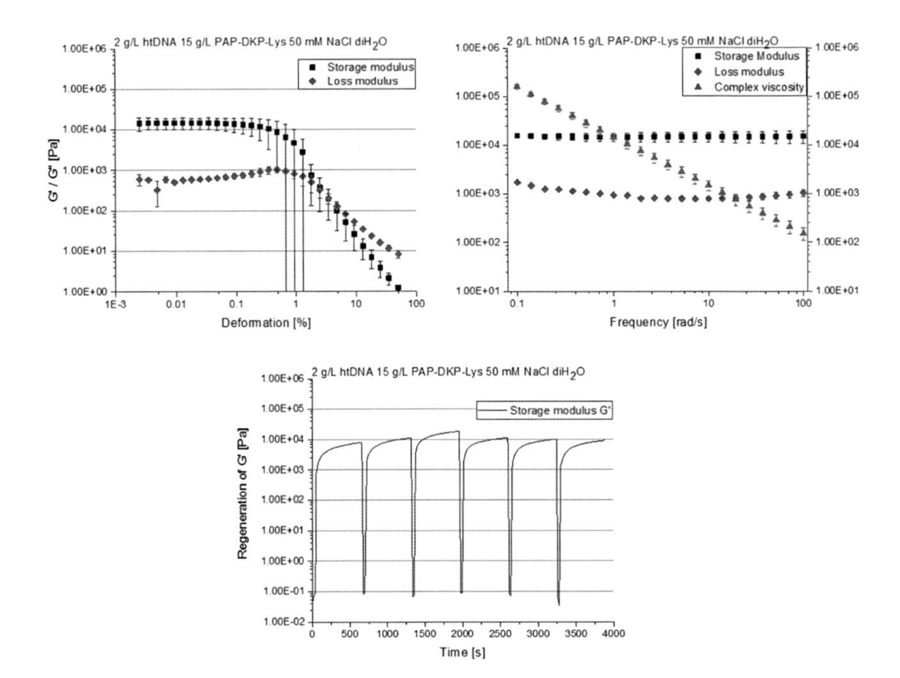

Figure 20. Gel D: 15 g/L (1.5%) of 1 and 2 g/L of htDNA in 50 mM aqueous NaCl; *S*train sweep experiment (top left); frequency sweep experiment (top right); regeneration of G' after shearing the gel for 30 sec at 100% deformation (bottom).

Thus, the storage modulus G' of the hydrogel after the addition of NaCl was measured by rheology for a 1.5% hydrogel (Figure 21). To exclude the influence of other variables, the formulation of gel A – D with DNA or a different concentration of gelator **1** was omitted. Based on this data, a consistent increase from 25 mM NaCl to 1 mol/L NaCl correlated with the storage modulus increase from

6.7×10^2 Pa to 9.5×10^4 Pa (Figure 53). This increasing rigidity (storage modulus G') is in agreement with the aforementioned enhanced gel-to-sol transition temperature by addition of NaCl (Table 9). Most likely, increasing the NaCl concentration leads to an increasing stability of the supramolecular assembly of the fibres,[119] because of the hydrophobic effect and a stabilisation of the hydrophobic core of the azobenzenes.

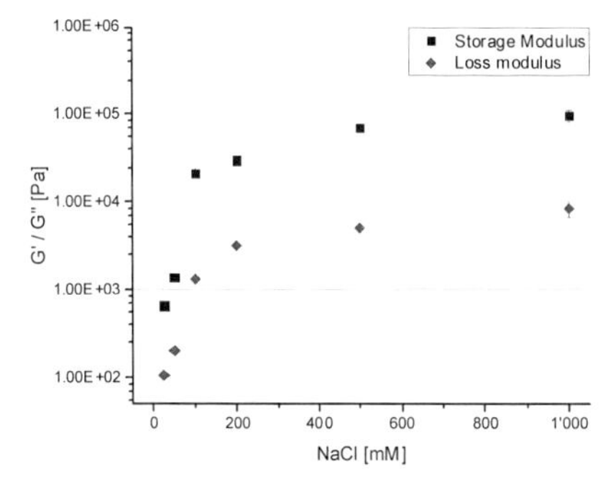

Figure 21. Logarithmic plot of the G' values vs. NaCl concentration (for the 15 g/L of 1), deformation 0.1%.

3.1.4 Drug release from hydrogels triggered by light

The hydrogels containing gelator **1** can reversibly dissipate upon irradiation. We decided to explore this property for encapsulation and the subsequent light-driven release of a drug or a bioactive molecule. To investigate the light-dependent drug delivery, two different cargo molecules were chosen. The fluorescent anticancer drug doxorubicin, which is used against various types of cancer,[178] and htDNA as an example of a biomolecule.

For the experiment, samples of 1.5% hydrogel in 50 mM NaCl were prepared in triplicates with 2 mg/mL doxorubicin. The light-triggered release is compared with the diffusion rate of the molecule itself in darkness over the duration of irradiation (30 min). To quantify the released doxorubicin, the buffer on top of the hydrogel was exchanged every 5 min and measured by UV-Vis at λ=485 nm. Within 30 min, the irradiated hydrogel dissipated, doxorubicin was released and a total recovery of 75% was measured (Figure 22). Without any irradiation, doxorubicin can diffuse, and this passive diffusion was overall below 20% of the encapsulated material within the same time period. The concentration of doxorubicin was measured by UV-Vis absorption in comparison to a calibration curve of a stock solution.

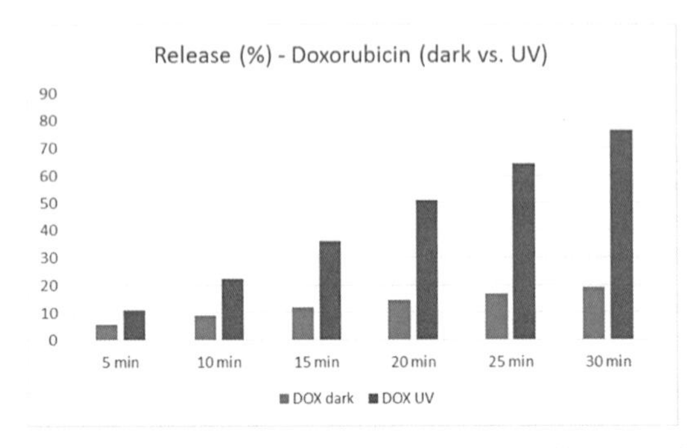

Figure 22. Release of htDNA to 1 mL of 50 mM NaCl solution applied on top of hydrogels composed of 15 g/L 1 and 2 g/L DOX in 50 mM NaCl; by diffusion in absence of light (blue bars on the left side) and by dissipation of the supramolecular structure (red bars on the right side) within the timescale of complete dissipation of gel 1 under UV light (30 min).

The experiments with htDNA were performed under the same conditions with loading of 2 mg/mL. As the UV-Vis absorption spectra of gelator **1** and DNA overlap, separation of the released mixture was necessary for detection and it was carried out by size-exclusion chromatography. The recovered amount of htDNA was quantified at 41% and the diffusion of htDNA in darkness was measured at below 3%. In conclusion, the cargo molecules can be released efficiently and there is a significant difference between diffusion and light-triggered release in both examples.

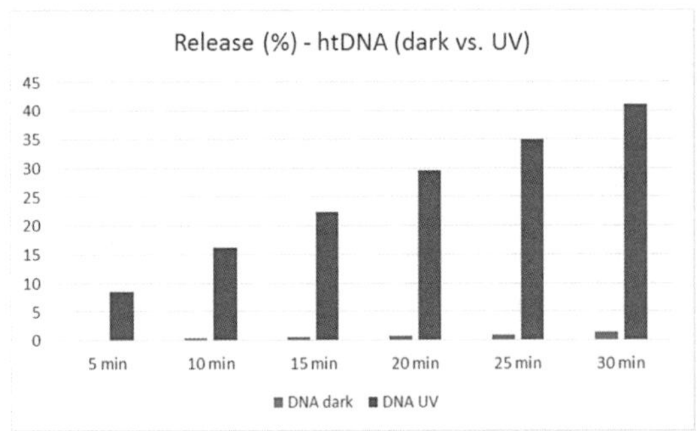

Figure 23. Release of htDNA to 1 mL of 50 mM NaCl solution applied on top of the hydrogel composed from 15 g/L 1 and 2 g/L DOX in 50 mM NaCl; by diffusion in absence of light (blue bars on the left side) and by degradation of the gel structure (red bars on the right side) within the timescale of complete gel erosion under UV light (30 min).

The ratio between the diffusion of doxorubicin and htDNA is presumed to be dependent on the individual set of interactions between cargo and hydrogel. As mentioned before, gelator 1 was designed to have an affinity towards DNA by the salt bridges between the lysine residue and the phosphate backbone. Doxorubicin is known to intercalate into DNA,[179] but it is unlikely that doxorubicin can intercalate into the supramolecular structure of gelator **1**. Otherwise, a lower diffusion would have been expected – comparable to the leaking rate of htDNA in the absence of light.

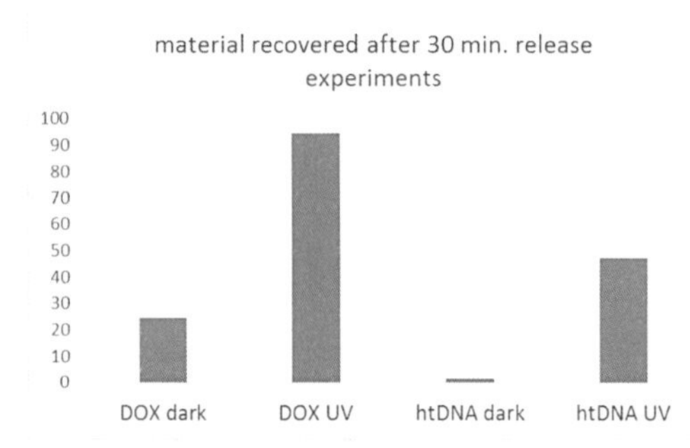

Figure 24. Total recovery of encapsulated material; Summary quantification in all fluid fractions isolated upon diffusion in darkness/under UV light, initial washing of gels in darkness and the remaining liquid after taking all described aliquots for analysis.

3.1.5 Hypothesis of supramolecular structure

In summary, gelator PAP-DKP-Lys can form hydrogels. The rigidity of the gel is dependent on several factors, including pH and concentrations of salts. The influence of salts can be predicted by the Hofmeister series[176] and adjusted to some extent for the desired application, here the drug release by light. It is presumed that the supramolecular structure in water is stabilized primarily by hydrogen bonding between 2,5-diketopiperazine units.[171] In addition, π-π stacking of azobenzenes can form a hydrophobic core. This hydrophobic core may be stabilized by an addition of NaCl, which means that the hydrophobic effect[180] is presumably another driving force in addition to π-π stacking (Figure 25).[181] The predicted arrangement of Tyr-DKP-Lys in a xerogel by Xie *et al.*[172a] indicates a bended lysine side chain ending inside this hydrophobic core. In the release experiment, an additional stabilizing interaction between the htDNA and gelator **1** (presumably between the protonated lysine amine groups and the DNA's phosphates) is indicated by the low leaking rate. Therefore, the bending of lysine residues inside the hydrophobic core does not have to apply to all of them and could rather be a dynamic arrangement in water. In analogy to simulations of β-peptides[182] or lipid bilayers[183] the structure is not fixed and several lysine residues can be charged to form dynamic salt bridges with the phosphate backbone of DNA. The increased gel-to-sol transition temperature $\Delta T_{g\text{-}s}$=13 °C of gelator **1** after the addition of 0.2% htDNA is an indication of this interaction and the low leaking rate in case of the light-triggered release of htDNA confirms the hypothesis.

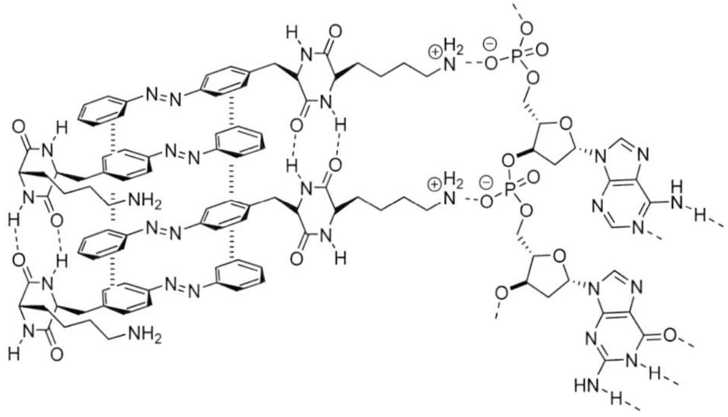

Figure 25. Hypothesis of the arrangement of the molecules in the hydrogel 1; cf.[184]

In relation to peptides, the tertiary structure of gelator **1** is probably a cross-linked network of multiple supramolecular fibres.[185] Other common hydrogels like gelatine, agar, collagen, chitosan, hyaluronic acid or alginate are amphiphilic biopolymers with a closely related 3D structure because of their long fibrils as a general characteristic. These fibrils in LMWG are detectable, for example, by TEM microscopy and were observed in gelator **1**, too (5.2.3.7, Figure 18). In contrast to these examples, gelator **1** is not a covalent polymer and forms these fibres exclusively by supramolecular interactions. The isomerization of azobenzene from *E* to the *Z*-isomer leads to distinct changes in conformation and polarity, and thus strongly affects these interactions, and in consequence, the entire 3D structure of the material.

The hydrogen bonding of the 2,5-diketopiperazine unit is putatively affected by the different conformation of the *Z*-isomer, which supposedly weakens them. The change in polarity directly affects its solubility in water[186] and hypothetically weakens the hydrophobic effect of the hydrophobic core of gelator **1**. The versatility of the supramolecular assembly and its modular synthesis to access multiple derivatives is a major advantage in comparison to common hydrogels based on polymers or biopolymers.[187] With the photochromic feature to reversibly change the gel into sol by irradiation with UV light, several possible applications can be accessed without contamination of the system by additional chemical reagents. In particular, the high spatial and temporal control of light can be an advantage in comparison to redox reactions, ultrasound, pH or other triggers. In regard to the dynamic chemistry introduced by J.-M. LEHN, supramolecular hydrogels are dynamic by nature,[10] but the photomodulation of the self-assembly leads to a dynamic control by intent.[188] This supramolecular assembly of diketopiperazines and their photomodulation is a robust concept and might serve for other supramolecular architectures like capsids for encapsulation, micelles for transfection or artificial muscles.

In most biological applications, cytotoxic UV light is a substantial disadvantage of gelator **1** due to poor penetration into cells and tissues with this light frequency. As such, it was the starting point for the development of the next-generation photochromic systems capable of overcoming this limitation. The redesign of gelator **1** by an exchange from the azobenzene unit to a spiropyran photoswitch would still require UV light irradiation – at least to switch the chromophore to the merocyanine form.[189] Additionally, it could hypothetically compromise the gel-forming properties of the system, particularly using derivatives of spiropyran switchable with visible light (> 390 nm),[190] because of the significant difference of the structure between azobenzene and spiropyrans. Moreover, kinetic studies of spiropyrans revealed that the merocyanine form can decompose by hydrolytic cleavage in aqueous buffer.[43] This could be prevented by replacement of the 6-NO_2 group with 8-carboxylate.[43] Despite such issues, spiropyrans had been successfully applied previously as photoswitches for supramolecular gels based on dipeptide *D*-Ala–*D*-Ala,[191] or a combination of cyclohexane trishydrazide and aldehyde,[192] or as a co-assembly with dipeptide Phe-Phe.[107] In our case, however, the limits of spiropyran, particularly these regarding UV light triggering, steric issues and fatigue in water, led to the fine tuning of the existing azobenzene switch, which ultimately resulted in the application of *ortho*-substituted azobenzenes.[30a, 30c]

3.2 Drug release from hydrogels with green light

For applications of photoswitchable supramolecular hydrogels in the biological environment, orthogonal conditions are essential. For cell cultures these conditions might include isotonic buffers, the operational temperature of 37 °C, low toxicity of the material and preferential restrain from UV light due to its cytotoxicity and low penetration ability into biological subjects. UV light can lead to cytotoxic effects, e.g. due to formation of mutagenic thymine dimers,[193] which is also relevant for the non-covalent interaction with DNA *in vitro* and, in addition, the penetration depth of UV light (300-400 nm) into common tissues is confined to a few fractions of a millimetre.[194]

These constraints are mostly overcome with our next system – the first photochromic low-molecular-weight gelator triggered by visible light. Here, we have demonstrated that the green light-induced release of a variety of drugs or peptides is possible under physiological conditions. In combination with an encapsulated antibiotic, the bacterial growth can be inhibited by green light irradiation, while normally sterilization processes require highly energetic UV light irradiation. The hydrogel described below exhibits high stability and suitable properties under physiological conditions, including low cytotoxicity.

3.2.1 Supramolecular gelator F_2-PAP-DKP-Lys

The structure of supramolecular gelator F_2-PAP-DKP-Lys **2** is based on a rational design and the evolution of the previously developed gelator **1** (Figure 14). To enhance photophysical properties of the gelator for biological applications, azobenzene as the photochromic unit has to be changed to a visible light-triggered photoswitch. Among interesting modifications of this scaffold, we have initially decided for *ortho*-fluoro substituted azobenzene derivatives introduced by HECHT *et al.*[30a] Properties of red light-triggered *ortho*-chloro or *ortho*-methoxy substitution, introduced by WOOLLEY *et al.*,[30c] were also of interest, but they were initially given lower priority because of the steric demands and somewhat less convincing synthetic routes.

The size of methoxy groups could potentially compromise π-π stacking when compared to the predicted supramolecular structure of Tyr-DKP-Lys,[172a] although a red light-triggered hydrogel based on interactions of methoxy-substituted azobenzene with β-cyclodextrin was reported recently.[116] The first published synthesis of *ortho*-chloro-azobenzenes[30c] did not work in our hands, and upon several attempts was proven to be unreliable [195] The same problem was observed by D. B. KONRAD from the TRAUNER group (*personal communication*), which ultimately led to the development of an alternative synthetic strategy for this useful chromophore.[36] Among *ortho*-fluorinated azobenzenes suitable for our system, several substitution patterns varying in the number of fluorine atoms, measured photostationary states, and thermal stability of *cis*-isomers were reported. The most optimal switch had all four *ortho*-positions saturated with fluorine atoms. However, the synthesis of this tetra-*ortho*-fluoro azobenzene requires more steps (Scheme 7) and therefore the shorter synthetic route was investigated first towards derivatives with two or three fluorine-substitutions at one side of the azobenzene. Hereby, the substitution pattern of 2,6-difluoro of **12** was compared to 2,4,6-trifluoro-azobenzene of photochromic amino acid derivative **13** (**12** 5.3.2.3, **13** 5.3.2.2). The 2,6-difluoro substitution was chosen because of the higher PSS ratio of 83% *Z*-isomer at 523 nm in ACN (Table 12 and 13). Scheme 6 illustrates an overview of the synthesis of gelator **2**, which is based on the modular approach of the previous gelator **1** (Scheme 5).

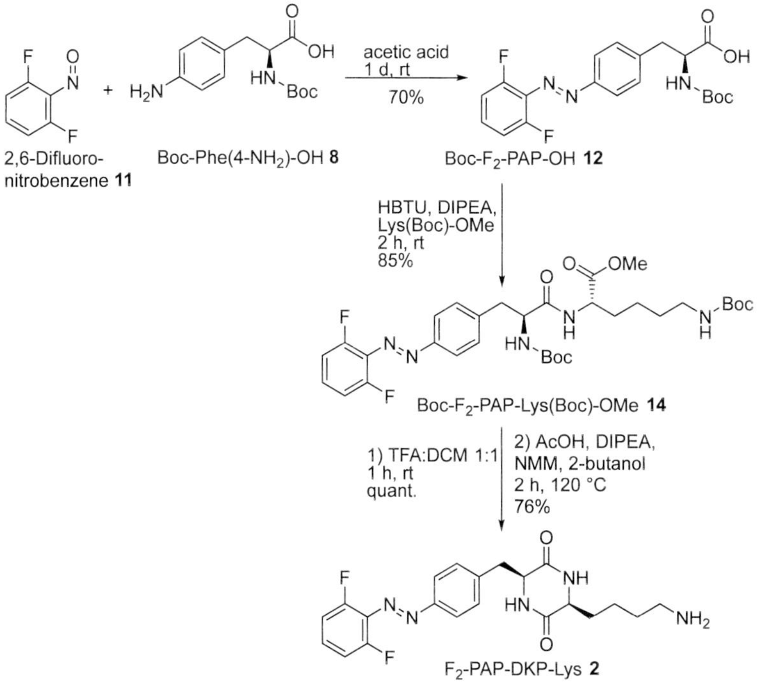

Scheme 6. Synthesis of the molecular gelator F_2-PAP-DKP-Lys 2.

The synthesis route towards gelator **2** starts with the MILLS reaction of 2,6-difluoronitrosobenzene **11** with commercially available Boc-Phe(4-NH_2)-OH **8** yielding photochromic amino acid derivative Boc-F_2-PAP-OH **12**. Then the amide bond is formed with the coupling reagent HBTU and H-Lys(Boc)-OMe·HCl as the coupling substrate. In the last synthesis step, the Boc protecting groups were removed first with a TFA:CH_2Cl_2 1:1 mixture and then dipeptide **14** was cyclized to form the 2,5-diketopiperazine unit. The product F_2-PAP-DKP-Lys **2** was obtained at the multi-gram scale with a yield of 45% over four steps. This route provided us then with adequate material supply for all further experiments with **2**.

3.2.1.1 Photochromic gelator triggered with green light

The photochromism of gelator **2** was analysed in Dulbecco's PBS buffer and DMSO for comparison. The UV-Vis spectra of **2** was measured (Figure 55) and the isosbestic point was determined at λ=224 nm for DMSO (Table 11). The photostationary state (PSS) was reached upon irradiation with 10 W LED at 523 nm and contains 66% *Z*-isomer, as determined by HPLC (Table 11). The back-isomerization with 410 nm leads ultimately to the PSS of 80% of *E*-isomer in DMSO. The PSS in PBS was similar with a slightly lower ratio and overall not as good as expected in comparison to the previous PSS of photochromic amino acids **12** and **13**. Bis-fluorinated azobenzene might be not ideal in terms of photophysical properties, particularly the relatively low discrimination between photoisomers at the respective PSS, but as supramolecular hydrogel formation is a collective behaviour through which gelation and dissipation of the gel can be described in terms of a phase transition, the difference between *E*-to-*Z* ratios under various irradiation regimes appears to be sufficient (Table 24).

In a further experiment, CD spectra of isolated *E* and *Z*-isomers were measured in 13.7 mm NaCl (Figure 26). A shift of the minimum from 215 nm to 209 nm and a reduction of the maxima at 240 nm was observed from *E* to *Z*-isomer. Figures 56 and 26 show no significant difference between the 13.7 mM NaCl or the 137 mM NaF samples. The change from *E*-isomer to *Z*-isomer shifts the minimum from 215 nm to 209 nm and the maximum at 240 nm is reduced to 0 mdeg. In the literature a similar CD spectrum was found for *L*-alanyl-p-hydroxy-*L*-phenylglycine diketopiperazine.[196] This indicates a chiral supramolecular assembly and photomodulation into another supramolecular arrangement, to which the following NOESY spectra gave us even more insight into its assembly (5.3.3.5).

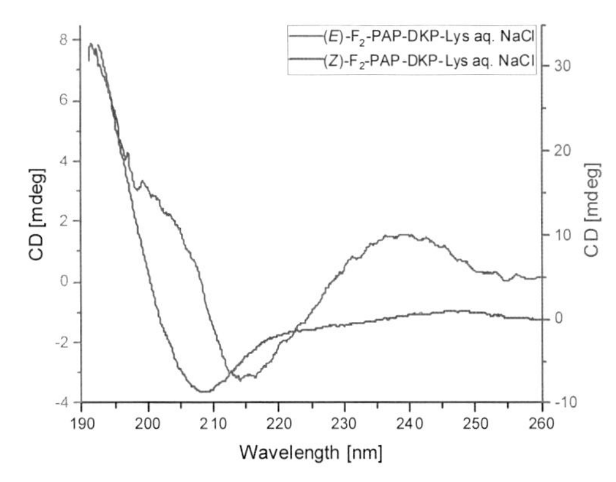

Figure 26. CD spectra of *E* and *Z*-F_2-PAP-DKP-Lys in 13.7 mM NaCl.

For the half-life measurement, the isolated *Z*-isomer was analyzed by HPLC and a half-life of 60 ± 1 day was determined in acetonitrile at rt, whereas in PBS this value was determined to be 109 ± 1 day, both by exponential fit (5.3.2.5). The photophysical properties are in accordance with the literature. Altogether, molecule **2** occurred to be suitable as a photochromic component of the reversible gelation system.[2, 30a]

Gelator **2** forms stable hydrogels at concentrations between 3% to 7% in PBS. The gel-to-sol transition temperatures of 82 ± 8 °C for the 4% hydrogel and 91 ± 3 °C for the 5% hydrogel were determined (Table 14). These $T_{g\text{-}s}$ are significantly higher compared to the $T_{g\text{-}s}$ of 2% gelator **1** in 50 mM NaCl with 70 ± 1 °C or 5% Phe-DKP-Lys·TFA **5** in 20 mM NaOH in PBS with 53 ± 4 °C. Gratifyingly, the stability of gelator **2** at 37 °C in Dulbecco's PBS buffer, as a necessary property for following assays, could be confirmed. The high $T_{g\text{-}s}$ of hydrogels produced from gelator **2** in combination with the efficient photoisomerization with visible (green) light are two major advantages over similar supramolecular systems, including the ones based on gelator **1.** In addition, the 5 mM solution of gelator **2** is stable against a 5 mM concentration of reduced glutathione in PBS for 6 days – a non-significant degradation of the molecule has been observed under these conditions, in contrast to more electron-rich azobenzene-based switches (e.g. *ortho*-methoxy substituted molecules) (5.3.2.4).

3.2.1.2 Supramolecular hydrogels dissipating with green light

The dissipation of hydrogels composed of gelator **2** with green light was investigated afterwards as the macroscopic effect of the photomodulation of the supramolecular assembly. The speed of the gel-to-sol transition is dependent on the concentration and samples of **2** with concentrations of 3% and 4% turned into sol upon irradiation for 30 min. The gel-to-sol transition for 5% hydrogel required 180 min under standard conditions with 10 W LED with 523 nm (Figure 27, 5.3.3). This formulation was chosen for all further experiments because of the good mechanical stability of the hydrogel with a storage modulus of $G'=1.3\times10^5$ Pa (Figure 62).

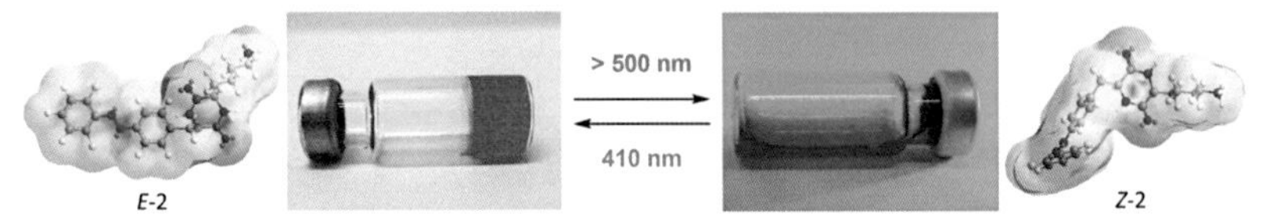

Figure 27. Reversible light-driven gelation of its aqueous solutions and minimal energy conformations (MM2) for *E*- and *Z*- isomers of gelator 2.

Gelator **2** was further characterized by TEM and a fibrous structure of the material was observed (Figure 28). By substitution of NaCl with NaI the same characteristics were observed and pictures with a higher contrast were recorded (Figure 63, Figure 64, Figure 65). The supramolecular structure was qualitatively analyzed by NMR. For these measurements, the PBS buffer (for allegedly interference of phosphate ions in such experiments) was substituted with 8 mg/mL NaCl solution in D_2O, which results in a similar and transparent hydrogel. For a comparison, other common deuterated solvents were screened and in deuterated acetic acid their solubility is good (> 50 g/L) and their peaks in ^1H NMR are narrow (5.3.3.5).

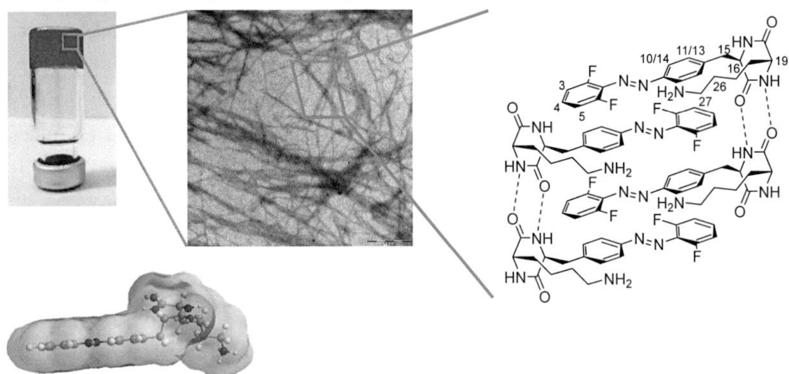

Figure 28. Nanometre-sized fibres visualized in the hydrogel samples and the hypothetical molecular structure of the fibres.[197]

By comparison of the NOESY-NMR of the solution of **2** in NaCl/D_2O with the solution in acetic acid, a qualitative prediction of the supramolecular structure can be made (Figure 28, Figure 69). In aqueous environments, the resonance of aromatic protons of the 2,6-difluro-residue (3, 4, 5) has a cross-signal with those of unsubstituted aromatic residue (10, 14, 11, 13), which is in accordance with the predicted π-π-stacking. The CH_2 group (27) has a cross-signal with all aromatic CH groups (3, 4, 5, 10, 14, 11, 13) and the CH_2 group (26) and the CH groups (16, 19) only with the unsubstituted aromatic part (10, 14, 11, 13), which means the lysine residue likely lays within the hydrophobic core. In addition, a small cross-signal was found and assigned between the CH_2 group (16) and CH_2 group (27). All these cross-signals were not found in the COSY-NMR of the aqueous solution, nor in the NOESY-NMR in acetic acid. The deuterated acetic acid likely inhibits the hydrogen bonding between the 2,5-diketopiperazine units, which results in good solubility and narrow peaks in the ^1H NMR (Figure 69).

As mentioned before, this hypothesis is based on a prediction by Xie *et al.*[172a] and is supposedly a dynamic structure like a secondary β-sheet of peptides or proteins, which is only revealed in aqueous solutions.[182] A prediction about the thickness of the fibril by NMR is not possible by the NOESY spectra alone. But the results and observations do not restrict the potential arrangements to the demonstrated antiparallel stacking with only one molecule upon each other (Figure 28). A higher order of the assembly, like a tertiary structure, is hypothetically possible, including more molecules in one layer or other "ladder" assemblies.[177]

Upon photoisomerization of azobenzene, the stability of the hydrogel decreases and the material is ultimately converted into sol. On the molecular level, the transition to *Z*-isomer destabilises the supramolecular structure comprising hydrogen bonding of 2,5-diketopiperazine and the π-π stacking of the aromatic part by change of the steric configuration of azobenzene. Additionally, the higher dipole moment of *Z*-isomer can weaken the hydrophobic effect, which is a second non-covalent interaction together with π-π stacking. This hypothetic conclusion is based on the increasing storage modulus by addition of NaCl of gelator **1** (Figure 53) and a basic comparison of the weak rigidity of hydrogel prepared from gelator **2** in diH_2O in comparison to PBS buffer (containing 137 mM NaCl).

3.2.1.3 Drug release from hydrogels with green light

The release of encapsulated substances from a photochromic gel is an important application of such soft materials in the area of drug delivery. As hydrogels are, due to their mechanical properties, generally biocompatible even with sensitive cell types and tissues, their loading with therapeutically relevant cargo and its triggered release can find practical applications, e.g. in the form of smart medical implants. The selectivity of such therapeutic materials is defined as the ratio between triggered and non-triggered release of the cargo. Maximizing this value is one of the crucial aims of the development. In the case of non-covalent (physical) encapsulation of the cargo inside the hydrogel matrix, the rate of non-triggered diffusion can indicate the strength of interactions (affinity) between the pharmaceutical cargo and the supramolecular network. To investigate this important parameter, hydrogels (5%) were prepared from gelator **2** with various encapsulated pharmaceutically important drugs: antibiotic ciprofloxacin, anti-inflammatory drug naproxen and diclofenac, anticancer drug actinomycin D and as the protein cytochrome c (Figure 29). The substances were selected to cover possibly diverse modes of action, and the final selection was done according to a reliable detection of the cargo by HPLC.

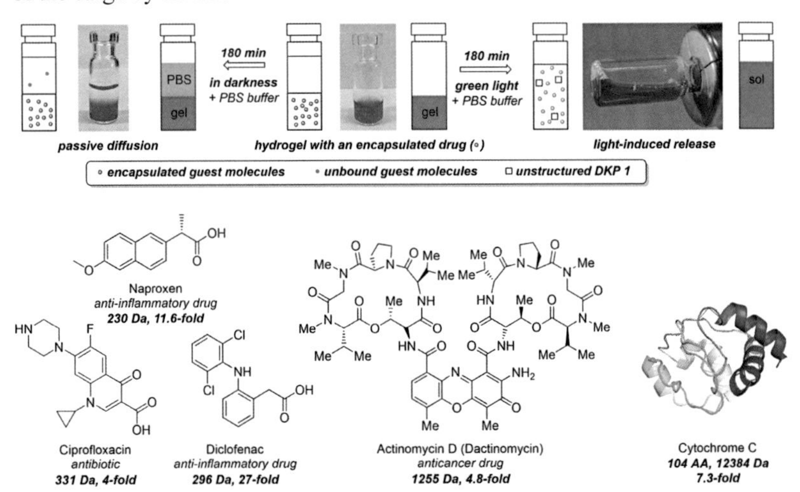

Figure 29. Green light-induced drug release from hydrogels. (top) Schematic representation of cargo encapsulation and release process in the irradiated and non-irradiated gel sample; (bottom) pharmaceuticals and a protein encapsulated as cargo inside hydrogels. The ratio (4-27-fold) between the light-induced release and passive diffusion in darkness is indicated below the structures.

Each hydrogel (500 μL) was covered with an aliquot of PBS buffer (500 μL) and irradiated for 180 min at rt with green light (523 nm, 10 W LED). After 30 min, the aliquot was exchanged with a fresh volume of the PBS buffer. All seven aliquots were weighed and quantified by HPLC. The rate of light-triggered release was compared to the diffusion of the cargo without irradiation over the same period. Their ratio fluctuates heavily for different cargos, as presented in Figure 29 (add. Figure 71 to Figure 77), spanning from a 4-fold difference for ciprofloxacin to a 27-fold one in the case of diclofenac – molecules with similar molar masses.

Thus, the molecular weight of the drug does not seem to be a significant factor for this ratio. The concentration of the drug in the hydrogel was adjusted for an optimal detection by HPLC. The release of peptide cytochrome c was compared with a release of 50 μg vs a 500 μg and a similar ratio of 7.6-fold for the lower loading and 6.3-fold for the higher loading was measured. Therefore, the loading does not lead to a huge difference of ratio between release and diffusion either, at least in the cases investigated. By comparison of the physical parameters of the guest molecules, the dependence of pK_a and the logP of the cargo fit most convincingly as the general explanation. The strength of non-covalent interactions between the cargo and gelator **2** correlates with the decreased leak of the cargo in absence of irradiation. This consideration does not lead to an exact value and instead a tendency can be estimated. Acidic cargos, e.g. the ones with a carboxylic acid, could likely form salt bridges with the basic lysine. Alternatively, more lipophilic drugs could intercalate more strongly within the hydrophobic core of the supramolecular structure of gelator **2**.

To verify this theory, an anchor was designed inspired by the photochromic antibiotic recently demonstrated by Feringa *et al.*[42, 198] We hypothesized that a structural motif comprising the photoswitch itself, or in combination with the DKP residue, will be incorporated into the supramolecular network forming the gel framework, which will reduce the unwanted diffusion. In fact, the diffusion is reduced to the two such designed derivatives: CiproFAzo **15** with photochromic amino acid F_2-PAP-OH **12**, and CiproDKP **16** including the complete gelator **2**. The 9-fold active release/passive diffusion ratio was measured for CiproFAzo **15** derivative, and this increased to 22-fold (background leaking below 5% of the triggered release rate) for CiproDKP **16**. The derivative with the photochromic amino acid supposedly has an additional interaction in the hydrophobic core by π-π stacking. Derivative CiproDKP **16** supposedly has the same interaction by π-π stacking. Additionally, the binding is enhanced because of the hydrogen bridges of the 2,5-diketopiperazine unit of CiproDKP **16** with gelator **2**. This resulted in a significantly increased ratio of light-triggered release vs. diffusion in darkness. Both modified derivatives of ciprofloxacin had a reduced antibacterial activity – from the nanomolar to the micromolar level (Figure 81, Figure 82). A similar reduction of activity was reported previously[42, 198] and can be explained with the increased MW over 500 g/mol, thus breaking the rule-of-five of LIPINSKI,[199] or by steric incompatibility with receptors of the original antibiotic.

In summary, both ciprofloxacin derivatives **15** and **16** have a lower diffusion because of an increased interaction with gelator **2**, although by modifying original ciprofloxacin structure the activity was reduced. This observation illustrates a common problem in photopharmacology, where the covalent integration of a photoswitch into the drug structure usually reduces its activity.[38] Therefore, the light-triggered release of an unmodified drug with high selectivity constitutes the most definite advantage of our system versus classical photopharmacological approaches. However, we demonstrated that the modification of a "leaky" drug with a linker can decrease its diffusion significantly. In such cases, the covalent modification of the drug with a biodegradable linker[200] appended to the gelator motif or its fragment, which is cleavable in the cellular environment, can be a reasonable compromise as it will circumvent the reduced activity problem and keep diffusion low. Especially, neutral or basic pharmaceuticals with a low-MW similar to ciprofloxacin may be prone to diffuse in absence of irradiation.

Despite the rather low ratio between the triggered release and background diffusion, we have used encapsulated non-modified ciprofloxacin as a model antibiotic to demonstrate that we can achieve significant selectivity in photocontrol of a simple biological system with visible light.

3.2.1.4 Photomodulation of bacterial growth with green light

Photomodulation of biological systems by controlled release of physically entrapped bioactive compounds is a complementary strategy[201] to photopharmacology, where the drug is covalently integrated with a photoswitch.[39, 170] As mentioned before, hydrogel **2** can be used for an *in vitro* release of ciprofloxacin by irradiation. Like other fluoroquinolones, ciprofloxacin is active in the nanomolar range against *E. coli*. This activity was reproduced by us with the MIC=24 nM against *E. coli K12 wt.* (Figure 80).[202] In contrast, gelator **2** is not active against bacterial cultures of *E. coli K12 wt.* below millimolar concentrations (with an EC_{50} of 5.4 ± 0.1 mM) (Figure 81). The structure of gelator **2** was designed in a rational manner for applications under physiological conditions. Thus, for the formulation of hydrogel **2** the isotonic PBS buffer was used. Also, the gelator **2** concentration of 5% has been selected as an optimal concentration to achieve the suitable mechanical stability of the material at 37 °C.

In this regard, the hydrogel has to withstand the shaking of the vial inside of a bacterial culture flask for the duration of 24 h (Figure 83). The hydrogel filled with ciprofloxacin in darkness does not release a sufficient amount of the antibiotic. Thus, bacteria can grow as fast as the growth control without the antibiotic. Even though ciprofloxacin leaks slowly by diffusion (Figure 76), its concentration is insufficient to inhibit the growth of *E.coli* effectively (Figure 30).

Upon irradiation, the hydrogel loaded with ciprofloxacin inhibits the growth of the bacteria culture at a similar level compared to the effect of unbound ciprofloxacin at the same final concentration (Figure 30, Figure 84, Figure 85). In this experiment, properties of hydrogel **2** were shown to be suitable for photomodulation of bacterial cultures. This includes the stability and encapsulation of unmodified drugs, and their selective release with green light as well as the low toxicity of gelator **2** alone.

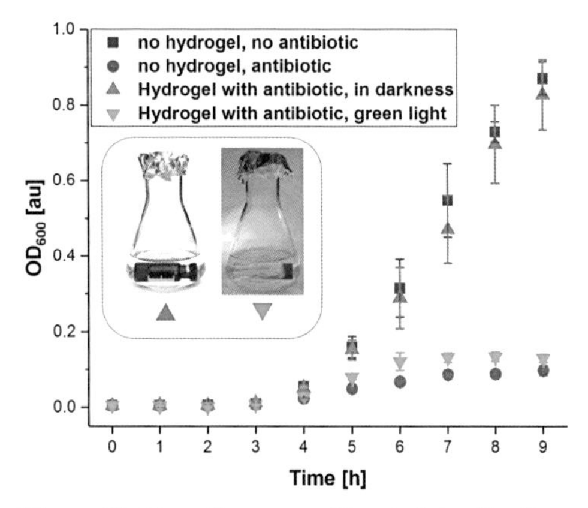

Figure 30. Growth curves of bacterial cell cultures exposed to ciprofloxacin-loaded hydrogel samples. The growth rate in darkness (*blue upward pointing triangles*) is not inhibited, whereas the growth rate under green light irradiation (*green downward pointing triangles*) is strongly inhibited. Control experiments also show the bacterial growth in absence of hydrogels (*dark-blue squares*), and in presence of free ciprofloxacin at a concentration equal to the MIC value (red circles).

3.2.1.5 Cytotoxicity against mammalian cells

2,5-diketopiperazine (DKP), which in our experiments is a scaffold of gelator **2** and other LMWG,[172a, 203] is at the same time a lead structure with broad pharmaceutical importance. For example, the drug Tadalafil is used to treat erectile dysfunction or pulmonary hypertension.[204] Retosiban was developed for the treatment of preterm labour and has a high affinity for the oxytocin receptor.[173, 205] Derivatives of Phenylahistin, such as plinabulin, act as a cytotoxic agent by inhibiting depolymerization of tubulin and are at the moment undergoing clinical trials for anticancer treatment.[206] Considering potential pharmaceutical applications of hydrogels based on **2**, we were as well interested in the cytotoxicity level that it exhibits. Either result could have practical consequences: low toxicity is desired for drug-delivering implants; but high cytotoxicity can also be interesting for photopharmacology applications as long as a sufficient toxicity difference between both mutually switchable photoisomers exists. In this manner, the toxicity of photoswitchable gelator **2** and derivatives CiproFAzo and CiproDKP was tested (5.3.4.8). The rapid colorimetric MTT assay with HeLa cells results in a EC_{50} value of the gelator above 0.5 mM (Figure 87-Figure 89). Toxicity against human cells is lower than ciprofloxacin alone, with an $EC_{50} = 0.43 \pm 0.06$ mM, (considered inactive against mammalian cells), (Figure 86) or that of a cytotoxic anticancer agent doxorubicin with $EC_{50} = 0.6 \pm 0.2$ µM (Figure 97), which we measured in accordance with the literature.[207] Photochomic ciprofloxacin derivative CiproFAzo is up to 10-fold more toxic ($EC_{50} > 50$ µM, Figure 90-Figure 92) than gelator **2** and conjugate CiproDKP with the full gelator scaffold is not toxic at concentrations below $EC_{50} > 2$ mM (Figure 93-Figure 96). Unfortunately, the therapeutic application of gelator **2** and derivatives CiproFAzo and CiproDKP as photopharmacology agents has to be excluded mainly because of the low or insignificant cytotoxicity difference between photoisomers. Nevertheless, if we consider hydrogels based on **2** as drug-delivery systems for the human organism, even with a moderate loading of 2 mg/mL doxorubicin in a 5% hydrogel, the reduction of cell viability would be only depended on the anticancer agent and not the gel (which is less cytotoxic with a 50-fold difference). Thus, the photomodulation of mammalian cells by encapsulation and release of a drug could be performed with hydrogel **2** in a way analogous to the experiment performed on prokaryotic cells.

3.2.1.6 Conclusions F_2-PAP-DKP-Lys

The combination of a photoswitch with the LMWG motif was designed rationally and resulted in a visible-light-triggered supramolecular gelator **2**, which forms hydrogels capable of releasing encapsulated pharmaceuticals upon irradiation with green light. Gelator **2** is stable under physiological conditions and suitable for applications on both prokaryotic and eukaryotic living cells because of the low toxicity and a sufficient therapeutic window. It was shown that the antibiotic ciprofloxacin can be encapsulated efficiently into the supramolecular network of gelator **2**. Upon irradiation with green light, hydrogel **2** dissipates into a fluid (sol) and, within this transformation, the drug is released. With a hydrogel bearing the encapsulated antibiotic ciprofloxacin, the growth of the *E. coli* culture was selectively inhibited by green light irradiation, but not in darkness. The most significant advantage of this system, in contrast to photopharmacology,[42, 198b] is that the encapsulated pharmaceuticals are unmodified and thus their biological activity is not diminished. Although structural modifications of many pharmaceuticals are not required in our system, in some cases a significant leak of the cargo by diffusion can be reduced considerably by covalent (but optionally bio-cleavable) attachment to the gelator molecule, as we have demonstrated with CiproFAzo and CiproDKP.

3.2.2 Supramolecular gelator F_4-PAP-DKP-Lys

In order to improve the photophysical properties of the photochromic unit, two additional fluorine-substituents were introduced in the *ortho* position of the azobenzene of gelator **2**. Reversible isomerization of tetra-*ortho*-fluoro-azobenzenes (TFAB) with green light to *Z*-isomer and with violet light back to *E*-isomer has been reported recently.[2] This state-of-the-art photoswitch can enhance photomodulation of the supramolecular assembly, resulting in improved properties of the hydrogels based on gelator **3**. The bidirectional photostationary states with the *E*/*Z* ratios over 85%, the enhanced thermal stability of *Z*-isomer up to 2 years at rt, and the reversible bidirectional isomerization with visible light are improvements in comparison to an unsubstituted azobenzene, but also, to some extent, to the chromophore of gelator **2**.[2, 30a] These properties can be a significant advantage from a biochemical viewpoint to develop novel tools or methods.

3.2.2.1 Synthesis of the TFAB-containing supramolecular gelator 3

Synthesis of photochromic amino acid Boc-F_4-PAP-OMe **19** is based on a two-step sequence of a NEGISHI coupling with amino acid precursor Boc-Ala(I)-OMe **18** and a subsequent MILLS reaction to form tetra-*ortho*-fluoro-azobenzene **20** (Scheme 7).The synthetic route over the NEGISHI cross-coupling was initially inspired by the procedure of ROSS *et al.* with iodoalanine **18** derived zinc reagents with various aryl halides.[208] The procedure was adapted and the novel artificial amino acid Boc-Phe(3,5-F_2-4-NH_2)OMe **19** was synthesized according to the optimised procedure of Susanne Kirchner (chapter 5.4.1).[209] After MILLS reaction, the photochromic amino acid Boc-F_4-PAP-OMe **20** was obtained, which is a versatile building block[210] that can be used for various coupling reactions.

The UV-Vis spectra and photophysical properties, e.g. isomerization, are in accordance with other TFAB derivatives reported in the literature (Figure 98).[2] After the saponification of methyl ester **20** with aqueous LiOH, enantiopure (*ee*=96% by chiral HPLC, Figure 108) Boc-F_4-PAP-OH **21** was obtained in multigram quantities. The photochromic building block **21** was coupled with H-Lys(Boc)-OMe·HCl with HBTU under standard conditions. After deprotection of the amino group with TFA, dipeptide **22** was cyclized to form the 2,5-diketopiperazine moiety, and gelator F_4-PAP-DKP-Lys **3** was obtained with the overall yield of 30% over 6 steps.

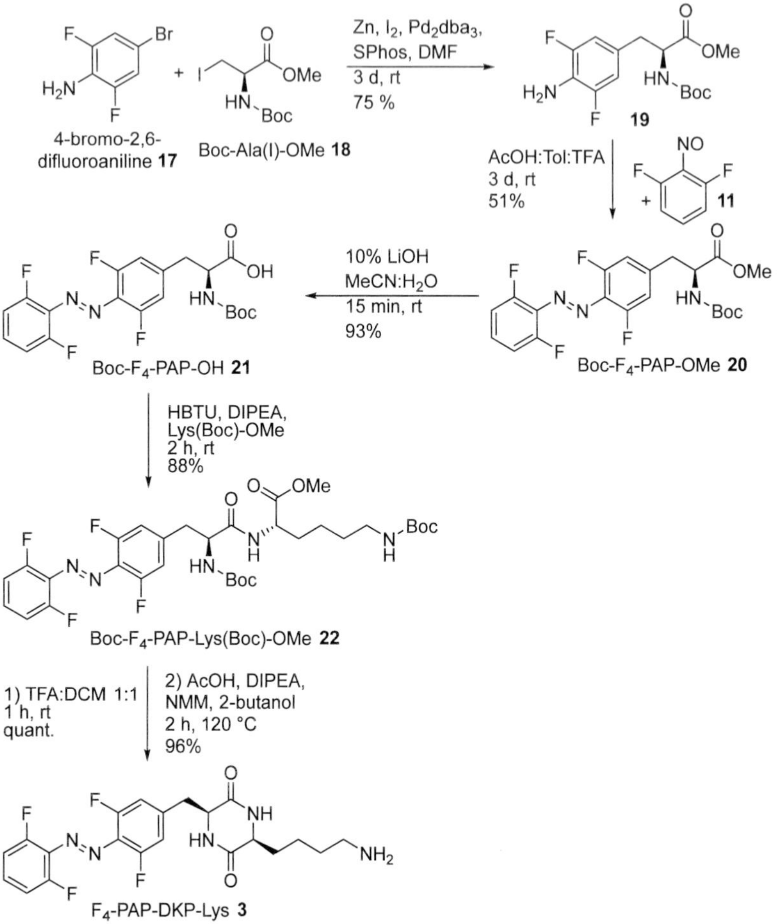

Scheme 7. Synthesis of the supramolecular gelator F_4-PAP-DKP-Lys 3.

3.2.2.2 Photophysical characterization and stability of the gelator

Gelator **3** can isomerize reversibly: with green light (530 nm), the PSS is established with 89% *Z*-isomer; with violet light (410 nm) 77% *E*-isomer in PBS is obtained at the respective PSS. The isomers were separated and stored at rt in the dark without significant thermal back-isomerization, which can be otherwise monitored by HPLC with the detection of the isosbestic point at 254 nm. The lifetime of *Z*-isomer **3** was determined as 70.9 ± 0.6 h at 60 ± 2 °C in ACN (Figure 101) because of the slow thermal isomerization at rt. This lifetime is in accordance with other TFAB derivatives described in the literature.[2] It was observed that the sol of the irradiated hydrogel of gelator **3** (2% in PBS, n=5) did not form a hydrogel for at least 7 days, which demonstrates the long lifetime macroscopically. The chemical stability of gelator **3** in a reducing environment, for example inside of mammalian cells, had to be determined in order to envision *in vivo* applications. To simulate the reducing conditions inside a cell, a solution of 0.5 mM gelator **3** (1.0 eq.) was prepared in 5.0 mM reduced glutathione (10 eq.) in PBS and was stable at rt for 3 d to 94% and after 6 d 92% of the molecules were still unaffected (Figure 102). The combination of these properties with the modular gram-scale synthesis constitutes optimal conditions for a broad spectrum of use in chemical biology and other applications. However, the stability limit of gelator **3** was found in boiling PBS buffer, which degrades the molecule slowly with an estimated half-life of 2.7 h ± 0.2 h (5.4.3). Overall, gelator **3** is stable under physiological conditions in PBS at ambient temperatures and against glutathione, and therefore suitable for preparation of hydrogels and their use under standard conditions like biochemical assays.

3.2.2.3 Gelation properties of gelator 3

Gelator **3** forms stable hydrogels under physiological conditions in PBS buffer in the range of 1.5% to 5% (Table 17). The gel-to-sol transition temperature of 1.7% hydrogel **3** is at moderate 57 ± 3 °C, but can be increased to 82 ± 1 °C with 2% of gelator **3**. The stability assessed by melting is in correlation with rheology measurements. 1.7% hydrogel **3** has a storage modulus of 1.3×10^4 Pa and can regenerate by self-healing within 10 min after 100% deformation reproducibly (Figure 104). Within the range of 1.5% to 2%, hydrogel **3** can reversibly dissipate into fluid (sol) upon irradiation for 30 min with green light at rt (523 nm, 5.4.3.2). With 2% hydrogel **3** the reversible isomerization of the sol was possible (n=5) after irradiation with violet light (410 nm, 3 W) for 60 min and incubation at rt overnight (Figure 31). The gelation properties are a significant improvement in comparison to those of gelator **2**, which required between 4-5% of the gelator to form a photochromic hydrogel. With a comparable stability, a storage modulus of ca. 1×10^4 Pa and same setup the gel-to-sol transition is reduced from 180 min to 30 min with green light. The reversible isomerization of gelator **2** was successfully carried out in solution (Figure 54), but macroscopically it was not possible to convert sol **2** back into a hydrogel **2** solely by light, without thermal equilibration at elevated temperatures. The reversible isomerization on a molecular level and a macroscopic scale was performed with gelator **3** in PBS. By the exchange of the photochromic core for tetra-*ortho*-fluoro-azobenzene, the improved photophysical properties enabled a reversible macroscopic sol to gel transformation (Figure 99 and 5.4.3.2).

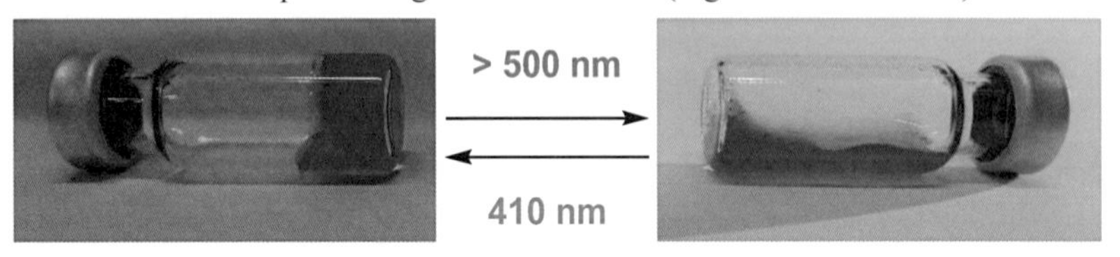

Figure 31. Photochromic 1.7% hydrogels based on 3 reversibly dissipate into fluids upon irradiation with green light.

3.2.2.4 Encapsulation of drugs and their release triggered with green light

The encapsulation of drugs or biomolecules in hydrogels and their retention is a common application, for example, for alginate-based drug delivery systems of pilocarpine in the eye.[211] Encapsulation properties of hydrogels based on gelator **3** were tested with two different pharmaceuticals, ciprofloxacin and plinabulin **23,** using the methodology analogous to the experiments determining release of the encapsulated cargo from hydrogel **2**. The antibiotic ciprofloxacin was chosen for the direct comparison to the previously tested system based on hydrogel **2**.

The anticancer pharmaceutical plinabulin **23** has a similar structure to gelator **3**, which (as we had earlier realized in the case of CiproFAzo and CiproDKP) should decrease the non-triggered diffusion, in contrast to ciprofloxacin. The diffusion rate is specific for each drug and can be estimated. A low pK_a and high logP value is an indication of a low diffusion rate. As expected, the antibiotic ciprofloxacin has a higher diffusion rate of 14% (Figure 107) within 30 min and plinabulin a low diffusion rate of 1% within 30 min (Figure 37). Ciprofloxacin has apparently low affinity with gelator **3** and plinabulin can supposedly intercalate into the supramolecular structure because of the 2,5-diketopiperazine moiety (Figure 32).

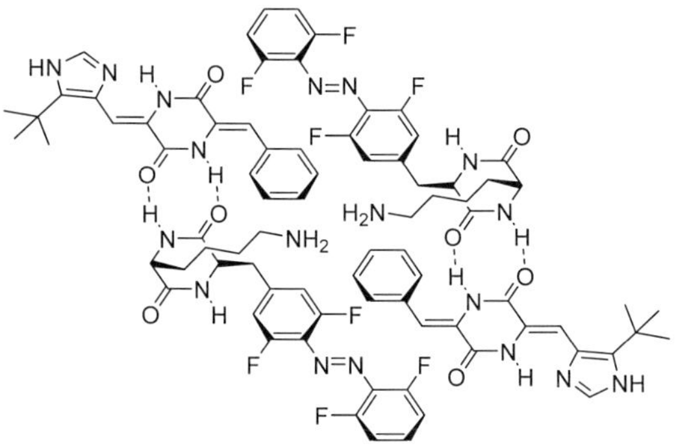

Figure 32. Hypothetic hydrogen bonds between plinabulin 23 and the gelator 3.

The release triggered with green light was measured for ciprofloxacin with a 3.4-fold ratio and for plinabulin with a 87-fold difference (Figure 33, 5.4.3.5). As shown before, plinabulin and gelator **3** can supposedly form mutual hydrogen bonds in solution or with the fibrils of hydrogel **3** between their 2,5-diketopiperazine substructure (Figure 32). This affinity might be too strong to release plinabulin and both molecules remain as a complex even in aqueous solution. The general concept of intercalation of a drug into supramolecular structures was shown, for example for fenbufen, in layered double hydroxides.[212] Despite these considerations, 77% of the unbound and solubilized plinabulin (free from the gelator) was quantified by analytic HPLC (Figure 33) and this recovery of plinabulin proves a release from gelator **3**. For this experiment the calibration curve of the TFA salt of plinabulin was determined and within this calibration a solubility of 7.5 ± 2 µmol/L in 1% DMSO in PBS was measured in order to quantify the released plinabulin. The solubility of plinabulin TFA **23** is higher in contrast to the solubility of plinabulin with < 3 µmol/L from the literature[213] because here the TFA salt and the 1% DMSO increased solubility.

Despite the low solubility of plinabulin TFA **23**, all hydrogels **3** (1.7% F_4-PAP-DKP-Lys·TFA, n=7) with 250 µmol/L plinabulin TFA **23** were transparent and remained transparent overnight. As a comparison, the same concentration of plinabulin TFA **23** forms visible precipitations after c.a. 1 h in the aqueous solution with 1% DMSO. This observation is relevant for the loading capacity of the hydrogel and encapsulation of plinabulin. An encapsulated suspension does not hamper the release of a substance in principle. But in this experiment a suspension would lead to a decreased diffusion that would convey a misleading impression of the encapsulation properties of hydrogel **3**. In addition, the solubility of the anticancer drug plinabulin might play a vital role in its approval as a drug. Hence, recently the solubility issue of plinabulin gave rise to further research, even though the low solubility of a drug does not necessarily lead to pharmacological effect loss.

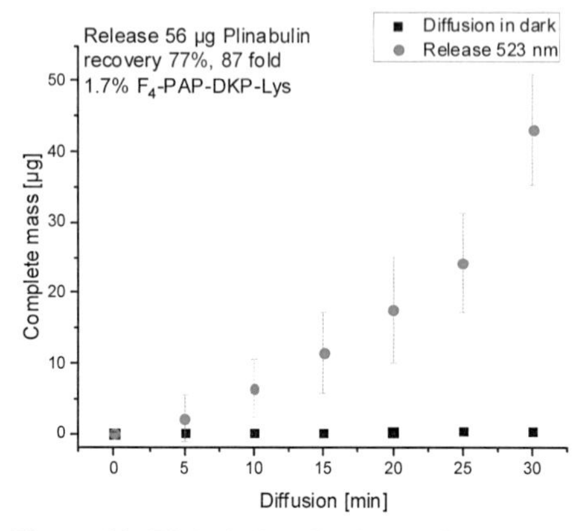

Figure 33. Light-induced release of encapsulated plinabulin (an anticancer drug candidate) from the hydrogel containing 1.7% 3 in comparison to the passive diffusion ("leaking") of the cargo in darkness. Significant decrease of "leaking" is attributed to the structural similarity between the gelator 3 and the cargo.

3.2.2.5 Potential for therapeutic applications

The low solubility of plinabulin was improved with prodrugs designed in the past[213-214] and other strategies like those of different formulations are common for other drugs in human medicine.[215] In the case of the anticancer drug plinabulin, an improved distribution could be achieved with an analogous strategy. The distribution of the drug should have the optimal concentration within the environment of the cancer tissue and a lower concentration in the rest of the body. In photopharmacology, the activity of the pharmaceutical can be switched on at any given time and in any given place, and in the rest of the body the pharmaceutical is not active, which results in fewer or less serious side-effects.

Here the local concentration in the tissue could be increased by a low diffusion, resulting in a continuous gradient that is matched to the clearance rate of the liver. The hysteresis of the diffusion could be adjusted by irradiation with light. One standard procedure for lung cancer is bronchoscopy and can include the removal of a sample from the cancer tissue, which serves mainly for analysis. During bronchoscopy the removal could be combined with an injection of hydrogel containing plinabulin. Hereby plinabulin would be delivered in proximity within the lung cancer, while previously encapsulated into a formulation with gelator **3**. The uptake by cancer cells might be increased with a further optimization of the structure. Regarding common transfection agents, gelator **3** is already positively charged and might be optimized with the substitution e.g. with quaternary amines.

3.3 *o*-Fluoroazobenzenes as photochromic building blocks

The modular syntheses of recently introduced tetra-*ortho*-fluoro azobenzenes[2, 30a] were extended to access achiral derivatives for other applications as hydrogels. As mentioned before, every photoswitch has a variety of properties and advantages which have to fit the desired application. Here, the robust synthesis of tetra-*ortho*-fluoroazobenzene and the improved photophysical properties were highly suited. In general, tetra-*ortho*-fluoro azobenzenes offer a high PSS of 90-95%, a lifetime of two years and reversible isomerization with visible light.[2] Nevertheless, by introduction into systems described below the anticipated properties are expected to have a major impact and provide significant advantage over the design based e.g. on the unmodified azobenzene switch.

3.3.1 Visible light activated catenanes as catalysts

Catenanes belong to the group of mechanically interlocked molecules (MIMs)[216] and are widely investigated because of their applications in guest release and catalysis.[217] They consist, like other MIMs, of two or more mechanically interlocked macrocycles which cannot be dissociated from each other.[218] In a recent publication, an asymmetric organocatalyst based on catenated bifunctional Brønsted acid was introduced. This catenated catalyst shows high stereoselectivities of up to 98% *ee* in the transfer hydrogenation of 2-aryl-quinolines in comparison to the non-interlocked derivative.[217b]

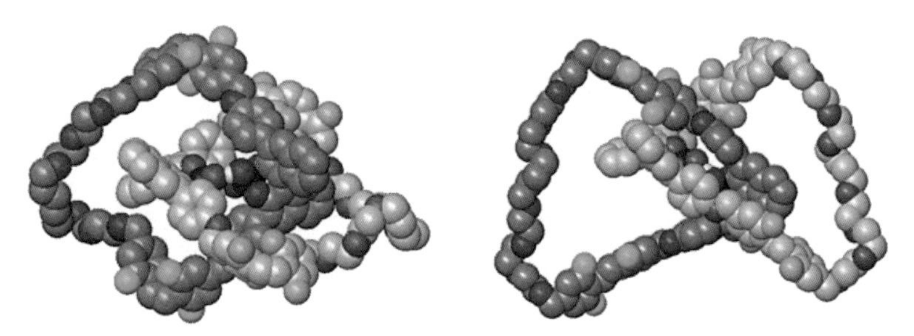

Figure 34. Calculated structures of the catenanes containing azobenzene units, left all *Z*, right all *E*.[217b, 219]

The linker consists of two hexa-ethylene glycols with a carbon-carbon double bond in-between. The but-2-ene-1,4-diol moiety can be replaced with azobenzene to enable a controlled isomerization. Tetra-*ortho*-fluoroazobenzene was chosen in this particular example because of the reversible isomerization with visible light and the high *E*/*Z* ratio. The structure of catenanes, including azobenzene, was calculated to evaluate the rational design and their potential performance as a photomodulated catalyst. In this process, the group of NIEMEYER *et al.* favoured a design with two azobenzenes and one hexa-ethylene glycol linker in between (Figure 34).

3.3.1.1 Synthesis of (*E*)-phenyl-*ortho*-F_4-azobenzyl alcohol

The symmetric phenyl-azobenzyl alcohol **29** with its hydroxy groups is the required building block for catenanes. In the literature, the analogous non-substituted 4,4′-bis(hydroxymethyl)-azobenzene was used for various other applications including the photomodulation of DNA hybridization.[24] The gram-scale synthesis of phenyl-*ortho*-F_4-azobenzyl alcohol **29** is summarized below (Scheme 8). The Rosenmund-von Braun reaction to 4-amino-3,5-difluorobenzonitrile **24** was scaled up in this process to 49 g. Following, the hydrolysis of nitrile **24** with sodium hydroxide and the reduction of benzoic acid **25** with $LiAlH_4$ are robust reactions, which required minor optimizations for the scale-up.

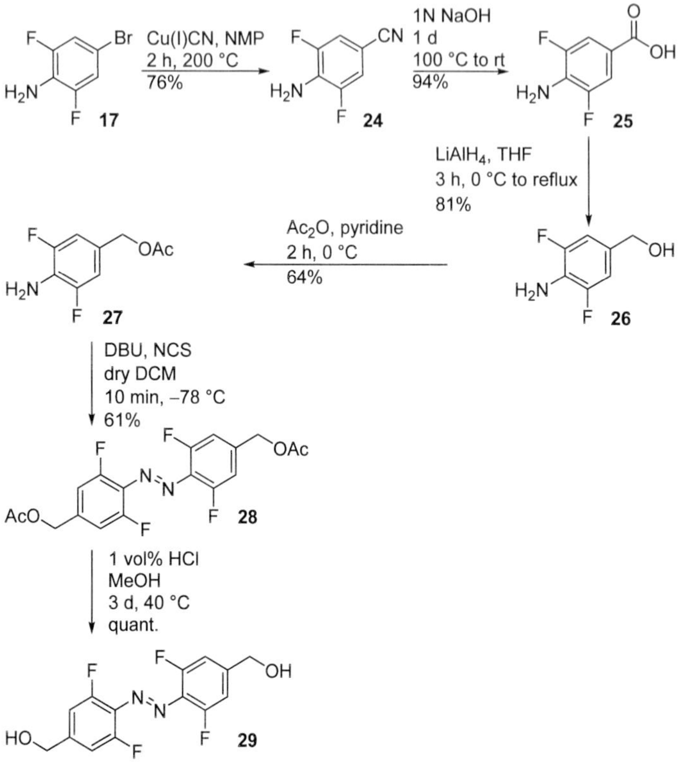

Scheme 8. Synthesis of (*E*)-phenyl-*ortho*-F_4-azobenzyl alcohol.

The direct coupling of (4-amino-3,5-difluorophenyl)methanol **26** *via* Mills reaction or oxidative azo coupling with $KMnO_4$ and $FeSO_4$ is possible, but low yields below 18% and tedious purification limited the synthesis pathway and hereby synthesis of the desired product **29** in gram quantities (chapter 5.5.1).[220] In order to supress side-reactions, the acetyl protection group for the primary alcohol was initially introduced, but the same limitations remained with a laborious workup and a low yield. The oxidative coupling procedure with $KMnO_4$ and $FeSO_4$ yielded 27%. An alternative pathway *via* the Mills reaction yielded 79% of the main coupling product, but side-products were difficult to separate even after several attempts to purify product **28** by recrystallization, silica gel column chromatography or preparative HPLC.

A recently published method of a symmetric azo coupling with DBU and NCS[221] provided a yield of 61% and product **28** can be purified by silica gel column chromatography. Then, an acid-catalyzed deprotection of acetyl protection groups was chosen because of recent reports of decomposition of tetra-*ortho*-fluoro azobenzenes with various bases including NaOH, DIPEA, ET_3N or Bu_3N.[222] The product (*E*)-phenyl-*ortho*-F_4-azobenzyl alcohol **29** was obtained with the yield of 23% over 6 steps in gram quantities and will be investigated further by the NIEMEYER group as a visible light photoswitch in asymmetric organocatalysis.

3.3.2 Photochromic peptides

The photomodulation of peptides can be achieved by different strategies, for example, by the introduction of photochromic amino acids derivatives, like Boc-F_4-PAP-OH **21**. This requires an enantiopure synthesis of the amino acid derivative to obtain a natural analogue of a peptide. A racemic derivative of photochromic amino acid derivative was introduced by genetic encoding using an engineered pyrrolysyl-tRNA/tRNA synthetase pair.[210] This procedure usually requires trial and error mutagenesis and has a low efficiency, and only a low amount of the protein can be extracted after a tedious purification. However, this method enables a photoswitchable control over enzymatic activity[210], fluorescence of GFP[223] and activity of other complex proteins that are not accessible by solid phase synthesis.[175a, 224]

For the chemical synthesis of photochromic peptides[225] a side-chain cross-linked azobenzene[165a-e] or the modification of the backbone are common strategies.[222, 226] The modular synthesis was adapted to access the tetra-*ortho*-fluoroazobenzene derivative[209, 227] of the literature known as 4-(*N*-Fmoc-aminomethylphenylazo)benzoic acid.[228] The photochromic building block was designed for further cooperation with the ULRICH group for the investigation of the membrane interaction of the antibiotic gramicidin S **30** (Figure 35).[229]

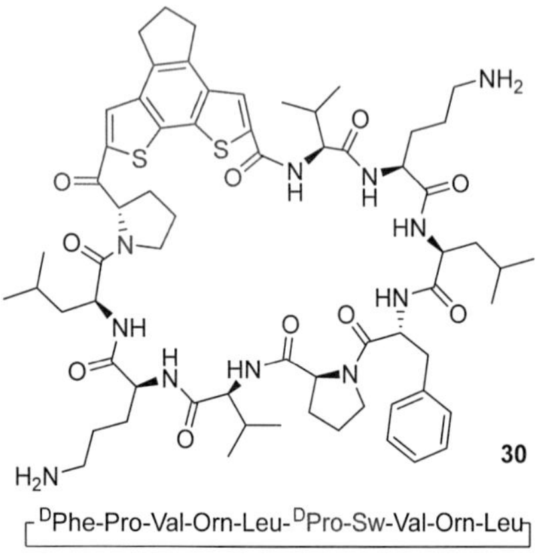

30

^{D}Phe-Pro-Val-Orn-Leu-^{D}Pro-Sw-Val-Orn-Leu

Figure 35. Photoswitchable gramicidin S analogue.[230]

The photocontrol of gramicidin S was achieved with diarylethene by its integration into the cyclic peptide, which can prevent the formation of two hydrogen bonds in its closed form.[231] In this case, the introduction of fluorine atoms enables studies of the interaction between the lipid membrane and peptidomimetic by solid-state ^{19}F NMR techniques.[231c, 232] The reversible photoisomerization of tetra-*ortho*-fluoroazobenzenes can be an advantage because diarylethenes require UV light to switch from the open to the closed form. Hypothetically, the azobenzene moiety can be synthetically modified by meta substitution at one or both sides. In addition, by the introduction of CH_2 groups or other linkers the rigidity of the whole cyclic system can be changed and further fine-tuned.[228c]

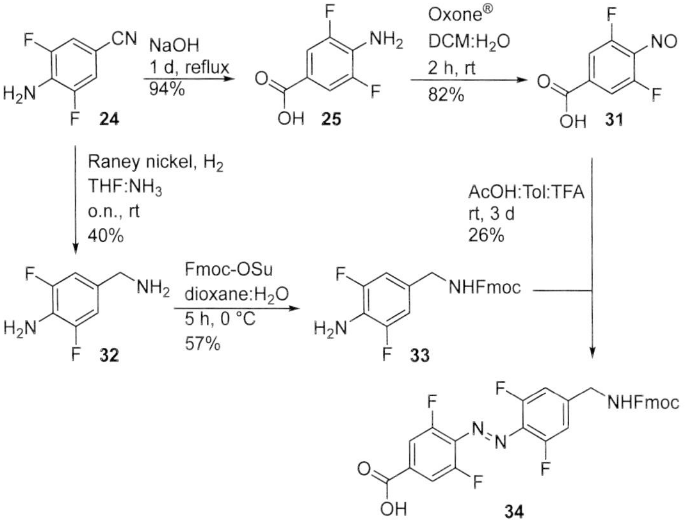

Scheme 9. **Synthesis of an Fmoc-protected achiral α,ω-amino acid containing the TFAB chromophore.**

4-amino-3,5-difluorobenzonitrile **24** can be used for both parts of the synthetic pathway and is accessible by the aforementioned ROSENMUND-VON BRAUN reaction in multi-gram quantities (Scheme 8, Scheme 9). This nitrile **24** is a versatile substrate for both aromatic parts of **34** and was either hydrolyzed into benzoic acid **25** with aqueous NaOH or reduced to benzylamine **32** with Raney nickel. The benzoic acid **25** derivative was oxidized with Oxone® to obtain the nitroso derivative **31**. Benzylamine **32** was protected with Fmoc-OSu at the primary amine. Then the Fmoc protected photochromic amino acid **34** was synthesized *via* the MILLS reaction under modified conditions.[209, 227] This photochromic amino acid **34** was further investigated under common reaction conditions for the removal of the Fmoc protection group. It included 20% piperidine, 2% DBU, 2% morpholine or 2% hydrazine hydrate or 0.1 M TBAF and no deprotection condition was suitable for a standard solid phase protocol. With a further reduction of the concentration to 5.0 eq. of the bases and a reaction time of 20 min the substrate was not completely degraded, but these conditions would be incompatible with standard SPPS (5.5.3). Deprotection conditions were reported for another Fmoc protected tetra-*ortho*-fluoroazobenzene with potassium fluoride (6.5 eq.) in DMF with catalytic amounts of 18-crown-6.[222, 233] The incompatibility of tetra-*ortho*-fluoroazobenzenes with standard Fmoc-based SPPS is a disadvantage for the synthesis of peptides in general. But the use of piperidine can often be circumvented in the synthesis of peptides on the solid support, for example, by introduction of tetra-*ortho*-fluoroazobenzene in the last step followed by the general acidic cleavage.[222, 233] Another possibility is the use of other protection groups like triphenylmethyl chloride or similar derivatives. This might add a minor complication for the synthesis of peptides, but does not diminish the general advantages of tetra-*ortho*-fluoroazobenzenes.

3.3.2.1 Peptide nucleic acid (PNA)

Peptide nucleic acids are artificial analogues of DNA with natural base pairs and consist of the backbone of aminoethylglycyl units instead of the phosphate backbone.[234] PNA-DNA hybrids are extraordinarily stable in comparison with B-DNA and can be synthesized by standard SPPS.[235] Furthermore, PNA-PNA dimers can mimic Watson-Crick complementary duplexes and a seeding of preferred chirality can be induced e.g. with lysine residues.[236] This type of artificial genetic polymer is biocompatible and resistant to nucleases and proteases. It hybridizes sequence-specifically with both complementary DNA or RNA chains and, as such, it has become a promising therapeutic agent for antigen and antisense therapy.[237] Moreover, PNA is also used as a diagnostic tool by fluorescence *in situ* hybridization to detect infectious diseases, for example, against staphylococcus aureus and other bacteria.[238] Inspired by the research conducted by the HILVERT group and their modulation of PNA-DNA hybridization by light,[132] we shared their optimism about using this scaffold for the development of stable photochromic antisense agents as a tool for chemical biology. In the published work, photocontrol of the transcription by T7 RNA polymerase with 360 and 425 nm provided by the unsubstituted azobenzene was reported.[132] In this regard, the chemical stability is compromised under physiological conditions because of the reduction of the azobenzene, for example by liver quinone reductase[239] or by adducts with glutathione.[240] There are even cell penetrating peptide–peptide nucleic acids reported, which use this reduction of azobenzenes for sequence-specific silencing of the luciferase expression *in vitro*.[241] A second limitation is the use of cytotoxic UV light for a photochromic antisense agent, which particularly restricts this application by formation of thymine dimers or other UV-induced DNA damages.[193, 242] In this regard, the high isomerization selectivity and the departure from using UV light are necessary steps towards a practically useful photochromic antisense agent for chemical biology. Therefore, a tetra-*ortho*-fluoroazobenzene functionalized with a CO_2H group in form of a substituted acetate was required and synthesized (Scheme 10).[209, 243]

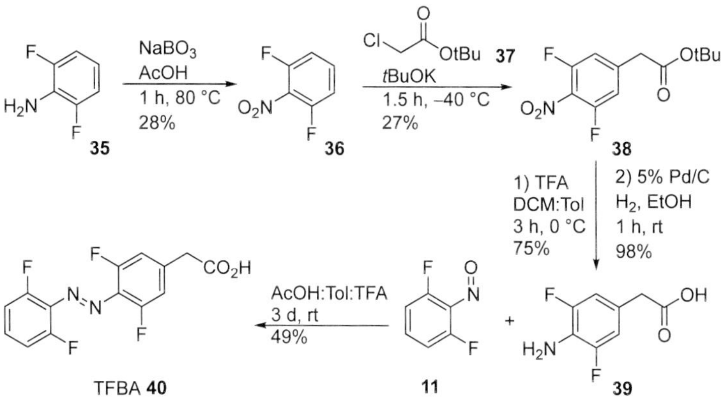

Scheme 10. Synthesis of (*E*)-phenyl-*ortho*-F_4-*p*-azobenzoic acid (TFBA); cf.[132]

2,6-difluoroaniline **35** is commercially available and the amine group can be oxidized with sodium perborate to the nitro group **36**. The key step is the C-C formation by the vicarious nucleophilic substitution between nitroarene **36** and carbanion, which is generated by the deprotonation *tert*-butyl 2-chloroacetate **37** with *t*-BuOK.[244] Subsequently, the *tert*-butyl protection group was removed with TFA and the nitro-group was hydrogenated with palladium on activated charcoal to amine **39**. The azo bond is formed *via* a Mills reaction under optimized conditions with nitrosobenzene **11** and tetra-*ortho*-fluoroazobenzene **40** was obtained in gram quantities. This photochromic building block is part of one strategy in the synthesis and development of the aforementioned photochromic antisense agent based on PNA. But this state-of-the-art photoswitch is not limited to this purpose alone and can be integrated in other structures to achieve reversible isomerization with visible light.

3.3.3 Photochromic building block for metal-organic framework

The concept of metal-organic frameworks – MOFs – was introduced by the Group of HOSKINS and ROBSON in 1989 upon observation of polymeric frameworks of $Cu^{I}(CH_3CN)_4^+$ ligand and 4,4',4",4'"-tetracyanotetraphenylmethane. The crystal structure showed cavities with a van der Waals surface of ca. 700 $Å^3$, which is a common property of MOFs.[245] After the introduction of the first 2D porphyrin network with 20 Å channels[246] and the definition of the term MOF with the introduction of a porous network of $CoC_6H_3(CO_2H_{1/3})_3(NC_5H_5)_2 \cdot 2/3NC_5H_5$[247] the field did grow with a broad variety of MOFs. The principles are based on the coordination between organic ligands and a metal coordination centre.[248] These crystalline materials typically have a high porosity of up to 90% and a high degree of variability of both the organic linker and the inorganic salts.[249] One promising application among others is energy storage, in particular by absorption of hydrogen,[250] or as a separator in lithium-sulfur batteries.[251] Both types of energy storage are potential key technologies to overcome oil dependency in the future.[252]

For a precise control of functions with high spatial and temporal control, light is the trigger of choice. Thus, azobenzenes were introduced into MOFs to enable pore size control[253] or cargo release.[254] Also here fluorinated azobenzenes can offer a higher *E/Z* isomerization rate and improved stability, but in addition, like other fluorinated materials, they can have superhydrophobic properties.[31, 255]

Here we report the synthesis of a photochromic dimethyl 4-aminonaphthalene-2,6-dicarboxylate derivative **46** with a tetra-*ortho*-fluoroazobenzene moiety (Scheme 11). (*E*)-phenyl-*ortho*-F_4-azobenzoic acid **44** can be synthesized from 4-amino-3,5-difluorobenzoic acid **25**, which is available from the aforementioned modular synthesis (Scheme 8). First, nitroso **31** is obtained by the oxidation with Oxone® and then tetra-*ortho*-fluoroazobenzene **44** is synthesized *via* the Mills reaction. The condition for the nitration of naphthalene **41** was optimized to supress double nitration. After the hydrogenation with palladium on carbon, the separation of the side-products is easier, which is a prerequisite for gram-scale synthesis. Various coupling conditions were screened to obtain the amide and the coupling was only possible *via* the highly reactive benzoyl chloride. This photochromic naphthalene derivative **46** and the asymmetric [4,4'-bipyridine]-3,5-dicarbonitrile **47** will be used for the preparations of photochromic MOFs by the Heinke group.

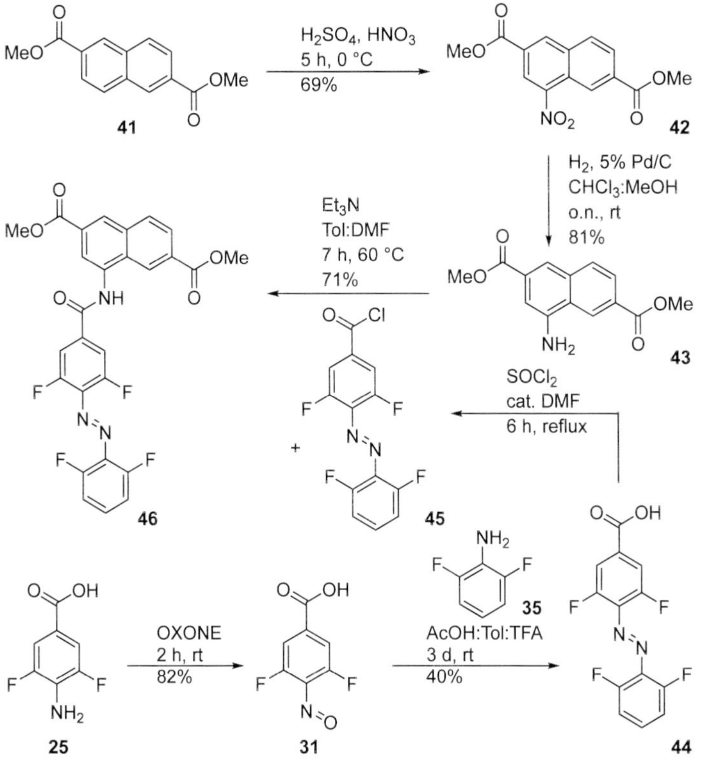

Scheme 11. Synthesis of the photochromic dimethyl 4-aminonaphthalene-2,6-dicarboxylate 46.

3.4 Supramolecular red-light-responsive supergelator

Photomodulation of supramolecular hydrogels with red light is optimal for therapeutic applications and would be a very useful feature in our cyclic dipeptide system. In contrast to other wavelengths, red light can penetrate tissues deeply because of its reduced absorption in comparison to UV, blue, or even green light. Therefore, it is located in the so-called therapeutic window.[116] In this work, the desired isomerization with red light was achieved with the integration of a cyclic dipeptide with tetra-*ortho*-chloroazobenzene. The resulting hydrogel **4** has an enhanced stability and does not dissipate even in a boiling water bath.

3.4.1 Synthesis of tetra-*ortho*-chloroazobenzene and Cl_4-PAP-DKP-Lys_2

The first synthesis of tetra-*ortho*-chloroazobenzenes was reported in 1969,[256] but only since 2013 advantages of an application *in vivo* and the isomerization with red light have been discovered by the WOOLLEY group.[30c] However, the synthesis of tetra-*ortho*-chloroazobenzenes *via* the initially reported diazonium-coupling remained challenging.[195] The demonstrated potential of tetra-*ortho*-chloroazobenzenes initiated the research for other synthesis routes. The most obvious route – MILLS reaction – is not suitable for the preparation of photochromic amino acid derivatives with tetra-*ortho*-chloro substitution due to the disfavourable combination of electron-withdrawing substituents with high steric hindrance.[227]

The emerging alternative coupling procedures leading to tetra-*ortho*-chloroazobenzene were the starting point for screening for a synthetic route towards red-light-responsive hydrogel **4**. Hereby the synthesis of a photochromic amino acid **49** requires an asymmetric azobenzene coupling. The coupling *via* direct palladium(II)-catalysed C–H activation[36] was reproduced, but was not suitable for photochromic amino acids of *L*-phenylalanine-4'-azobenzene with acetyl **48** or Boc-protection group **9** either. The synthesis by *ortho*-lithiation of the aromatic substrate and the followed coupling reaction with an aryldiazonium salt[37] was also tested and a red coloured reaction mixture was obtained. But the detection or isolation of Cl_4-Boc-PAP-OMe **49** was not achieved.[227] In general, the harsh conditions and, in particular, the strong base *n*-BuLi could potentially lead to the racemization of the amino acid and was not further investigated.

However, the recently introduced symmetric azobenzene coupling method with NCS and DBU at −78°C under argon[221] was applied successfully to the synthesis of the aforementioned tetra-*ortho*-F_4-4,4’-bis(hydroxymethyl)azobenzene. As expected, the synthesis was not suitable for the asymmetric photochromic amino acid Cl_4-Boc-PAP-OMe **49**, which would be required for a hydrogel that is structurally analogous to the previously developed hydrogel F_4-PAP-DKP-Lys **3**. But the symmetric photochromic amino acid Cl_4-$(Boc)_2$-PAP-OMe **52** was obtained with a good yield of 70% in multi-gram quantities. In general, the synthesis of a symmetric gelator seems to be a risk, which can compromise the hydrogel properties at first glance and initiate a rational consideration prior to the synthesis. In the literature, there is a report on a multi-responsive organogel with a symmetric phenylalanine-azobenzene subunit within two tripeptides of benzyl-protected aspartic acids.[118] The supramolecular architecture is also broadly related to a para-substituted bis-cysteine hydrogel with an azobenzene as its core that enables the π-π stacking of these azobenzene moieties[122] or the self-assembly of a symmetric dichromonyl compound.[121]

Besides that, the predicted supramolecular arrangement of Tyr-DKP-Lys suggested a point symmetric orientation and did show a similar length for the lysine and the tyrosine residue.[172a] Therefore, two hypothetically fused Tyr-O-O-Tyr residues of the gelator would have a similar length and would fit approximately in the length of the lysine residues and a packing should not be disturbed. In this hypothetical supramolecular arrangement of F_2-PAP-DKP-Lys, the two 2,5-diketopiperazines strands were not parallel, which is also the case for the energy minimised 3D model (MM2) of a symmetric Cl_4-PAP-DKP-Lys_2 with a predicted angle of 69° between the two 2,5-diketopiperazines.

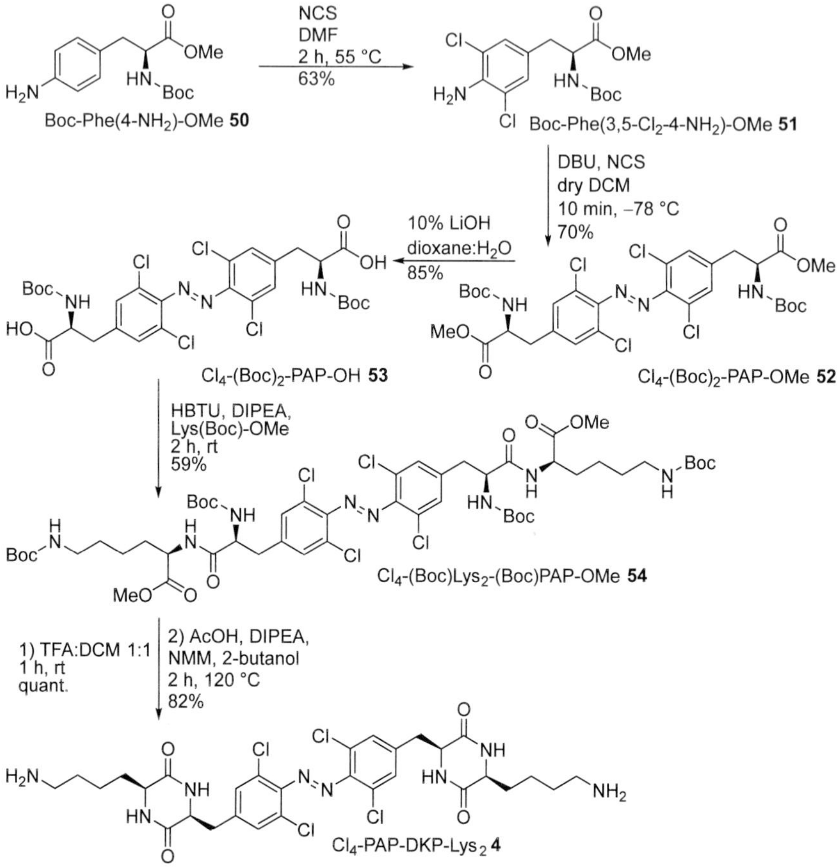

Boc-Phe(4-NH$_2$)-OMe **50**

NCS
DMF
2 h, 55 °C
63%

Boc-Phe(3,5-Cl$_2$-4-NH$_2$)-OMe **51**

DBU, NCS
dry DCM
10 min, −78 °C
70%

Cl$_4$-(Boc)$_2$-PAP-OMe **52**

10% LiOH
dioxane:H$_2$O
85%

Cl$_4$-(Boc)$_2$-PAP-OH **53**

HBTU, DIPEA,
Lys(Boc)-OMe
2 h, rt
59%

Cl$_4$-(Boc)Lys$_2$-(Boc)PAP-OMe **54**

1) TFA:DCM 1:1
1 h, rt
quant.

2) AcOH, DIPEA,
NMM, 2-butanol
2 h, 120 °C
82%

Cl$_4$-PAP-DKP-Lys$_2$ **4**

Scheme 12. Synthesis of the supramolecular gelator Cl$_4$-PAP-DKP-Lys$_2$ 4.

The synthesis of Cl$_4$-PAP-DKP-Lys$_2$ **4** starts with the commercially available Boc-Phe(4-NH$_2$)-OMe **50**, which was synthesized in multi-gram quantities over four steps with a total yield of 32% (chapter 5.6.1). This amino acid was chlorinated in the *ortho*-position with NCS with a yield of 63% of **51**. Then the symmetric tetra-*ortho*-chloroazobenzene **52** was obtained with the DBU/NCS coupling conditions at −78 °C under argon.[221] The resulting methyl esters were deprotected under mild basic conditions with 10% aqueous LiOH in comparison to NaOH, and both resulting free carboxylic acids of **53** were coupled with H-Lys(Boc)-OMe·HCl and HBTU. Subsequently, the protected amine groups with Boc were deprotected and the amino acids were cyclized to form the 2,5-diketopiperazine subunit of **4** (Scheme 12).

3.4.2 Photophysical properties Cl_4-PAP-DKP-Lys_2

Then photophysical properties Cl_4-PAP-DKP-Lys_2 were analyzed regarding physiological conditions. Thus, the isomerization of tetra-*ortho*-chloroazobenzene was measured in aqueous PBS buffer and MeOH (Figure 111-Figure 113). With red light at 623 nm, the azobenzene isomerized to *Z*-isomer and with violet light at 410 nm the azobenzene isomerized back to *E*-isomer. The ratio of *E*/*Z*-isomer was quantified with HPLC at the isosbestic point at 274 nm. With red light (623 nm) irradiation 78% *Z*-isomer (Figure 115) and with violet light (410 nm) 91% *E*-isomer were obtained at the respective photostationary states in water (Table 20). The relative absorption of Cl_4-PAP-DKP-Lys_2 is < 1% at > 600 nm and even for diluted samples of 1 mM gelator **4** a high-power LED at 623 nm with 10 W and 700 lumen is necessary to reach the PSS after 10 min in water. In the organic solvent DMSO, a higher PSS of 86% *Z*-isomer was reached (Figure 115). The reversible isomerization of gelator **4** with red and violet light shows favourable properties for a photoswitch in an aqueous environment, but in order to reach the PSS quickly, a sufficient light source is conceivable for a hydrogel.

The thermal stability of gelator **4** was quantified with HPLC in DMSO and water at rt. In water the lifetime was $t_{1/2}$=60.2 ± 0.4 days, and in DMSO $t_{1/2}$=23.7 ± 0.1 days (Figure 115). This observation is positively surprising, as the previously reported lifetime for unsubstituted tetra-*ortho*-chloroazobenzene was approximately 3.5 hours at ambient conditions.[30c] In respect to the lifetime, these photophysical properties of **4** are suitable not only for a hydrogel, but also for other applications in chemical biology with a duration of several weeks. Regarding the stability of gelator **4**, Dulbecco's PBS buffer can lead to a partial degradation of 33% upon boiling the 1 mmol/L samples and this degradation is particularly pronounced in similarly diluted samples. In another isotonic buffer with 0.9% NaCl or in Ringer's solution, as well as in diH_2O, the degradation upon boiling remained below < 5% by HPLC. Only for hydrogels below 0.5% gelator **4** in PBS do the repeated boiling and re-gelation lead to a macroscopic effect, a less stable hydrogel or a sol. Here the stability of the hydrogel is not compromised in general and instead gelator **4** is remarkably stable in an isotonic buffer like 0.9% NaCl or in Ringer's solution.

3.4.3 Supramolecular hydrogels dissipating with red light

In analogy to other hydrogels, the gelator and aqueous solution have to be heated up for a short time to initiate gelation upon cooling the sol. Here, in the same way gelator **4** formed hydrogels in the range between 0.2% and 2% in PBS, counting as a supergelator because of the efficient gelation below 1%. The stability of hydrogels depends on the concentration and the gel-to-sol transition temperature starts at 66 ± 9 °C for the 0.2% hydrogel. The $T_{g\text{-}s}$ 0.5% hydrogel reaches an impressive 92 ± 2 °C. The hydrogel with 1% gelator **4** in PBS does not turn into a sol at 100 °C and remains a hydrogel after temperature equilibration in the boiling water bath for 5 min, and upon inversion of the vial the hydrogel remains intact. The hydrogels with 1.5% and 2% gelator **4** are stable upon shaking the vial and remain a hydrogel after temperature equilibration at 100 °C. This remarkably high gel-to-sol transition temperature is surprising for gelator **4,** which is based formally on a tetra-peptide MW = 712 g/mol and notably high in comparison with other supramolecular hydrogels based on LMWGs. In the literature, a LMWG with 1.4% based on *D*-sorbitol with MW = 270 g/mol reached the same $T_{g\text{-}s}$ in 6 M KOH aqueous solution.[257] Another amphiphilic tris(urea) hydrogel with three β-D-glucopyranose residues, a MW = 1598 g/mol and 0.8% reached a thermal stability of $T_{g\text{-}s} > 100$ °C.[258] To our best knowledge, a supramolecular photochromic hydrogel based on peptides with a $T_{g\text{-}s} \approx 100$ °C has not been reported so far. Nevertheless, all hydrogels with ≥ 1% of **4** turn into a sol upon boiling the hydrogel itself with a heat gun inside a closed vial. The gelation after the preparation of the hydrogel depends on the concentration and requires up to 7 d for the 0.2%, typically 1 h for the 0.5% hydrogel and ca. 15 min for the ≥ 1% hydrogels.

The gel-to-sol transition with red light was achieved for the 0.2% hydrogel in isotonic Ringer's solution or 0.9% NaCl, after irradiation with a 10 W LED (623 nm) for 30 min at rt. For the 0.5% hydrogel the gel-to-sol transition was possible after irradiation with 120 W (623 nm) within 10 min without cooling the sample. For concentration of ≥ 0.5% gelator **4** the gel-to-sol transition by irradiation with red light at rt has not been possible so far with our irradiation setup. The technical limitation restricted further investigations because a continuous irradiation (ca. 100 W) inside a reflecting chamber, for example, aluminium foil (reflexion ca. 80-85%) at rt, was not acquired. The relative absorption of red light with < 1% is rather low and a combination of sufficient light energy is required, and with the higher reflexion of aluminium foil in comparison with brushed steel the efficiency can increase 3-fold (Figure 114). Also, an increased irradiation time or lower concentration can speed up the isomerization rate. The technical effort or the low absorption of tetra-*ortho*-chloroazobenzenes seems to be a disadvantage at first glance, but in fact, a real life application could be compromised by high absorption in the visible range because daylight would isomerize the photoswitch and only in the dark would the material remain a gel. Ambient daylight did not in fact lead to a collapse of gel **4** to the sol macroscopically.

In conclusion, gelator **4** forms hydrogels in isotonic buffers with a high thermal stability. Tetra-*ortho*-chloroazobenzene enables the reversible isomerization with red and violet light and thereby a gel-to-sol dissipation within the therapeutic window. The supramolecular arrangement might serve as well for other photochromic assemblies for different purposes. Together with enhanced properties and a presumably similar orthogonal biocompatibility, like hydrogel **2,** further applications can be envisioned from regenerative medicine to electronic devices,[3] besides the future vision towards photopharmacology by release of a co-assembled and unmodified drug. In detail, the formulation of the hydrogel as a composite with other gelators, or as a co-assembly, other LMWGs based on cyclic dipeptides, in particular Phe-DKP-Glu **55**, might improve the gel properties or reduce the amount of the photochromic gelator further while retaining similar properties. This formulation could form an injectable vehicle[259] or an implant for a systemic release within a certain timeframe.[260]

3.4.4 Outlook Cl_4-PAP-DKP-Lys_2

The photomodulation of drugs is envisioned in mammals *in vivo* with gelator **4** and drugs as a cargo. A comparable spatial and temporal control of the encapsulated drug might be possible in comparison to photoswitchable drugs of an evolving field of photopharmacology.[39] However, in photopharmacology, the development of photoswitchable derivatives requires a re-design of the drug, which can decrease the activity or the necessity of other impaired like stability or solubility.[42] Every modification of a drug would force the re-evaluation of toxicity measurements and all clinical trials. This necessity is not the case for a release of unmodified drugs from a photochromic hydrogel. In contrast, the encapsulation into a photochromic releasing system like a photochromic hydrogel does not require covalent modifications of the cargo and is therefore an advantage when releasing unmodified drugs.

The design of the gel could be an injectable vehicle[259] or an implant for a systemic release within a certain timeframe.[260] The combination of a photochromic gelator **4** with another hydrogel can reduce the amount and sustain or even enhance desired properties. As an example, Phe-DKP-Glu **55** is a straightforward choice because of the related structure of a cyclized dipeptide and likely a similar supramolecular structure based on 2,5-diketopiperazine. A composite hydrogel of the basic lysine and the acidic glutamic acid derivatives have already been investigated and inspired this idea.[171] The nano- to microgel particles encapsulate the drugs and can be injected into the bloodstream. It is reasonable to investigate nano/microgel formulations of our materials in the near future. For gelators **2** and **3** the treatment of eyes with retina cancer or skin cancer with localized melanomas could be then possible. The encapsulated cargo in gelator **4** can be released by irradiation with red light, which is in the therapeutic window and can penetrate tissues more deeply. Analogously to photopharmacology, the drug can accumulate in the irradiated area and the overall dose in the rest of the body will be lower. Therefore, the negative systemic effects for a patient might result in significantly reduced side effects. An example of formulation for nano- and microgel[261] is a poly(ethylene glycol)-based adhesive with biodegradable gelatin.[262] The composite of gelatin and gelator **4** can lead to biodegradable microgels with a controlled and slow release of diclofenac.[263]

Gelator **4** may also be formulated as an implant under the skin, which can release drugs on demand, for example, strong painkillers, by irradiation with red light. Patients with chronic rheumatoid disease could dose themselves with painkillers or *via* an electronic circuit for the release of the optimal drug concentration can be adjusted. Although hydrogel typically requires a strong light source of 10 W, this could be reduced due a continuous irradiation and the increased timeframe of a couple of days, and only the hysteresis of the drug concentration has to be stabilized in some cases. Throughout the night, the painkiller concentration can be regulated to the necessary minimum, which potentially reduces long term side effects.

3.5 3-Arylazopyridinium derivatives switchable with green light

Arylazopyridinium salts are versatile molecules first described in 1959[264] and developed as dyes, which is a common application for azobenzene derivatives. The cationic and colourful colorants were subjected to several patents as hairdye [265], keratin fibres in general[266], and polyester and acrylic fibres.[267] Recently, another area of medicinal applications was introduced. For example, a patent for azoaryl was granted as a reversible modulation of tubulin inhibitors with the focus on chemotherapy, which can also contain the 3-arylazopyridinium moieties.[268] Another medicinal patent describes the treatment of inflammatory diseases, e.g. arthritis, by the allosteric inhibition of P38 mapk and one of the hits included a 3-arylazopyridinium derivative, which was active in the micromolar range.[269]

In materials science, azopyridinium salt was used to create supramolecular self-assembly conductive nanofibres, which consist of one[270] or two other amphiphilic salts.[271] Based on a related supramolecular architecture an amphoteric azopyridine carboxylic acid was designed based on the π- π stacking of azobenzenes, the function of pyridine as a hydrogen acceptor and carboxylic acid as a hydrogen donor. This molecule is able to self-assemble into macroscopic microfibres that are influenced by heat, pH change and light.[272] The self-assembly characteristic of the molecules was also used for functionalized surfaces for a second-order non-linear optical material[273] and for the formation of a [2]catenane by the strongly π-electron deficient azopyridinium unit.[274]

In a biological context, azopyridinium salts were used as capping of silica mesoporous material, which can release rhodamine B upon cleavage by reductases or esterases. These nanoparticles were also applied for the delivery of cytotoxic camptothecin into HeLa cells through internalization, which reduced cell viability.[275] Azopyridinium salts were used as a fluorescent quencher with the advantage of a fast elimination, hydrophilicity and the absorption from 450 to 600 nm.[276] In addition to these properties, azopyridinium salts can isomerize between *E* and the *Z*-isomer like other azobenzenes.

For the design of fast push-pull azoaryl switches (with thermal relaxation times below 1 s), azopyridinium salt constitutes an interesting building block for the electron deficient part, as it is one of few available designs which makes the resulting azoaryl switch hydrophilic.[277] For biological applications the photoswitch is typically immersed in an aqueous environment, like peptides or other biopolymers. Azobenzenes can be derivatized with hydrophilic substituents, which was previously performed with sulfonate groups.[278] Azopyridinium salts are also more soluble in water than unsubstituted azobenzenes. The phenyl ring of a photo-pharmacophore DENAQ, reported recently as a photochromic ion channel blocker, was isosterically substituted by the pyridinium group (Figure 36). However, the resulting azopyridinium derivative "AFM2-10" did not exhibit the desired biological effect, most likely because of the insufficient lifetime of *Z*-photoisomer (Figure 36).[279] Comparable 4-azopyridinium salts have a lifetime in the nanosecond timescale,[280] which is presumably too fast for meaningful interactions with the targeted ion channel.

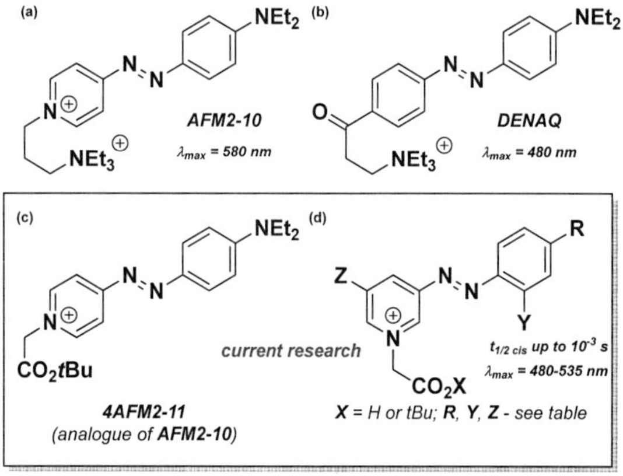

Figure 36. Arylazopyridinium salts as potential molecular photoswitches. (a) AFM2-10, a potential photochromic blocker of voltage-gated potassium (K_V) ion channels based on the 4-arylazopyridinium scaffold;[279] (b) DENAQ, a carbocyclic photochromic blocker of K_V ion channels;[279] (c) 4-substituted pyridinium derivative 55 with short *cis* lifetime (nanoseconds); (d) novel 3-arylazopyridinium photoswitches 56-66 with extended lifetimes of the *cis*-isomer (substituent combinations given in Table 2).

These considerations as well as the versatility of pyridinium azobenzenes led us to investigate hydrophilic photoswitches based on 3-arylazo-, rather than the extremely short-living 4-arylazo-pyridinium salts.[281] Briefly, we identified several compounds switchable with green light and soluble in water. Lifetimes of the respective *Z*-isomers ranged from the microsecond to the millisecond timescale, thus increasing by several orders of magnitude.

These properties may allow more reliable photomodulation of fast biological processes. For an initial screening of the bathochromic shift, 10 derivatives of 3-arylazopyridinium salts **56-66** were prepared (Table 2) and one 4-arylazopyridinium salt **4AFM2-11** as a reference (Figure 36).[282] For the 4-arylazopyridinium analogue, the absorption maximum of 580 nm was determined, which is identical to

AFM2-10 (Figure 36) with the same chromophore. The lifetime of these 4-substituted derivatives was reported in the millisecond timescale for aromatic ethers,[277] for hydroxy groups in the microsecond timescale[283] and for primary or tertiary amines in the nanosecond timescale.[280] The bathochromic shift of these aromatic ethers is significantly lower with a maximum at 415 nm in comparison with other 4-arylazopyridinium derivatives, e.g. with amines.[277] This bathochromic shift with the same maximum at 415 nm can only be achieved with a diethylamino substitution of an azobenzene without the azopyridinium moiety, both dissolved in ethanol.[284]

Hypothetically, these modifications, e.g. with hydroxy groups, can slow down the electron flow, which is common in para-substituted push-pull systems.[277] This leads to a rapid thermal relaxation and can be interrupted by breaking the linear conjugation with a *meta*-substituted pyridinium moiety. However, 3-arylazopyridinium should retain a comparable bathochromic shift caused by the presence of a charged heterocycle.

The bathochromic shift of the 3-arylazopyridinium derivatives in the visible range is listed for different pH values of 2.1, 7 and 12 (Table 2) and is within a range of 365 up to 530 nm. The absorption maxima of the 3-arylazopyridinium derivatives with para substituted diethylamino **56**, hydroxy **58** and acetyl-piperazine (AcPip) **60** are shifted to a moderate 15-25 nm with a bromine substitution **57**, **59**, **61** and **63** in position 5 in comparison without this halogen substitution. In the literature, a comparable shift was described for an analogue azobenzene with diethylamino substitution.[284] The change of thermal isomerization rates by substitution of azobenzenes in correlation with their bathochromic shift has been intensively studied,[285] including push-pull azobenzenes.[286] However, for azopyridinium derivatives no consistent literature reports on the substitution influence on the lifetime of *Z*-isomer nor on the bathochromic shift have been published so far.

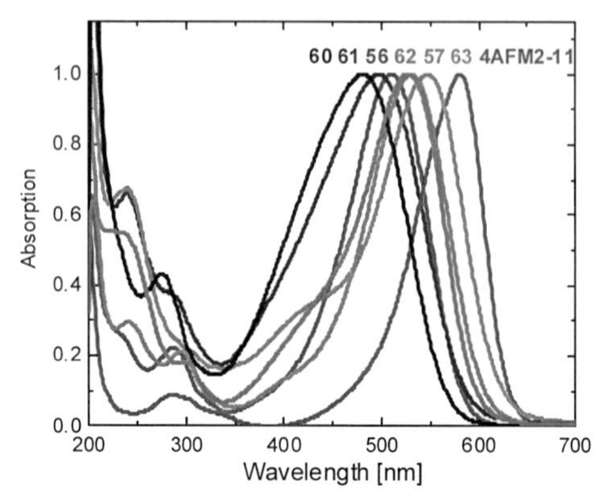

Figure 37. Absorption spectra of molecules 60, 61, 56, 62, 57, 63 and 4AFM2-11 (from left to right) in PBS buffer, pH 7.4; All spectra are normalized to the same peak amplitude in the visible range.

Table 2. Substitution pattern and photophysical properties of the newly synthesized 3-(arylazo)-pyridinium salts at pH 2.1, 7 and 12; Values in the right column refer to the absorption maxima (in nm) of the UV-Vis bands.

Compound	R	Y	Z	pH 2.1 / pH 7 / pH 12
56[a]	-NEt_2	-H	-H	385 / 510 / 510
57[a]	-NEt_2	-H	-Br	480 / 530 / 400
58	-OH	-H	-H	370 / 480 / 480
59	-OH	-H	-Br	380 / 505 / 415
60	-AcPip[b]	-H	-H	485 / 480 / 380+500
61	-AcPip[b]	-H	-Br	495 / 495 / 390
62[a]	-NMe_2	-NMe_2	-H	430+510 / 530 / 530
63	-NMe_2	-NMe_2	-Br	430+515 / 535 / ns[c]
64	-OH	-OH	-H	390 / 405 / 420+505
65	-$NHCOCH_3$	-H	-H	365 / 365 / 365

[a] For compounds 56, 57, and 62: X = H; for the remaining compounds: X = *t*Bu. [b] "AcPip"–(4'-*N*-acetylpiperazinyl). [c] ns – not soluble. Aqueous buffers used were 0.1 M citric acid (pH 2.1), 50 mM NaH_2PO_4/Na_2HPO_4 (pH 7.0), 50 mM Na_2HPO_4/Na_3PO_4 (pH 12.0).

Derivatives with an electron-rich carbocyclic aren have an absorption maximum at 480 nm or higher. By an anilide substitution **65**, the absorption maximum is shifted to the UV region and compromises the visible light isomerization. The influence of the deprotection of the *tert*-butyl protected carboxylic acid was compared to the protected derivative, e.g. for AcPip derivatives **60**, but no significant change in the UV-Vis spectra was observed. On the other hand, the pH of the aqueous solution did influence the UV-Vis spectra significantly, but in a predictable manner depending on the protonation or deprotonation of the functional groups. For a detailed photophysical characterization six compounds with various substituents were chosen and the synthesis of AcPip derivative **60** was optimized. Additionally, the synthesis of oligopeptides **67** and **68** and the functionalisation with aspartic acid at the piperazine side **60D** were shown. Hydroxy derivatives **58**, **59** and **64** were not selected because of the limited potential for derivatization and the significant photophysical changes by derivatization to aromatic ethers.[277]

3.5.1 Synthesis of 3-arylazopyridinium salts

3-arylazopyridinium salts were synthesized partly by common azo coupling with sodium nitrite in acidic aqueous solution at 0 °C. The yield and versatility of the scope were increased by the preparation of stable diazonium salts with tosylates **70**[287] as the anion (Scheme 13). The preparation of diazonium tetrafluoroborates is also possible,[288] but they are not as safe as a dry substance in comparison with stable diazonium tosylates. Presumably, only the meta-substituted diazonium tosylates of pyridine are stable and can be stored under argon at −20 °C for several days. Azopyridine was coupled with *tert*-butyl bromoacetate to introduce a protected carboxylic acid in multi-gram quantities. In the same structure, it was also possible to introduce a Fmoc-protected aspartic acid at the piperazine **60D** as an example of a further derivatization of this substructure of **60** (Figure 17).

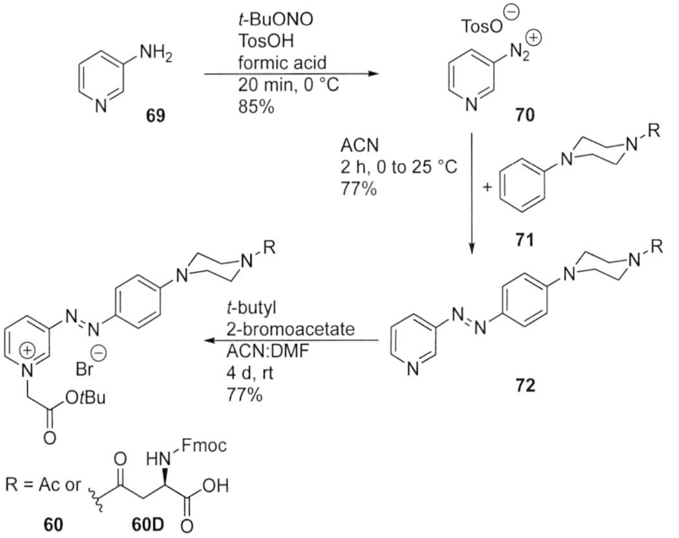

Scheme 13. Synthesis of 3-arylazopyridinium derivatives 60 and 60D.

3.5.2 Physical properties of *Z*-isomers of selected 3-arylazopyridinium salts

The lifetimes of *Z*-isomers were measured with a nanosecond flash photolysis system in PBS buffer with a pH of 7.4. Azobenzene was isomerized with a 6 ns Nd:YAG laser pulse at 532 nm. The influence of aqueous buffer was compared to a mixture of PBS:glycerol 1:4, which increased the lifetime significantly (Table 3).

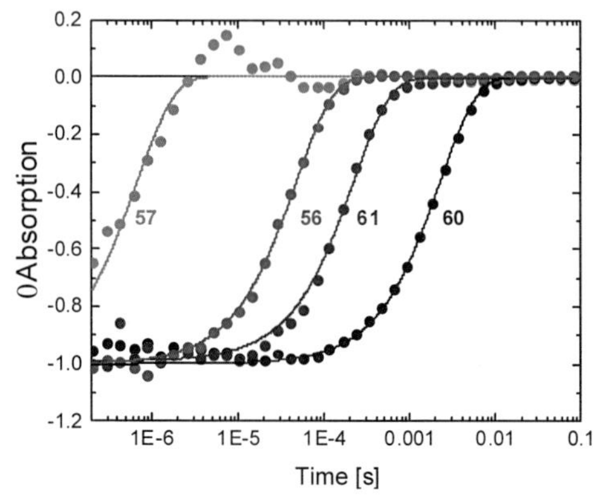

Figure 38. *Z-E* relaxation kinetics of molecules 56 and 60 (non-brominated) as well as 57 and 61 (brominated) in PBS buffer, pH 7.4, recorded after light-induced *trans-cis* isomerization. (X = CO_2H for 56 and 57; X = tBu for 60 and 61). Lines represent mono-exponential fits of the data.

Acyclic amine derivative **56** has a higher bathochromic shift in comparison with cyclic derivatives **60**, but only cyclic derivative **5** exhibited a lifetime of its *Z*-isomer in the millisecond range (Table 3). By bromo-substitution the lifetimes of both derivatives were reduced significantly by one order of magnitude, and with a moderate bathochromic shift of 15-25 nm. This might be an unfavourable trade-off for a photoswitch for fast biological processes. Nevertheless, the introduction of bromine is synthetically not challenging and can be a strategy to tune the thermal isomerization of bis-azo photoswitches for a temporal resolution.[289]

Table 3. Lifetimes of the *E*-isomers of molecules 56, 57, 60 and 61 in PBS, pH 7.4 and 75%/25% (vol./vol.) glycerol/PBS.

Compound	τ (s) (in PBS)	τ (s) (in glycerol/PBS)
56	$(4.4 \pm 0.3)\times10^{-5}$	$(4.87 \pm 0.04)\times10^{-4}$
57	$(1.1 \pm 0.2)\times10^{-6}$	$(1.5 \pm 0.1)\times10^{-5}$
60	$(2.47 \pm 0.05)\times10^{-3}$	$(9.4 \pm 0.3)\times10^{-3}$
61	$(2.2 \pm 0.1)\times10^{-4}$	$(4.5 \pm 0.2)\times10^{-4}$

Most biological and some supramolecular systems require the accurate control of pH and functional groups can be deprotonated or protonated under the respective conditions. Therefore, the lifetime of *Z*-AcPip-derivative **60** was measured at different pH values and fitted to the Henderson-Hasselbalch equation with a $pK_a = 7.4 \pm 0.1$. At a low pH of 2.7 the lifetime is significantly reduced by two orders of magnitude and at a neutral and basic pH the lifetime remains in the range of milliseconds (Table 4 and Figure 121).

Table 4, Lifetimes of the *Z*-isomer of compound 60 at different pH values

pH	τ (s)
2.7	$(5.7 \pm 0.47)\times10^{-5}$
5.4	$(1.1 \pm 0.022)\times10^{-4}$
6.2	$(5.6 \pm 0.28)\times10^{-4}$
7.4	$(2.0 \pm 0.01)\times10^{-3}$
10.2	$(3.5 \pm 0.032)\times10^{-3}$

The lifetime of other derivatives, for example, that of compound **64**, was only determined under basic conditions (pH = 12) with a lifetime of $\tau = (3.7 \pm 0.1)\times10^{-5}$ s and measurements at other pHs did not lead to conclusive results. In another example, the photoisomerization kinetics of compound **4AFM2-11** in PBS leads to a very small absorption difference of $\Delta A < 0.01$, which hinders the accurate determination of the lifetime. Overall, not all lifetimes can be determined with the experimental setup described. This can also occur with extremely short lifetimes below nanoseconds, which is outside of the experimental time window, e.g. the analogue-to-digital converter. This observation is in line with the reported lifetime of *Z*-isomer of AFM2-10, which was too short for the intended use as a photoswitchable ion channel blocker.[12] An exchange of the solvent with a viscous solvent like ethylene glycol can slow down thermal isomerization[290], and this was applied to compounds **56, 57, 60** and **61** (Table 3).

Additionally, the polarity of the solvent can also influence its thermal isomerization[291] and can be used to enhance the lifetime above the detection limit. However, the use of 75% or 95% glycerol, ACN or dimethoxyethane did not lead to any sign of photoinduced isomerization of compound **4AFM2-11**. Another explanation can be similar UV-Vis spectra of *E*- and *Z*-isomer, which overlap at the chosen wavelength of the monochromator, e.g. at 480 nm. In the literature and in the present setup, the wavelength of the laser for isomerization is green at $\lambda = 532$ nm[289] or as an alternative UV light at $\lambda = 355$ nm.[277, 280, 283]

In comparison with tetra-methoxy or tetra-chloro-azobenzenes,[30c] which can be isomerized with red light, several azopyridinium salts absorb light above 600 nm sufficiently, e.g. compounds **56** and **60**. For example, an LED with a wavelength of 623 nm was chosen for gelator **4** with the tetra-chloro-azobenzene substructure at the edge of the absorption maxima with a relative absorption below 1% and this resulted in the highest ratio of *Z*:*E*-isomer (Figure 115). In the same way, azopyridinium salts can isomerize hypothetically with wavelengths above 600 nm in the therapeutic window. Due to the technical restrictions of a Nd:YAG laser the range between 600-700 nm was not accessible.

3.5.3 Photo-oxidation of compounds 62 and 63

Compounds **62** and **63** have two electron-donating substituents on the aryl substituent, respectively (Figure 39). Both compounds have absorbance spectra with two maxima in the visible range, for example, for compound **62** at 430 and 530 nm (Figure 40, Figure 128). By increasing the pH, the band at 430 nm decreases and the band at 530 nm increases concomitantly, which may indicate the deprotonation of a *bis*-dimethylamine substituent. Its absorption can be fitted to the Henderson-Hasselbalch relation and a $pK_a = 5.5 \pm 0.1$ for compound **62** and a $pK_a = 4.9 \pm 0.1$ of compound **63** was determined (Figure 40).

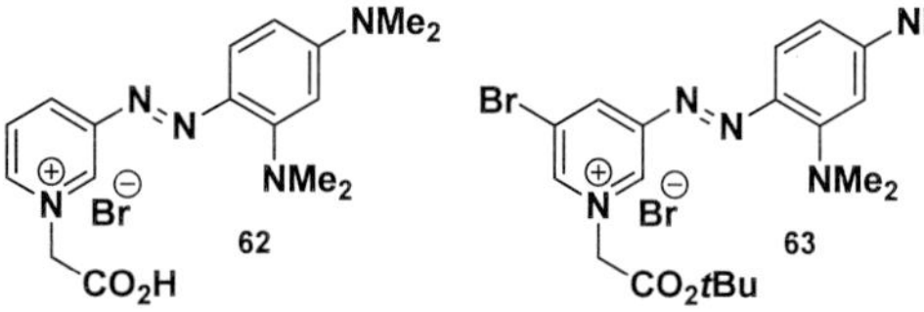

Figure 39. Structures of pyridinium salts 62 and 63 with *bis*-dimethylamino-substituted carbocycles.

An opposite effect was reported recently from the group of WOOLLEY: *ortho*-substituted azobenzenes can form azonium ion at the neutral pH range[292] and this is a significantly higher pH in comparison with the usual pK_a value from 1.5-3.5. But, in contrast to these findings, the absorption maximum at a low pH is blue-shifted in comparison with that at high pH values. Another unforeseen property of compounds **62** and **63** is the irreversible, light-induced structural change that occurs at pH 7.4 (Figure 40). Upon irradiation with laser at 532 nm, intermediates **62X** and **63X** are formed with absorption maxima at 450 nm and 501 nm, respectively. In comparison with the substrate, the absorption is slightly blue-shifted for both compounds and, due to the bromo-substitution, compound **63** is slightly red-shifted by 21 nm (Figure 119). To reach the complete formation of compounds **62X** and **63X** it takes up to 15 h in the dark and the reaction does not take place in the absence of dioxygen, e.g. in degassed PBS buffer. These two molecules can isomerize with lifetimes of *Z*-form of **62X** $\tau = (4.3 \pm 0.1)\times10^{-4}$ s and for **63X** $\tau = (1.33 \pm 0.09)\times10^{-5}$ s (Figure 130).

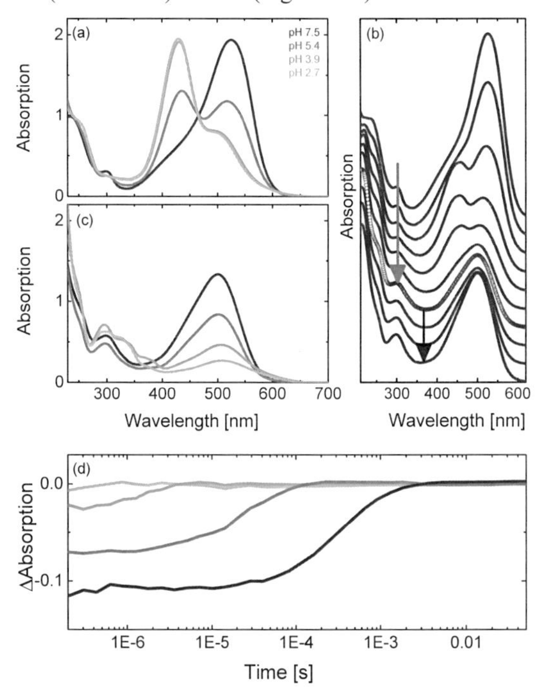

Figure 40. Photophysical characteristics of pyridinium salt 62. a) Absorption spectra at four different pH values, colour-coded as indicated in the panel; b) In air-saturated buffer, light irradiation leads to an irreversible transformation to a new compound 62X, as indicated by the changing UV-Vis spectra at pH 7.4; c) absorption spectra of 62X (*trans*-isomer); pH values are colour-coded as in panel a; d) *Trans*-

cis **photoisomerization kinetics of 62X, at various pH values (colour coded as in panel a), revealing that 62X exhibits reversible photoisomerization.**

The mass spectrum of an irradiated sample of compound **62** with an LED with 523 nm in PBS buffer indicates a full consumption of the substrate and a clean formation of compound **62X** with a molecular mass increased by 14 Da (Figure 131) . A further analysis by NMR did not reveal the oxidation site and attempts to grow crystals for X-ray analysis failed. In the literature, the 1st and 2nd reductions of a bi-pyridinium salt lead to a comparable change in the UV-Vis spectra, although the electrochemical reduction and re-oxidation process was reversible.[293]

3.5.4 Comparison of 3-arylazopyridinium salts and azobenzene in oligopeptides

The photomodulation of biopolymers with UV light is one common application of azobenzenes besides numerous others to investigate the dynamic of the protein folding.[294] Isomerization with unsubstituted azobenzenes requires UV light that can induce artefacts in an *in vitro* experiment.[193] This might not be suitable for applications with tissues or even in living organisms because of the poor penetration of UV light through the skin or thicker cell layers.[194] Additionally, the nonpolar nature of unsubstituted azobenzene can also limit its use due to aggregation in aqueous solution. With 3-arylazopyridinium salts both limitations of hydrophobicity and UV light can be overcome and, additionally, *Z*-isomer undergoes a fast thermal back-switching in the millisecond range, which can be an advantage for the investigation of fast biological processes.

The applicability of 3-arylazopyridinium salts as a photochromic peptide component was proven with a solid-phase synthesis of a hydrophobic tetrapeptide also containing alanine and phenylalanine. The unsubstituted azobenzene was used as the negative control in comparison with compound **60**. Both, the heterocyclic ring of pyridinium with the positively charged nitrogen and the piperazine moiety, enhanced the overall hydrophilicity of the peptide. Piperazine is a frequently used moiety in drug design[295] to achieve, for example, better solubility, e.g. ciprofloxacin, to form a hydrochloride salt. It can also constitute a rigid linker[296] for further derivatization.

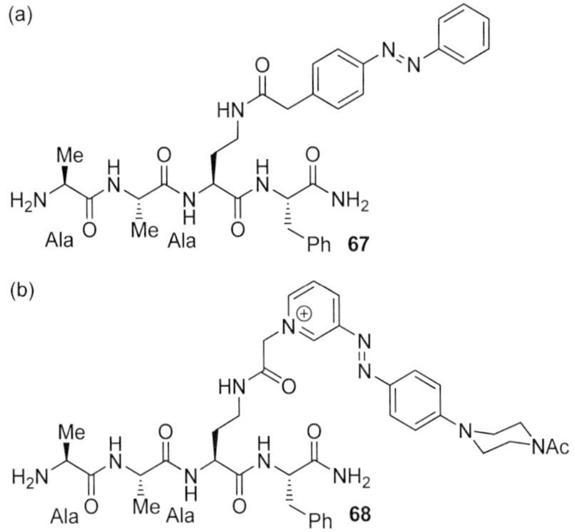

Figure 41. Structures of tetrapeptides containing: (a) an azobenzene photoswitch 67 (b) a 3-arylazopyridinium photoswitch 68.

In this proof-of-principle, the nonpolar nature of *E*-azobenzene together with the hydrophobic residues of other amino acids should render oligopeptide **67** (Figure 41**b**) insoluble, or to aggregate in water at least. The same oligopeptide with compound **60** (compound **68**, Figure 41**a**) should exhibit enhanced water solubility under the same conditions due to permanently charged pyridinium salt. Both oligopeptides were dissolved in diluted (≤50 mM) aqueous PBS buffer (pH 7.4) and passed through a 3 kDa size-exclusion filter.

The remaining retention of the filtrate was assessed with absorption of the UV-Vis spectra. As expected, the differences between both oligopeptides are significant and less than 30% at 20-50 mM of compound **67** and over 95% of compound **67** were retained by the size exclusion filter (Figure 42). The molar masses of 755 Da for **68** and 628 Da for **68** are significantly below the cut-off threshold of the filter (3000 Da), therefore the observed retention should be only correlated to the aggregation tendency of the given molecule in water. In a longer peptide, this effect might be reduced due to the increased influence of other side chains of the amino acids and can depend, e.g. on the orientation of the photoswitch on the secondary structure of the peptides. Nevertheless, the influence can be significant and highly increase water solubility of a designed photochromic peptide.

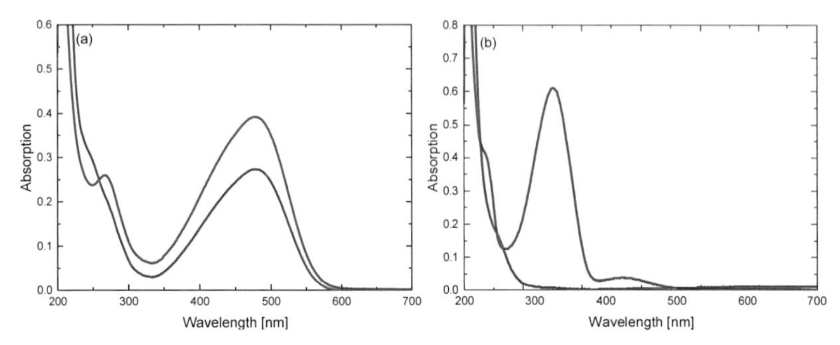

Figure 42. Absorption spectra of solutions (50 μM in PBS, pH 7.4) of tetrapeptides (a) 67 and (b) 68 before (red line) and after (blue line) size-exclusion spin filtration.

In summary, the investigated 3-arylazopyridinium salts are green-light-triggered photoswitches with good solubility in water and polar solvents, and with tuneable thermal isomerization of *Z*-isomer in the range from milliseconds to microseconds. Depending on substituents on the carbocyclic ring, the photoisomerization frequency can vary from the UV to the green light. The lifetime of *Z*-isomer decreases in general with an increasing bathochromic shift with electron-donating substituents. The influence of this effect is not as pronounced in their 4-arylazopyridinium analogues because of the asymmetry of their electron distribution.

Their use in oligopeptides can reduce hydrophobicity significantly and prevent aggregation in comparison with an unsubstituted carboxylic azobenzene. The synthesis of the preferred structure of compound **60** with the piperazine moiety was upscaled to a gram-scale, and further amide coupling with amino acids at the piperazine and pyridinium sides was both demonstrated. In this respect, 3-arylazopyridinium salts are anticipated for photochomic applications, like the folding of peptides and supramolecular assemblies based on their amphiphilic character in the future.

4 Conclusion and outlook

4.1 Photoresponsive self-healing supramolecular hydrogels

Molecular photoswitches enable sophisticated control of macroscopic properties of biochemical and advanced materials. In this regard, azobenzenes have been established as a diverse and robust photoswitch moiety particularly for biomolecules,[28] photopharmacology[38] and supramolecular chemistry.[117] By converting the energy of light into molecular movement, contamination is avoided while precise temporal and spatial control is possible. By integrating a photoswitch into supramolecular hydrogels, photoresponsive self-healing materials can be created.[123] To investigate photoresponsive low-molecular-weight gelators, an azobenzene moiety was introduced to the known Phe-DKP-Lys **5** gelator[171-172] *via* rational design (Figure 43), as this class of gelators exhibits transparency, as well as considerable thermal and mechanical stability.[101]

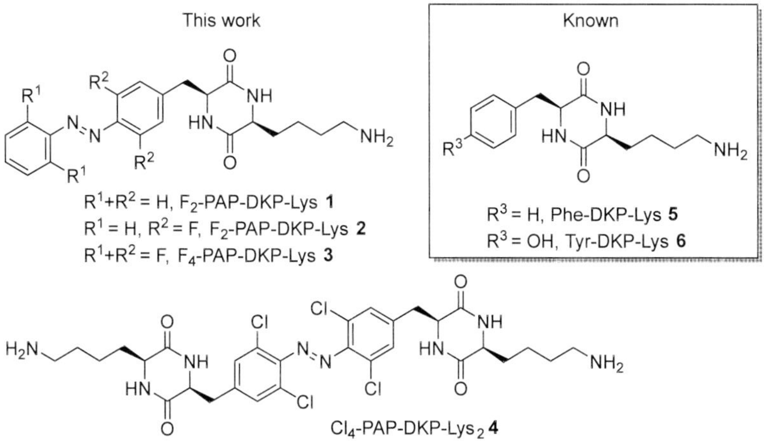

This work

R^1+R^2 = H, F_2-PAP-DKP-Lys **1**
R^1 = H, R^2 = F, F_2-PAP-DKP-Lys **2**
R^1+R^2 = F, F_4-PAP-DKP-Lys **3**

Known

R^3 = H, Phe-DKP-Lys **5**
R^3 = OH, Tyr-DKP-Lys **6**

Cl_4-PAP-DKP-Lys_2 **4**

Figure 43. Structures of photochromic supramolecular hydrogels PAP-DKP-Lys 1, F_2-PAP-DKP-Lys 2, F_4-PAP-DKP-Lys 3, Cl_4-PAP-DKP-Lys_2 4, and Phe-DKP-Lys 5,[171] Tyr-DKP-Lys 6.[172a]

To accomplish this, the synthesis of photochromic amino acid derivatives of phenylalanine is an essential element of this work. The first gelator PAP-DKP-Lys **1** was synthesized with unsubstituted azobenzene. As expected, a multi-responsive hydrogel was formed, which can be triggered by a change in pH, temperature and ionic strength, and which is, most importantly, responsive to UV-light. Additionally, it was demonstrated that encapsulated cargo molecules, such as dsDNA or doxorubicin, can be released from the hydrogel in a light-dependent manner. Together with the self-healing behaviour caused by its robust supramolecular structure, the hydrogel made from **1** exhibits a potential for photodynamic therapies based on a temporally and spatially-controlled release of drugs and therapeutic oligonucleotides by light.

Based on these promising results, a second generation of gelators containing *ortho*-fluoro-substituted azobenzenes was prepared with the aim for bathochromic shift of the photoisomerization wavelength, to avoid the use of cytotoxic UV-light. The required photochromic amino acid derivatives were synthesized in multi-gram quantities and their scope of application is not restricted to hydrogels alone.

In the first synthetic attempt yielding the F_2-PAP-DKP-Lys hydrogel, two fluorine were substituted to create the gelator **2**. This gelator was synthesized over three steps with a 45% yield in multi-gram quantities. Despite the lower content of *Z*-**2** in the photostationary state, as compared with the UV-light-irradiated gelator **1** containing the unsubstituted azobenzene, the predicted bathochromic shift of **2** was achieved with efficient gel-to-sol transition using green light (523 nm). The supramolecular hydrogel composed of **2** shows similar multi-responsive properties to the gels prepared from **1**, but the thermal and mechanical resistance is significantly increased (Figure 44).

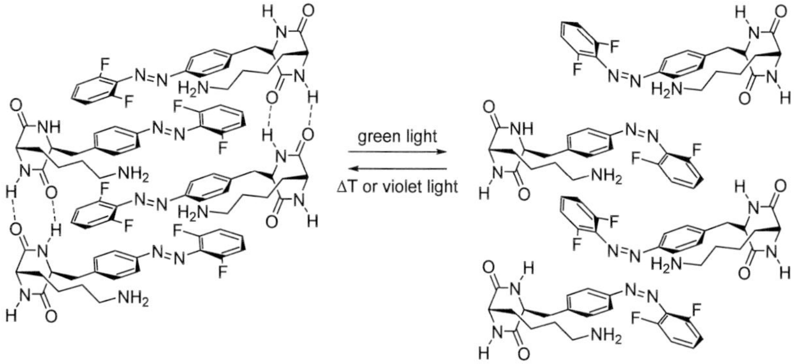

Figure 44. Supramolecular gel assembly of the *trans*-F_2-PAP-DKP-Lys 2 in aqueous solutions (left) and the partial disassembly caused by isomerization to the *Z*-isomer (right).[197]

Thus, hydrogels prepared from **2** can sustain shaking and elevated temperatures in bacterial cultures while no decrease of cell viability for *E. coli* is observed at ≤ 5 mM concentration of the gelator **2**. Moreover, cytotoxicity of **2** against mammalian cells is moderate (lower than a commonly used antibiotic). Overall, this material can be applied in biological experiments without eliciting unspecific toxicity. The photoresponsive release of bioactive guest molecules encapsulated in the hydrogels based on the gelator **2**, like the anti-inflammatory drugs naproxen and diclofenac, the antibiotic ciprofloxacin, the anticancer drug actinomycin D and the protein cytochrome C, were successfully shown *in vitro*. Importantly, these common drugs did not require any covalent structural modifications, as their encapsulation occurred in a purely physical manner. As the consequence, they were released directly in their physiologically active form. As a proof-of-principle in photomodulation of biological systems, the growth of *E. coli* was stopped *in vivo* by the phototriggered release of encapsulated ciprofloxacin. In contrast, the growth was not affected in the absence of light.

4.1.1 Outlook F_4-PAP-DKP-Lys 3

To further improve the photochromic properties, tetra-*ortho*-fluoro- and chloro-substituted azobenzene switches were integrated into the DKP-based gelator systems, in order to enhance the *Z/E*-isomer ratio at the respective PSS, and to further red-shift the photomodulation wavelength (Figure 41).

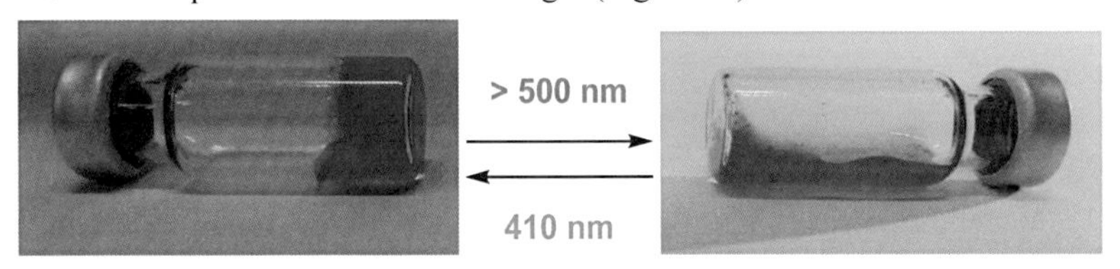

Figure 45. Photochromic hydrogels based on 3 (1.7% mass content) reversibly dissipate to fluids upon irradiation with green light.

Hydrogels prepared from **3** reversibly switch from gel-to-sol with green (523 nm) and violet light (410 nm) and exhibit a decreased critical gelation concentration of 1.5%, compared to 3% of the gelator **2**. An evaluation of drug release properties showed that the encapsulated anticancer drug plinabulin in hydrogel **3** can overcome its solubility limits in water, most likely due to supramolecular aggregation/intercalation effects enabled by their mutually related molecular structures. Upon irradiation with green light, plinabulin was efficiently released, as shown by an 87-fold difference between diffusion rate in the absence of light. and the rate of light-triggered release.

4.1.2 Supramolecular red-light-responsive supergelator Cl_4-PAP-DKP-Lys_2

Lastly, the symmetrical tetra-*ortho* chlorinated gelator **4** was prepared. A reversible and photoinduced gel-to-sol transition, using red (623 nm) and violet light (410 nm) in the range of 0.2-0.5%, in isotonic Ringer's solution was demonstrated. This further bathochromic shift improves the therapeutic applicability due to the increased potential for deep tissue penetration with red light. The supergelator forms hydrogels in several isotonic aqueous solutions, like 0.9% NaCl or DPBS buffer. At higher concentrations of $\geq$ 1%, hydrogel **4** remains stable, even in a boiling water bath. This corroborates the enhanced stability of the supramolecular assembly of **4,** in comparison with the previous asymmetric design.

4.1.3 Outlook Cl_4-PAP-DKP-Lys_2

Photoswitchable drug derivatives require molecular re-design, which can be accompanied by a decrease in activity, solubility or stability.[42] Every modification of a drug would force the reevaluation of toxicity properties and all clinical trials. To circumvent this necessity, photoresponsive drug-encapsulating hydrogels are used in the evolving field of encapsulation[101] in contrast to photopharmacology.[38] The present rationally designed photoresponsive hydrogels could be further investigated with the aim to integrate in one system several structurally similar gelators with orthogonal photoresponsive wavelengths.

With regards to applications, the lysine moiety of the diketopiperazine can be substituted with any other amino acid for catalysis, charge-charge interactions, or other light-triggered applications. The formulation of our materials as injectable micro/nanogels may provide reliable systems for spatiotemporally controlled release of pharmaceuticals, which in turn will reduce the therapeutic dosage and diminish systemic side effects, e.g. associated with existing anticancer therapies.

4.1.4 Tetra-*ortho*-fluoroazobenzene and 3-arylazopyridinium as photochromic building blocks

By incorporation of the photoswitches synthesized for the gelators **2-4**, , the photomodulation of a broad variety of molecules, not only peptides, is possible., As an illustrative example, modular synthesis of the tetra-*ortho*-fluoroazobenzenes was developed further to apply this photochromic core in artificial oligonucleotides, MOFs, and catenanes applied for asymmetric catalysis. Thus, the reversible isomerization with green and violet light renders these systems and materials with novel interesting properties. Particular incentives for such incorporations are the long lifetime of *Z*-photoisomer (up to two years at ambient temperature), its population at the photostationary state of 90-95%, and the NMR-active ^{19}F atoms for analytics.[2]

In addition to this bistable photoswitch based on tetra-*ortho*-fluoroazobenzenes, another class of fast switching 3-arylazopyridinium photoswitches was developed (Figure 46). Their isomerization can be controlled precisely with green laser and their high thermal relaxation rate is tuneable within the range of a microsecond to a millisecond.

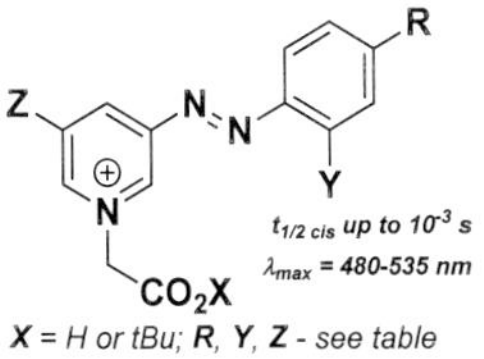

Figure 46. 3-Arylazopyridinium photoswitches 56-65 with extended lifetimes of the *cis*-isomer (substituent combinations given in Table 2).

Their synthesis was scaled up and improved by the use of stable diazonium salts. Non-symmetric functionalisation at both ends is possible with amino acids. The aggregation of a hydrophobic tetrapeptide in water could be significantly reduced by integration of this cationic photoswitch, as a proof-of-principle. Thus, e.g. photomodulation of the folding process in peptides could be achieved with 3-arylazopyridinium salts.

5 Experimental part

5.1 Laboratory standards

The following chapter about the work and laboratory standards are a translation of the internal regulations[297] and combined additionally with the supporting information.[298]

5.1.1 Solvents and reagents

All reagents and starting materials are commercially available (SIGMA-ALDRICH, FLUOROCHEM, CHEMPUR, ALFA AESAR or BEPHARM) and were used as supplied unless otherwise indicated. Solvents of technical quality were distilled prior to use. Absolute solvents were prepared by the methods shown in Table 5 and stored under argon atmosphere. All experiments were conducted in air and in deionized water (MILLIPORE) unless otherwise noted. In particular, htDNA (Deoxyribonucleic acid sodium salt from herring testes, Sigma-Aldrich, cat.#: D6898, containing polydisperse double-stranded DNA with the approximate length of c.a. 1300 bp) was used in all experiments that involved oligonucleotides.

In the experiments where physiological conditions were required, Dulbecco's Phosphate-Buffered Saline (DPBS) buffer pH 7.4, (-/-): no calcium, no magnesium, Gibco™ from THERMOFISHER, cat.#: 14190136, abbreviated below and in the thesis simply as "PBS buffer", was used.

All experiments with photoisomerizable molecules in the visible light spectrum, namely the azobenzene derivatives, were performed in absence of sunlight, for example with brown glassware, or colourless glassware wrapped with aluminium foil, working in a room with dimmed light. All experiments were conducted in air unless otherwise noted and all reactions containing air- and moisture-sensitive compounds were performed under argon using oven-dried glassware applying common Schlenk-techniques. Liquids were added *via* steel cannulas and solids were added directly in powdered shape. For the aqueous extraction tap water and for all analytic samples and for the HPLC deionized water (diH_2O) from MILLIPORE was used unless otherwise noted.

5.1.2 Nuclear magnetic resonance spectroscopy (NMR)

NMR spectra were recorded using the following devices: ^1H NMR: BRUKER 300 (300 MHz), BRUKER AVANCE 400 (400 MHz), BRUKER ASCEND 500 (500 MHz), ^{13}C NMR: BRUKER 300 (75 MHz), AVANCE 400 (101 MHz), ASCEND 500

(126 MHz), ^{19}F NMR: AVANCE 400 (377 MHz) or ASCEND 500 (471 MHz). The following solvents from EURISOTOP were used: chloroform-d_1, acetic acid-d_4, DMSO-d_6, acetonitrile-d_3, methanol-d_4, acetone-d_6 and D_2O. Chemical shifts δ were expressed in parts per million (ppm) and referenced to chloroform-d_1 (^{1}H: $\delta = 7.26$ ppm, ^{13}C: $\delta = 77.16$ ppm), acetic acid-d_4 (^{1}H: $\delta = 2.04$ ppm, ^{13}C: $\delta = 178.99$ ppm), DMSO-d_6 (^{1}H: $\delta = 2.50$ ppm, ^{13}C: $\delta = 39.52$ ppm) and D_2O (^{1}H: $\delta = 4.79$ ppm). methanol-d_4 (^{1}H: $\delta = 3.31$ ppm, ^{13}C: $\delta = 49.00$ ppm), acetone-d_6 (^{1}H: $\delta = 2.05$ ppm, ^{13}C: $\delta = 30.83$ ppm), and acetonitrile-d_3 (^{1}H: $\delta = 1.94$ ppm, ^{13}C: $\delta = 118.26$ ppm).[299] ^{19}F NMR were not referenced, except the ^{19}F NMR in acetic acid of F_2-PAP-DKP-Lys and its derivatives with ^{19}F: $\delta = -76.55$ ppm.[300][54] The signal structure is described as follows: s=singlet, d=doublet, t=triplet, q=quartet, quin=quintet, bs=broad singlet, m=multiplet, dt=doublet of triplets. The spectra were analysed according to the first order. All coupling constants are absolute values and expressed in Hertz (Hz).

5.1.3 UV-Vis spectrophotometer

The UV-Vis spectra were measured with the Lambda 750 from PERKINELMER at 20 °C, slit=2 nm. The volume was prepared with pipet4u® performance pipets 100-1000 µL with an accuracy of 2%-0.6% and 10-100 µL with an accuracy of 2%-0.8%. For the measurement of the DNA and doxorubicin concentration of the photo release experiment a NanoDrop™ 1000 spectrophotometer from THERMO SCIENTIFIC was used. The absorbance was measured from 200 nm to 800 nm and the average absorbance between 700 nm and 800 nm was abstracted from each measured value as a baseline correction.

5.1.4 Rheometry

For the rheometry measurements 8.0 mL freshly prepared hydrogel was heated up and filled in the rheometry plates. After the gelation overnight at rt the measurements were carried out at an ARES-G2 rheometer from TA INSTRUMENTS. The strain sweep experiment was performed at 10 rad/s and at 20 °C. That result allows the determination of the linear viscoelastic regime (LVR) and also the strength of the hydrogel.[171] The second experiment was a frequency sweep within the linear viscoelastic region at 0.1% deformation. The third experiment was carried out to test the regeneration of the hydrogel sample. Therefore, the hydrogel was exposed to 100% deformation for 30 s and a regeneration time for 10 min. This measurement was repeated 5 times.

5.1.5 High-power ultrasonic cleaning unit

To treat the samples with ultrasound the BANDELIN Sonorex RK 52 was used. This high-power ultrasonic cleaning unit was applied with a water bath at rt for 1 min per sample.

5.1.6 Infrared spectroscopy (IR)

IR spectra were recorded on a BRUKER IFS 88 using ATR (Attenuated total reflection). The absolute intensities of the peaks are given as follows: vs=very strong 0-9% T, s=strong 10-39% T, m=medium 40-69% T, w=weak 70-89% T and vw=very weak 90-100% T. All spectroscopy samples were taken at rt.

5.1.7 Analytic and preparative HPLC

Analytic HPLC was measured with the 1200 Series from AGILENT TECHNOLOGIES with a YMC C_{18}-column JH08S04-2546WT with 250×4.8 mm and 4 μm (column material). Preparative HPLC separation was performed with a JASCO LC-NetII / ADC HPLC system equipped with two PU-2087 Plus pumps, a CO-2060 Plus thermostat, an MD-2010 Plus diode array detector and a CHF-122SC fraction collector from ADVANTEC or PURIFLASH® 4125 from INTERCHIM, equipped with InterSoft V5.1.08 software and a UV diode array detector. The stationary phase was in both systems a VDSpher column with C_{18}-M-SE, 250×20 mm and 10 μm from VDSOPTILAB. For the elution, a linear gradient was used from 5% to 95% MeCN with 0.1% aqueous TFA in diH_2O with 0.1% aqueous TFA and with a flow rate of 10 mL/min at 25 °C.

5.1.8 Lyophilization

Substances purified by HPLC were dried by lyophilization. The used device was a CHRIST LDC-1 ALPHA 2-4 system.

5.1.9 Mass spectrometry (EI, FAB, GC-MS and ESI)

Mass spectra were recorded on a FINNIGAN MAT 95 mass spectrometer using electron ionization-mass spectrometry (EI-MS) or fast atom bombardment-mass spectroscopy (FAB-MS). For FAB measurements *m*-nitrobenzyl alcohol (3-NBA) was used as the matrix. The software of FAB and EI adds the mass of one electron. The molecular ion is abbreviated [M+] for EI-MS, the protonated molecular ion is abbreviated [M+H] for FAB-MS. Electrospray ionization-mass spectrometry (ESI-MS) spectra were recorded on a THERMO FISHER SCIENTIFIC Q EXACTIVE mass spectrometer. Calibration was carried out using premixed calibration solutions

(THERMO FISHER SCIENTIFIC). The molecular fragments are stated as ratio of mass per charge *m/z*. The GC measurements have been recorded with an AGILENT TECHNOLOGIES model 6890N (electron impact ionization), equipped with an AGILENT 19091S-433 column (5% phenyl methyl siloxane, 30 m, 0.25 μm) and a 5975B VL MSD detector with turbo pump. As a carrier gas helium was used.

5.1.10 Elemental analysis (EA)

Elemental Analyses were carried out on an ELEMENTAR VARIO MICRO device using a SARTORIUS M2P precision balance. The following abbreviations were used: calcd =calculated value, found =experimental value.

5.1.11 Thin layer chromatography (TLC)

Analytical thin layer chromatography was carried out using silica coated aluminium plates (silica 60, F_{254}, layer thickness: 0.25 mm) with fluorescence indicator by MERCK. The spots were detected by fluorescence quenching of UV-light at $\lambda = 254$ nm or $\lambda = 366$ nm and subsequent staining with phosphomolybdic acid solution (5% phosphomolybdic acid in ethanol, dip solution); potassium permanganate solution (1.00 g potassium permanganate, 2.00 g AcOH, 5.00 g sodium bicarbonate in 100 mL water, dip solution); Mostain solution (5% $(NH_4)6Mo_7O_{24}$ in 10% sulfuric acid with 0.03% $Ce(SO_4)$-2) or β-naphthol solution (10% β-naphthol, 10% AcOH and 80% ethanol, dip solution) followed by heating in a hot air stream. For amino acids the Seebach reagent (2.5% phosphomolybdic acid, 1.0% cerium(IV) sulfate tetrahydrate, 6.0% conc. sulfuric acid, 90.5% water; dipping solution) was used with subsequently heating in a hot-air flow. Amines were detected by dipping the TLC-plate in ninhydrin solution (0.2% ninhydrin in 0.1% AcOH in EtOH, dip solution) and subsequently heating in a hot-air stream. Alcohols were detected by dipping the TLC-plate in $KMnO_4$-solution (1.5 g of $KMnO_4$, 10 g K_2CO_3 and 10 mL 10% NaOH in 200 mL of water) and subsequently heating in hot air stream.[301]

5.1.12 Analytical balance

Used devices: SARTORIUS Basic (0.001 g), METTLER TOLEDO AE163 or RADWAG AS220.X2 (0.0001 g) and Sartorius M2P Micro Balance (0.001 mg).

5.1.13 TEM and SEM microscope

SEM images were obtained using a QUANTA 650-FEG scanning electron microscope from FEI Company, with an accelerating voltage of 10-20 kV and a spot size: 3.0. TEM measurements were performed using Zeiss 912 Omega microscope on 400 mesh carbon-coated copper grids, with the sample sprayed on the surface using an ultrasound device.

5.1.14 Preparative work

Table 5. Methods for the preparation of absolute solvents.

Solvent:	Method:
Dimethylformamide	drying of DMF (p.a.) using molecular sieves (4 Å) under an argon atmosphere.
Tetrahydrofuran	distilled over sodium (benzophenone as an indicator) or MB SPS5-system from MBRAUN
Dichloromethane	distilled over calcium hydride or MB SPS5-system from MBRAUN
Diethyl ether	distilled over sodium metal (benzophenone as an indicator) or MB SPS5-system from MBRAUN
Toluene	distilled over sodium metal (benzophenone as an indicator) or MB SPS5-system from MBRAUN

Reactions without further temperature indications were performed at rt. Reactions at low temperatures were cooled in flat Dewar flasks from ISOTHERM, Karlsruhe. The following cooling mixtures were used:

Table 6. Cooling mixtures.

Solvent:	Method:
−0 °C	Ice / water
−10 °C	Ice / water / sodium chloride
−20 °C	Methanol / dry ice
−35 °C	Stirred mixture of ethylene glycol:water 1:1 / dry ice
−42 °C	Stirred mixture of ACN / dry ice

In general, solvents were removed at preferably low temperatures (< 60 °C) under reduced pressure with a rotary evaporator. Solvents for solvent mixtures were separately volumetrically measured. If not indicated otherwise, solutions of inorganic salts are aqueous, saturated solutions.

5.1.15 Product purification

Crude products were purified according to literature procedure by column chromatography *via* flash-chromatography.[302] Silica gel 60 (0.063×0.200 mm, 70 – 230 mesh ASTM) (MERCK), Geduran® Silica gel 60 (0.040×0.063 mm, 230 – 400 mesh ASTM) (MERCK) or Celite® (FLUKA) and sea sand (calcined, purified with hydrochloric acid, Riedel-de Haën) were used as stationary phase. Column dimensions were indicated as follows: Height of silica gel × column diameter column. Eluents (mobile phase) were distilled and volumetrically measured. The flow of eluants were regulated with a pressure valve to typical 300 mbar with pore size of 2 (40 – 100 μM).

5.1.16 LED and isomerization

Sample irradiation (for photoisomerization of hydrogels, or measurements of photostationary states) was performed using LED diodes with following emission maxima: 10 W LED diodes: 365, 405, 460, 523, 590, 623 and 660 nm – from LED ENGIN or 3 W LED diodes: 365, 380, 400, 410, 420, 430, 450, 470, 490, 530, 590 and 605 nm – from LEDS-GLOBAL. For the time of irradiation, samples were maintained at constant temperature (22 ± 2 °C) using a metal cooling block unless otherwise noted.

5.2 PAP-DKP-Lys

5.2.1 Synthesis of PAP-DKP-Lys

Boc-PAP-OH, (*S,E*)-2-((*tert*-butoxycarbonyl)amino)-3-(4(phenyldiazenyl)phenyl)propanoic acid 9

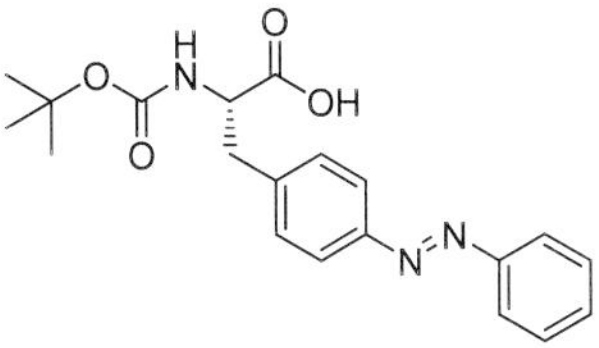

Nitrosobenzene **7** (4.31 g, 40.3 mmol, 1.10 eq.) was suspended in glacial AcOH (73 mL) and Boc-4-amino-*L*-phenylalanine **8** (10.3 g, 36.6 mmol, 1.00 eq.) was added. The resulting mixture was stirred at rt for 1 d in darkness. The solution was diluted with PhMe (100 mL) and concentrated under reduced pressure (60 °C, < 75 mbar, excess of the green nitrosobenzene was removed by this kind of steam distillation with a rotary evaporator). The crude product was purified by silica gel column chromatography: H=8 cm, start in pure CH_2Cl_2 $R_f = 0.0$ until flow through is colourless, stepwise gradient 1% MeOH; 2% MeOH; 10% MeOH in CH_2Cl_2 ($R_f = 0.1$-0.3 in 10% MeOH in CH_2Cl_2). Evaporation of combined fractions and drying under high vacuum overnight resulted in 10.1 g non-symmetrical azobenzene Boc-PAP-OH **9** (27.4 mmol, 75%) as an orange solid.[175b]

^{1}H NMR (400 MHz, DMSO): δ = 12.66 (s, 1H), 7.91 – 7.86 (m, 2H), 7.82 (d, J = 8.2 Hz, 2H), 7.63 – 7.55 (m, 3H), 7.48 (d, J = 8.1 Hz, 2H), 7.21 (d, J = 8.5 Hz, 1H), 4.18 (ddd, J = 10.4, 8.3, 4.5 Hz, 1H), 3.13 (dd, J = 13.7, 4.6 Hz, 1H), 2.94 (dd, J = 13.8, 10.4 Hz, 1H), 1.32 (s, 9H) ppm. **^{13}C NMR (101 MHz, DMSO):** δ = 173.4, 155.5, 152.0, 150.6, 142.2, 131.4, 130.2, 129.5, 122.5, 122.4, 78.1, 54.9, 36.3, 28.2 ppm. **^{1}H NMR (400 MHz, Acetone-*d*$_6$):** δ = 8.01 – 7.84 (m, 4H), 7.62 – 7.49 (m, 5H), 4.50 (dd, J = 9.1, 5.0 Hz, 1H), 3.32 (dd, J = 13.9, 5.0 Hz, 1H), 3.12 (dd, J = 13.9, 9.1 Hz, 1H), 1.36 (s, 9H) ppm. **^{13}C NMR (101 MHz, Acetone-*d*$_6$):** δ = 173.3, 156.2, 153.5, 152.3, 142.4, 132.0, 131.2, 130.1, 123.6, 123.5, 79.3, 55.4, 38.0, 28.5 ppm. **HRMS (FAB+):** *m/z* calcd for $C_{20}H_{24}O_4N_3$: 370.1767 Da [M+H], found: 370.1765 Da (Δ = 0.5 ppm). **IR (ATR):** $\tilde{\nu}$ = 2978 (vw), 1716 (w), 1690 (w), 1522 (w), 1445 (vw), 1393 (vw), 1367 (vw), 1250 (vw), 1158 (w), 1104 (vw), 1054 (vw), 1019 (vw), 927 (vw), 836 (vw), 768 (vw), 687 (w), 635 (vw), 561 (vw), 388 (vw) cm^{-1}. **UV-Vis (MeCN):** λ_{max} = 230, 324, 440 nm.

Boc-PAP-Lys(Boc)-OMe, methyl N^6-(*tert*-butoxycarbonyl)-N^2-((*S*)-2-((*tert*-butoxycarbonyl)amino]-3-(4-((*E*)-phenyldiazenyl)phenyl]propanoyl)-*L*-lysinate 10

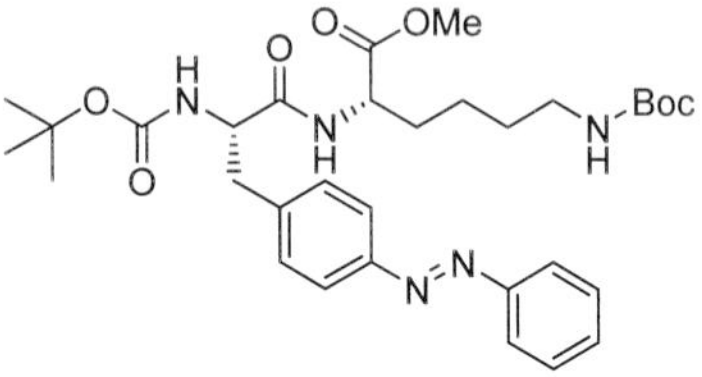

Boc-PAP-OH **9** (5.00 g, 13.5 mmol, 1.00 eq.), HBTU (5.13 g, 13.5 mmol, 1.00 eq.) and DIPEA (5.9 mL, 33.8 mmol, 2.50 eq.) were dissolved in anhydrous DMF (13 mL) and stirred for 10 min at rt under argon. The same amount of DIPEA (5.9 mL, 33.8 mmol, 2.50 eq.) was added together with *ω*-*N*-Boc-lysine methyl ester hydrochloride (H-Lys(Boc)-OMe·HCl, 4.02 g, 13.5 mmol, 1.00 eq.). The reaction mixture was stirred further at rt under argon and the reaction progress was followed by TLC. After 3 h full conversion of the starting material was observed. The reaction mixture was quenched with sat. aqueous NH_4Cl solution (200 mL) and extracted once with EtOAc (200 mL). The organic layer was washed with sat. aqueous NH_4Cl solution (3×200 mL), brine (1×200 mL), dried with anhydrous Na_2SO_4 and the solvent was evaporated under reduced pressure. The crude was purified by silica gel column chromatography. Initial elution with cH:EtOAc 3:1 washed out colourless non-polar impurities. Then the column was eluted with cH:EtOAc 1:1 (R_f = 0.45). Evaporation of combined orange fractions and drying under reduced pressure resulted in 5.92 g Boc-PAP-Lys(Boc)-OMe **10** (9.68 mmol, 72%) as an orange solid. During evaporation we observed formation of a gel in this solvent system before the product was completely dried.

^{1}H NMR (400 MHz, DMSO): δ = 8.33 (d, *J* = 7.6 Hz, 1H), 7.88 (dd, *J* = 8.2, 1.6 Hz, 2H), 7.82 (d, *J* = 8.3 Hz, 2H), 7.64 – 7.53 (m, 3H), 7.51 (d, *J* = 8.3 Hz, 2H), 7.00 (d, *J* = 8.7 Hz, 1H), 6.78 (t, *J* = 5.8 Hz, 1H), 4.34 – 4.20 (m, 2H), 3.62 (s, 3H), 3.06 (dd, *J* = 13.7, 4.1 Hz, 1H), 2.95 – 2.78 (m, 3H), 1.67 (dt, *J* = 33.9, 7.7 Hz, 2H), 1.33 (d, *J* = 29.4 Hz, 22H) ppm. **^{13}C NMR (101 MHz, DMSO):** δ = 172.5, 171.8, 155.6, 155.2, 152.0, 150.6, 142.3, 131.4, 130.3, 129.5, 122.4, 122.3, 78.1, 77.3, 55.3, 52.0, 51.9, 37.3, 30.7, 29.1, 28.3, 28.1, 27.9, 22.6 ppm. **HRMS (FAB+):** *m/z* calcd for $C_{32}H_{46}O_7N_5$: 612.3397 Da [M+H], found: 612.3396 Da (Δ = 0.3 ppm). **IR (ATR):** ṽ = 3335 (vw), 2934 (vw), 1742 (vw), 1684 (w), 1655 (w), 1517 (w), 1444 (vw), 1390 (vw), 1365 (vw), 1291 (vw), 1268 (w), 1249 (w), 1211 (vw), 1165 (w), 1044 (vw), 1017 (vw), 925 (vw), 849 (vw),

766 (vw), 687 (vw), 630 (vw), 569 (vw), 528 (vw), 429 (vw) cm^{-1}. **UV-Vis (MeCN):** λ_{max} = 230, 324, 440 nm.

PAP-Lys-OMe·TFA, methyl *N*^{2}-((*S*)-2-amino-3-(4-((*E*)-phenyldiazenyl)phenyl]propanoyl)-*L*-lysinate trifluoroacetate 10d

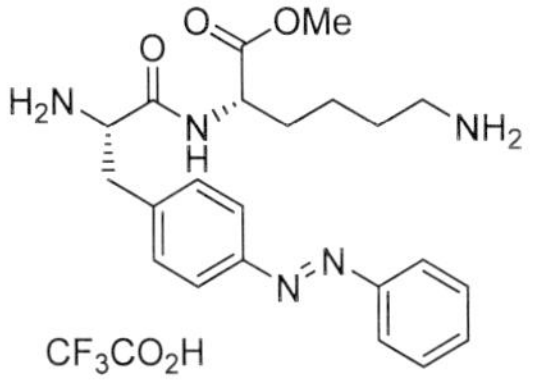

CF_3CO_2H

(Boc)PAP-(Boc)Lys-OMe **10** (4.90 g, 8.01 mmol, 1.00 eq.) was dissolved in CH_2Cl_2 (80 mL). TFA (80 mL) and 1 vol% Triisopropylsilane (1.6 mL) was added at rt. The mixture was stirred for 1 h at rt. The reaction mixture was diluted with PhMe (100 mL) and the solvents were evaporated under reduced pressure. After evaporation to dryness 5.00 g PAP-Lys-OMe·TFA **10d** (7.82 mmol, 98%) was obtained and was used without further purification in the next reaction.

^{1}H NMR (400 MHz, DMSO): δ = 8.97 (d, J = 7.5 Hz, 1H), 8.32 (s, 3H), 7.98 – 7.74 (m, 7H), 7.69 – 7.55 (m, 3H), 7.50 (d, J = 8.4 Hz, 2H), 4.31 (td, J = 8.2, 5.4 Hz, 1H), 4.15 (s, 1H), 3.62 (s, 3H), 3.15 (ddd, J = 55.0, 13.9, 6.8 Hz, 2H), 2.83 – 2.70 (m, 2H), 1.81 – 1.69 (m, 1H), 1.68 – 1.46 (m, 3H), 1.35 (dq, J = 14.4, 6.8 Hz, 2H) ppm. **^{13}C NMR (101 MHz, DMSO):** δ = 171.7, 168.1, 158.2 (q, J = 31.5 Hz), 152.0, 151.2, 138.6, 132.0, 130.7, 129.5, 122.7, 122.5, 117.1 (d, J = 299.2 Hz), 53.1, 52.0 (d, J = 17.4 Hz), 38.5, 36.7, 30.4, 26.6, 22.1 ppm. **HRMS (FAB+):** *m/z* calcd for $C_{22}H_{30}O_3N_5$: 412.2349 Da [M+H], found: 412.2350 Da (Δ = 0.4 ppm). **IR (ATR):** $\tilde{\nu}$ = 2953 (w), 1665 (m), 1435 (w), 1179 (m), 1129 (m), 836 (w), 798 (w), 767 (w), 721 (w), 687 (w), 560 (vw), 519 (vw), 409 (vw) cm^{-1}. **UV-Vis (MeCN):** λ_{max} =230, 322, 441 nm.

PAP-DKP-Lys, (3*S*,6*S*)-3-(4-aminobutyl)-6-(4-((*E*)-phenyldiazenyl]benzyl)piperazine-2,5-dione 1

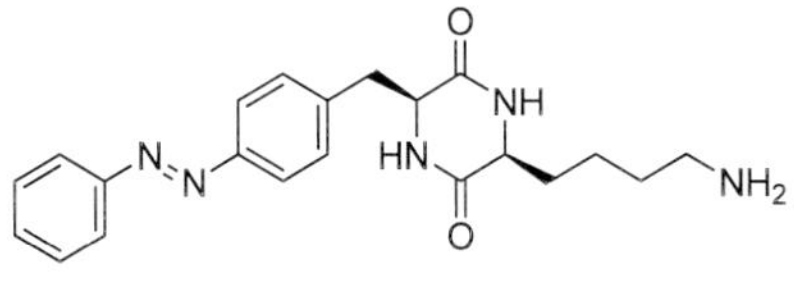

The crude compound PAP-Lys-OMe·TFA **10'd** (4.94 g, 7.73 mmol, 1.00 eq.) was dissolved in 2-butanol (327 mL). It was added subsequently glacial AcOH (1.63 mL, 28.5 mmol, 3.69 eq.), *N*-methylmorpholine (1.08 mL, 9.87 mmol, 1.28 eq.) and *N,N*-diisopropyl-*N*-ethylamine (DIPEA, 1.95 mL, 11.2 mmol, 1.45 eq.). The resulting mixture was refluxed for 2 h in an oil bath with external heating (120 °C), then cooled down. Next, approximately 60% of the solvent was removed under reduced pressure. Cooling down to rt resulted in orange gel-precipitation. The precipitate was filtered off, washed with small amounts of cold 2-butanol (2×10 mL) and dried *in vacuo*. Fractional recrystallization over three steps in 2-butanol is necessary to obtain analytic purity. The product was obtained in 3.54 g PAP-DKP-Lys **1** (7.18 mmol, 93% yield) as an orange solid.

^{1}H NMR (400 MHz, DMSO): δ = 8.27 (d, J = 2.1 Hz, 1H), 8.13 (d, J = 2.1 Hz, 1H), 7.88 (dd, J = 8.0, 1.8 Hz, 2H), 7.81 (d, J = 8.4 Hz, 2H), 7.65 (s, 3H), 7.64 – 7.54 (m, 3H), 7.39 (d, J = 8.4 Hz, 2H), 4.28 (s, 1H), 3.68 (s, 1H), 3.23 (dd, J = 13.5, 4.1 Hz, 1H), 2.99 (dd, J = 13.5, 5.1 Hz, 1H), 2.55 (s, 2H), 1.22 (dt, J = 11.2, 6.8 Hz, 3H), 1.01 – 0.73 (m, 3H) ppm. **^{13}C NMR (101 MHz, DMSO):** δ = 167.1, 166.3, 158.1 (d, J = 30.8 Hz), 152.0, 151.0, 140.4, 131.5, 131.4, 129.5, 122.6, 122.2, 55.2, 53.4, 38.4, 37.9, 32.5, 26.4, 20.5 ppm. **(*E*)-PAP-DKP-Lys – ^{1}H NMR (300 MHz, D_2O):** δ = 7.87 (d, 4 H, J =7.6 Hz), 7.64 (m_c, 3 H), 7.42 (d, 2 H, J = 6.8 Hz), 4.57 (bs, 1 H, CH) ppm. **(*Z*)-PAP-DKP-Lys – ^{1}H NMR (300 MHz, D_2O):** δ = 7.33 (d, 4 H, J = 7.8 Hz), 7.13 (m_c, 3 H, J = 8.2 Hz), 6.94 (d, 2 H, J = 8.0 Hz), 4.57 (bs, 1 H, CH) ppm. **HRMS (FAB+):** *m/z* calcd for $C_{21}H_{26}O_2N_5$: 380.2087 Da [M+H], found: 380.2086 Da (Δ = 0.2 ppm). **IR (ATR):** $\tilde{\nu}$ = 2891 (w), 1666 (m), 1555 (w), 1456 (w), 1334 (vw), 1204 (m), 1127 (m), 1014 (vw), 925 (vw), 836 (w), 800 (w), 768 (w), 722 (w), 688 (w), 574 (w), 525 (w), 487 (vw), 431 (w) cm^{-1}. **UV-Vis (MeCN):** λ_{max} = 230, 324, 443 nm.

5.2.2 Photophysical properties PAP-DKP-Lys

The following chapters about the properties of PAP-DKP-Lys are quoted verbatim from the supporting information.[298a] Photoisomerization of the compound **1** (PAP-DKP-Lys) in solution was examined by irradiating a sample (2 mL of 60 µM solution in deionized water) with 10 W UV-LED diode (365 nm). The photostationary state PSS_{365} with predominant *cis*-isomer was already achieved upon short (1 s) irradiation. Longer exposure on the UV-light did not result in any further changes of the UV-Vis spectrum. Short (1 s) irradiation of the second sample with blue light (10 W LED diode, 460 nm) did not cause significant *trans*-to-*cis* photoisomerization. Its spectrum was almost identical to the spectrum of a sample, which was not irradiated at all ("the dark state").

Although the isomer distribution in different photostationary states in solutions of the unmodified azobenzene was described in numerous publications[28] we wanted to verify the isomer ratio in our particular experimental system consisted of the molecule **1** (15 g/L) and htDNA (2 g/L) dissolved in 50 mM NaCl solution in D_2O. The *cis/trans* (*E/Z*) isomer ratio in darkness and upon irradiation with 365 nm UV-light was determined by ^{1}H NMR measurements and integration of peaks at the aromatic spectral region. The following peaks were identified as contributions from the respective isomer:

***(E)*-PAP-DKP-Lys – ^{1}H NMR (300 MHz, D_2O):** δ = 7.87 (d, 4 H, J = 7.6 Hz), 7.64 (m, 3 H), 7.42 (d, 2 H, J = 6.8 Hz), 4.57 (br s, 1 H, CH) ppm. ***(Z)*-PAP-DKP-Lys – ^{1}H NMR (300 MHz, D_2O):** δ = 7.33 (d, 4 H, J = 7.8 Hz), 7.13 (m, 3 H, J = 8.2 Hz), 6.94 (d, 2 H, J = 8.0 Hz), 4.57 (br s, 1 H, CH) ppm.

The peak at 7.87 ppm for the *trans*-isomer of **1** was clear shifted from the other ones. By comparing the integration of both peaks (7.87 ppm) to the sum of all peaks in the aromatic region, the ratio of the isomers was estimated of both isomers in darkness and upon 30 min irradiation with 365 nm (Table 7). The PSS is in agreement with common literature data for the unsubstituted azobenzene.[28]

Table 7. *E/Z* ratio of 15 g/L PAP-DKP-Lys, 2 g/L H-DNA in 50 mM NaCl in D_2O.

Procedure	*E/Z* ratio	Peaks for calculation [ppm]
Dark state	Approx. 100%	7.87 / all A_r
30 min 365 nm	23% / 77%	7.87 / all A_r
30 min 460 nm	66% / 34%	7.87 / all A_r
PSS_{365}[28]	20% / 80%	-

5.2.2.1 Stability of the PAP-DKP-Lys – photodegradation, multiple switching cycles

Photostability of the compound **1** (PAP-DKP-Lys) was investigated by irradiation of its 60 µM solution in diH_2O for 24 h with UV-light (10 W, 365 nm) at 20 ± 2 °C in a sealed cuvette and measurement of its UV-Vis spectra. No significant changes were observed in the spectrum after 24 h of irradiation with UV-light (Figure 47, empty triangles) in comparison to the same sample photoisomerized shortly (10 sec irradiation) with UV-light (Figure 47, empty squares, PSS_{365}). The sample after 24 h of UV-light irradiation was shortly (10 sec) irradiated with blue light (460 nm, Figure 47, black triangles). Again, no significant changes in comparison to the PSS_{460} (Figure 47, black dots) achieved after 10 sec irradiation of the fresh sample with blue light was observed. In conclusion the compound **1** does not undergo significant photodegradation on the timescale of at least 24 h.

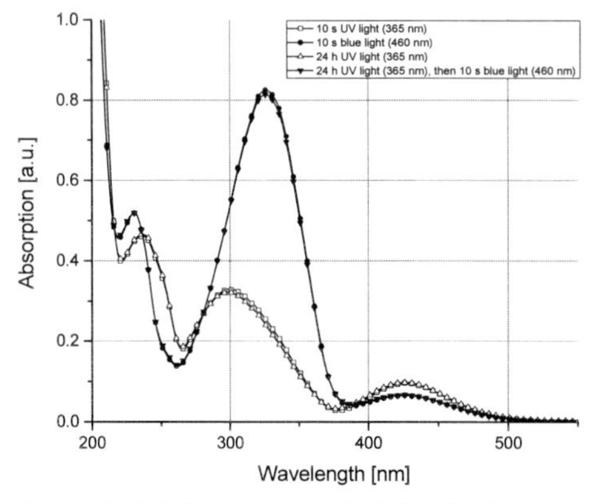

Figure 47. UV-Vis spectra: 60 µM PAP-DKP-Lys 1 in diH_2O. Photostationary states (10 sec irradiation) at 365 nm (10 W, empty squares) – and 460 nm (10 W, black dots). The same sample after 24 h irradiation with UV-light (365 nm, 10 W, empty triangles), and additionally 10 sec irradiation with blue light (460 nm, 10 W, black triangles).

The resistance of the material to multiple photoisomerization cycles was also investigated by UV-Vis spectrophotometry. A 60 µM solution of the compound **1** (PAP-DKP-Lys) was alternatively irradiated with UV (365 nm, 10 W) and blue (460 nm, 10 W) light 100 times each. At the end, the respective spectra were compared to the photostationary state achieved after single irradiation with the given wavelength. No meaningful differences in the behaviour of the freshly prepared material and the same material after 100 full photoisomerization cycles were found (Figure 19).

5.2.3 Gelation properties of PAP-DKP-Lys

5.2.3.1 Solubility of 1 in water and organic solvents

20 mg of **1** as finely crushed powder was mixed in a 10 mL glass vessel with 1 mL of solvent. **1** was readily dissolved in DMF, DMSO and methanol. Samples in water, acetonitrile and PhMe were shortly boiled and slowly cooled down to rt. **1** was not soluble in MeCN and PhMe. **1** was fully soluble in boiling water and this

solution formed a gel upon cooling. The following Table 8 summarizes solubility of the compound **1** in different solvents.

Table 8. Solubility of the compound 1 in various solvents (conc. of 2% in each case).

Solvent	water	MeOH	DMF	DMSO	MeCN	PhMe
Property	gel	soluble (gel)*	soluble	soluble	insoluble	insoluble

***) The 5% solution of 1 in MeOH forms a gel with gel-to-sol T. of c.a. −22 °C.**

5.2.3.2 Procedure for preparation of the hydrogels

In a 10 mL glass vessel was added the finely crushed powder of the PAP-DKP-Lys **1** (40.0 mg for 2% gels or proportionally less for gels of lower concentration) and deionized water (between 1.5 mL and 2.0 mL). The closed glass vessel was instantly treated with ultrasonic in a water bath for 1 min with gentle stirring. To this opaque suspension was added, if needed, stock solution (400 mM) of NaCl and/or htDNA stock solution (6.0 g/L) in deionized water to reach the desired concentration of the components in the total volume of 2.0 mL. The closed glass vessel was warmed up vertical in a water bath at 80 °C for 5 min. The yellow suspension was dissolved in the closed glass vessel after heating it up to the boiling point with a heat gun. The hot fluid turned to an orange solution and this fluid was gelated after 5 min at rt. In the case of incomplete gelation, the fluid was incubated overnight at rt in darkness.

5.2.3.3 Measuring the melting temperature of the hydrogels.

2.0 mL samples of the hydrogels in 10 mL cylindrical glass vials are swimming horizontally on the surface of water in a water bath stirred with 60 rpm at 25 °C, which is heated up with a heating rate between 1 °C/min and 2 °C/min. The hydrogel will start melting slowly before the hydrogel will abruptly flow down at a certain moment. The measurement was repeated 5 times and the average melting point was reported (cf. Figure 61).

Gels prepared from **1** in deionized water without any additives were stable and homogenous in the range between 20 g/L and 30 g/L (2-3 wt% of **1**). Below that concentration gelation was slow (overnight) or did not occur at all. Moreover, the gels containing less than 2 wt% of **1** were very sensitive on mechanical deformation and almost immediately turned into liquid upon slight shaking of the vial.

Addition of TFA (1 wt%, 10 g/L) or sodium chloride (to the final concentration of 50 mM) to the gel **A** resulted in a significant increase of the melting point as well as significantly improved the mechanical properties of the resulting materials, that could be examined both by optical observation of samples and verified by rheometry (vide infra). It can be intuitively understood by protonation of the lysine side chain in case of acids or increasing ionic strength of the medium – both can increase the strength of electrostatic and polar interactions that in turn stabilize fibres and the entire hydrogel structure. Addition of long dsDNA (htDNA 2 g/L, ca. 1300 bp) alone (gel **C**) did not significantly improve the mechanical properties, but increased melting temperature of resulting gels, although not to the same extent as in the case of TFA (DNA is much less acidic) or NaCl.

That could be explained as a combination of the ionic interactions between the amines from lysine side chains and phosphates of DNA together the additional templating effect of the covalent DNA backbone. As expected, the combination of NaCl and htDNA led to formation of hydrogels that were mechanically more stable and had higher melting temperatures in comparison with all other combinations with same concentration of **1** (PAP-DKP-Lys).

Table 9. Melting temperatures of hydrogels containing 1 and additives (TFA, NaCl, htDNA).

Composition of the gel	Gel to sol T. °C	Remarks
15 g/L **1** (1.5 wt%)	38 ± 4	
16 g/L **1**	41 ± 4	
17 g/L **1**	43 ± 4	
18 g/L **1**	46 ± 8	
19 g/L **1**	48 ± 6	
20 g/L **1** (2 wt%)	51 ± 4	gel **A**
30 g/L **1** (3 wt%)	69 ± 2	
20 g/L **1**, 2 g/L htDNA	64 ± 8	gel **C**
20 g/L **1**, 10 g/L TFA	68 ± 1	
20 g/L **1**, 50 mM NaCl	70 ± 1	gel **B**
20 g/L **1**, 2 g/L htDNA, 50 mM NaCl	71 ± 2	
15 g/L **1**, 2 g/L htDNA	44 ± 3	
15 g/L **1**, 50 mM NaCl	51 ± 4	
15 g/L **1**, 2 g/L htDNA, 50 mM NaCl	60 ± 3	gel **D**

5.2.3.4 Light-induced gel-to-sol transitions

Gel samples were irradiated at maximum power for 30 min at 20 ± 2 °C with two 10 W 365 nm LED diodes (LED Engin) in a water-cooled aluminium chamber. Immediately after irradiation the glass was turned upside down. The state of the sample was assigned as fluid, unstable gel (that liquefied after slewing), or stable gel (resistant on moderate shaking). The following Table 10 summarizes the results of gel irradiation with UV-light.

Table 10. Mechanical properties of various 1-based gels in darkness and upon UV irradiation.

Gel composition	dark	365 nm, 30 min
3% **1**	stable gel	stable gel
2% **1**	stable gel	unstable gel
2% **1**, 1% TFA	stable gel	stable gel
2% **1**, 50 mM NaCl	stable gel	stable gel
2% **1**, 0.2% htDNA	stable gel	unstable gel
2% **1**, 0.2% htDNA, 50 mM NaCl	stable gel	stable gel
1.5% **1**	unstable gel	fluid
1.5% **1**, 50 mM NaCl	stable gel	fluid
1.5% **1**, 0.2% htDNA	unstable gel	fluid
1.5% **1**, 0.2% htDNA, 50 mM NaCl	stable gel	fluid

The fluid samples containing 1.5% of the DKP **1** after UV-light irradiation were irradiated for 30 min at 20 ± 2 °C with a single 10 W 460 nm (blue) LED (LED Engin) and left overnight in darkness. Upon that time, they formed gels with mechanical properties, similar to the gels formed with the same composition and left without any irradiation. In comparison, UV-irradiated fluid samples incubated in darkness without previous blue light irradiation remained liquids for at least one week.

That behaviour can be explained, as follows: the existence and rigidity of hydrogels depends on the presence of fibres that constitute its inner structure. Upon UV-light irradiation and photoisomerization around three quarter of molecules (as judged by the NMR signals) loose the flat unpolar aromatic system capable of strong π-π stacking interactions. That causes fibres to break and dismantle. In case of 2% PAP-DKP-Lys gels, the remaining fraction of *trans* molecules can still sufficiently support the inner structure. For 1.5% PAP-DKP-Lys samples the remaining *trans* fraction is too low and the whole gel structure tends to collapse. Thermal back-isomerization (*cis*-to-*trans*) rate for azobenzene is in the range of 5-10 d and this corresponds well to the rate of gel regeneration in darkness. Irradiation of such samples with blue light speeds up this process because it increases the population of the *trans*-isomer. However, the process of fibre reconstitution is slow and therefore the gel structure is reconstituted after a few hours.

5.2.3.5 Rheology

The rheological characterization of hydrogels formed by the compound **1** in water or 50 mM aqueous NaCl was performed in presence or in absence of htDNA. The following results are in accordance with our macroscopic observations that indicate that both the addition of NaCl and DNA to hydrogels formed by **1** increase their mechanical stiffness and resistance on sample shaking.

Hydrogel samples for rheological measurements were generated by cooling their solutions directly on the rheometer plate from the boiling point to rt. Strain sweep experiments were performed at 10 rad/s to determine the linear viscoelastic regime and the mechanical strength of the hydrogel. Frequency experiments were performed at low strain within the linear viscoelastic region (LVR) of the sample. For regeneration experiments, the samples were exposed to a deformation of 100% for 30 sec to destroy the supramolecular network, afterwards the regeneration of G' was measured at low strain within the LVR.

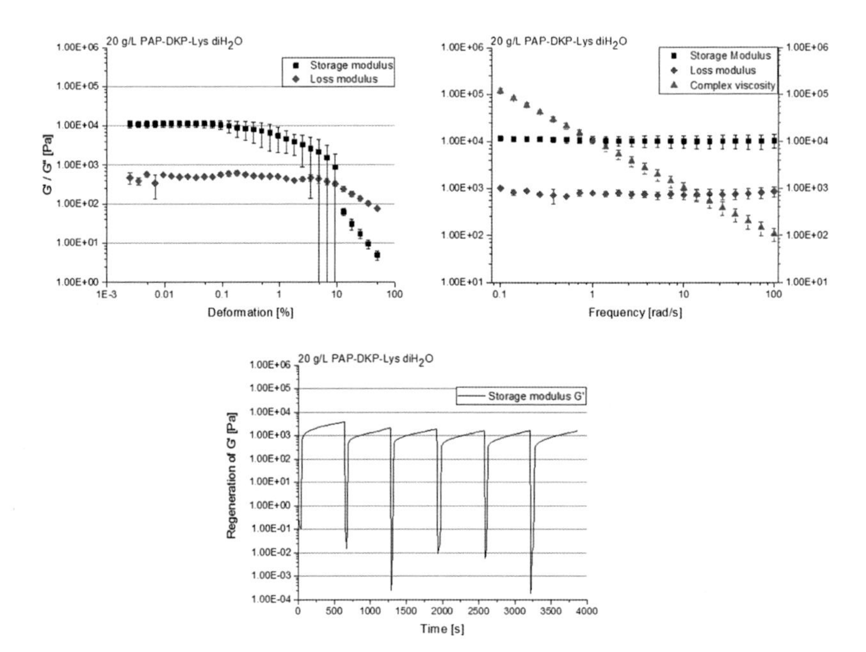

Figure 48. Gel A: 20 g/L (2%) of 1 in diH$_2$O. Strain sweep experiment (top left); frequency sweep experiment (top right); regeneration of G' after shearing the gel for 30 sec at 100% deformation (bottom).

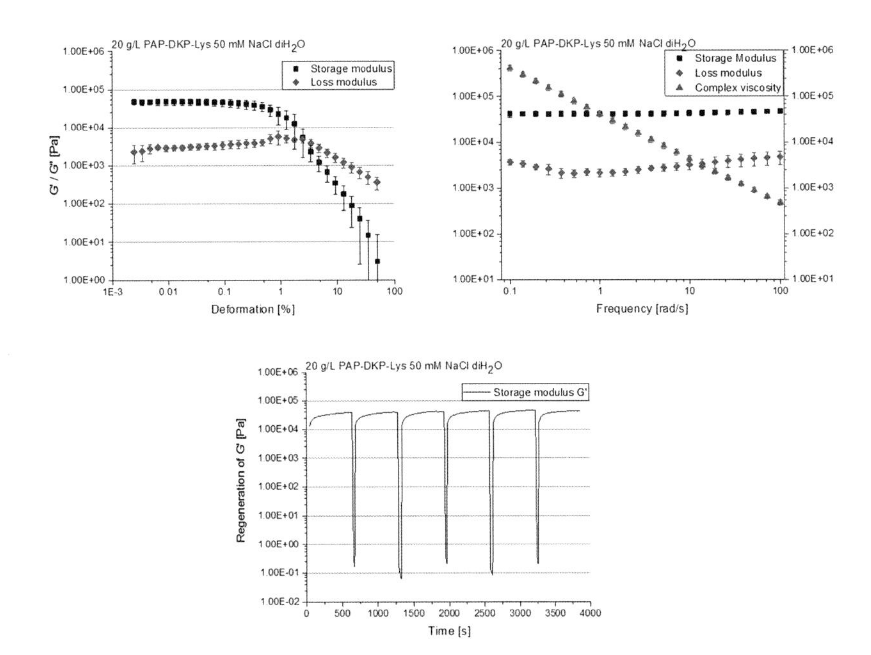

Figure 49. Gel B: 20 g/L (2%) of 1 in 50 mM aqueous NaCl. Strain sweep experiment (top left); frequency sweep experiment (top right); regeneration of G' after shearing the gel for 30 sec at 100% deformation (bottom).

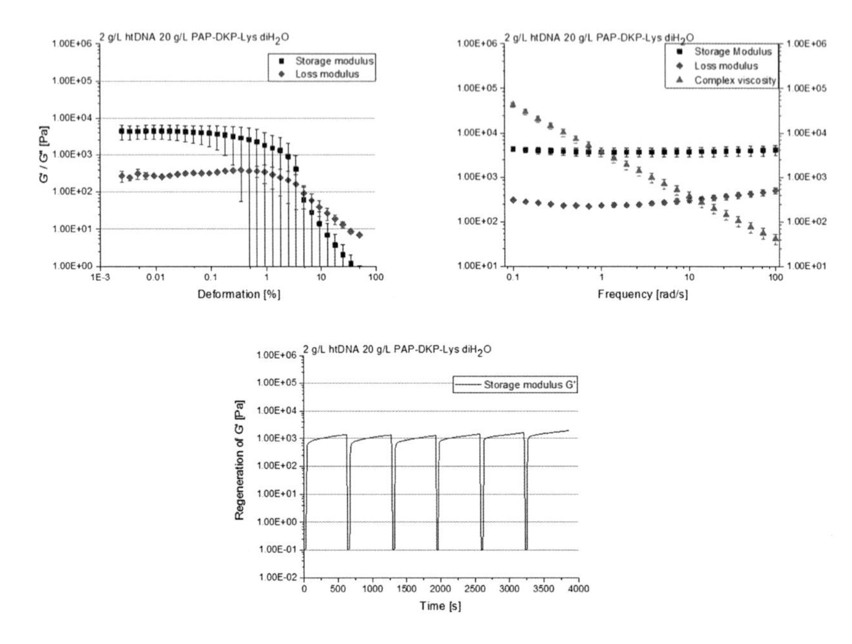

Figure 50. Gel C: 20 g/L (2%) of 1 and 2 g/L of htDNA in diH₂O. Strain sweep experiment (top left); frequency sweep experiment (top right); regeneration of G' after shearing the gel for 30 sec at 100% deformation (bottom).

5.2.3.6 SEM and TEM characterization

With scanning electron microscopy, the samples previously characterized by rheology were further characterized. Samples for high vacuum mode SEM were prepared by lyophilization of gels that were rapidly frozen with liquid nitrogen. The dried SEM images of the samples were obtained with a QUANTA 650-FEG scanning electron microscope from FEI Company (accelerating voltage: 10 kV, spot size: 3.0). Before imaging, all the samples were sputter-coated with platinum. For environmental SEM (ESEM) mode, a small piece of the hydrogel was placed on the sample holder without sputter-coating and the ESEM images were directly recorded using the same microscope (accelerating voltage: 20 kV, spot size: 3.0).

5.2.3.7 TEM characterization

Carbon-coated copper grids (400 mesh) were covered with diluted solution of the compound **1** in aqueous 50 mM NaI or NaCl by short exposure on spray generated from the samples with ultrasounds. The grids dried under atmospheric pressure and were examined using Zeiss 912 Omega transmission electron microscope. Under these conditions, fine fibrous structure of our material was revealed.

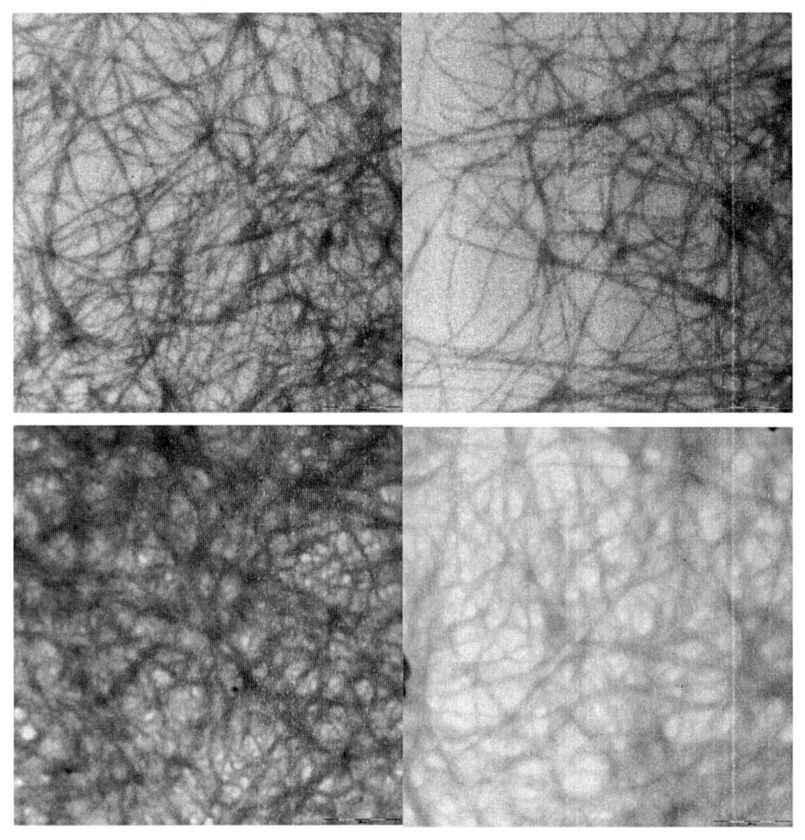

Figure 51. TEM of the sample 5 g/L PAP-DKP-Lys 1. Staining with 50 mM NaI (top), 50 mM NaCl (bottom left) and 500 mM NaCl (bottom right). Scale-bar = 200 nm.

5.2.3.8 Guest release from the gel matrix

Here we wanted to investigate how efficient hydrogels based on PAP-DKP-Lys **1** are in releasing various types of encapsulated guest molecules by means of diffusion (in darkness) or decomposition of the inner gel structure (upon UV irradiation). As the gel matrix we have chosen the following composition: 1.5% (15 g/L) of the DKP **1** in 50 mM NaCl. As described in the section 5 of this supporting information, it forms a stable gel in absence of light that is completely converted to fluid upon 30 min of UV-light irradiation.

For the release experiments we prepared (in the same way as described above – section 4) six samples of 1 mL volume containing 2 g/L htDNA and 15 g/L DKP **1** in 50 mM NaCl and 6 samples of 1 mL containing 2 g/L doxorubicin (DOX) and 15 g/L DKP **1** in 50 mM NaCl. The samples were warmed up until the stage of homogenous liquid and left in darkness overnight to produce uniformed gels upon slow cooling. All the release measurements (in darkness and upon UV-light irradiation) were performed in triplicates and the average values were taken for the final conclusions and result plots.

1 mL of 50 mM aqueous NaCl was slowly added on top of a gel sample (on the wall of the vial) and immediately removed with micropipette, to wash away unbound or loosely bound guest molecules from the surface. Addition of fresh 1 mL of 50 mM aqueous NaCl followed. The gel was incubated together with the liquid on the top in darkness for 5 min and 1 mL of the liquid was collected after 5 min by gently turning the vial sideways and pipetting off the liquid from the side wall of the vial. Then, fresh 1 mL of 50 mM aqueous NaCl was added on the side wall of the vial, removed after 5 min the same way as described above, and that process was repeated for the total duration of 30 min (by collecting six subsequent 1 mL-volume aliquots). After that time, both gel types (with DNA and DOX) remained visually unaffected. By measuring the remaining liquid volume after removal of the last 1 mL aliquot from the top of the gel we estimate that the total decay of the gel volume was lower that 15%.

To measure the release process upon UV-light irradiation, we exactly repeated the procedure described above, but after initial washing of the gel surface the sample was placed in an irradiation chamber and illuminated with two 10 W LED diodes. Short breaks in irradiation (< 30 sec) were taken for the replacement of 1 mL aliquot with fresh 1 mL of 50 mM NaCl every 5 min, but the summary irradiation time was 30 min. That period was sufficient to fully convert all the gel samples into liquid. In the case of htDNA, slight pale-yellowish precipitate was observed on the bottom of the sample and aliquots. That might correspond to partial precipitation or aggregation of some DNA material, as that precipitation was not observed for the gel samples containing DOX or the DKP **1** alone in 50 mM NaCl solutions.

5.2.3.9 Release of doxorubicin

The resulting 1 mL aliquots of solutions collected above the gels in the time course of experiments were analysed using UV-Vis absorption spectrophotometry (Figure 52, top left). The concentration of doxorubicin was measured by quantifying the sample absorbance at λ=485 nm. At that wavelength there is no spectral overlap with the PAP-DKP-Lys **1** molecule. Calibration was performed using solutions of doxorubicin at the concentration range of 10 mg/L to 1000 mg/L (the maximal possible DOX concentration in case if 100% of the material was instantly released from the gel) and a mixture of DOX and DKP **1** in the ratio 1:7.5 (identical as the ratio in the gel) at the same concentration range of doxorubicin.

5.2.3.10 Release of htDNA

Quantification of the released DNA was more complicated, as the DKP **1** molecule strongly absorbs in the UV-light range that is normally used to quantify DNA. After comparing various techniques (HPLC alone or in combination with chemical reduction of azobenzene or acetylation of the lysine side chain) to separate the two molecules from our samples we obtained the best results using size-exclusion chromatography (SEC). The 1 mL sample (yellow solution, the colour comes from **1**) directly from the release experiment was applied on the DextraSEC PRO10 size-exclusion cartridge (AppliChem, cat.#: A8822,0050), previously equilibrated with 50 mM aqueous NaCl, and eluted according to the attached protocol to obtain 1.5 mL of the filtrate as colourless liquid. The yellow band containing DKP **1** remained on the column and could be fully removed with further 15 mL of the eluent. The amount of DNA in the filtrate was quantified using Nanodrop-1000 with the standard procedure for dsDNA quantification. Every fraction was additionally controlled with standard UV-Vis spectrum. The DKP **1** has a strong absorption maximum at λ =327 nm, but no peak at that wavelength was detected in the filtrates from the size-exclusion column, which confirms good separation of both components. As the size-exclusion filtration leads to certain material losses due to various factors (e.g. incomplete material separation, aggregation, incomplete collection of the material in the fractions collected according to standard protocols), to measure the material recovery after size-exclusion filtration we prepared (in triplicate) a 1 mL sample of 100 mg/L htDNA and 750 mg/L DKP **1** in 50 mM

NaCl. Those amounts exactly corresponded to 10% of the total material that could be released after full degradation of our gel sample. The sample was applied on the SEC column and eluted in the identical way as described above to the final volume of 1.5 mL. The average value of DNA concentration measured for three output samples was multiplied by the factor of 10 and the resulting absorbance value was used as the reference: "100% material release".

The 1 mL aliquots from the htDNA release experiments performed in absence of light or under UV irradiation were purified on the SEC columns. The DNA concentration was measured and compared with the reference absorbance "100% material release" described above. In the samples incubated in darkness, the DNA release was well within the experimental error, so the total 1.42% material release reported on the plot below is the maximal value, but it rather reflects the sum of baseline fluctuations during the absorbance measurements. Even that result is, however, in the striking contrast with the amount of htDNA recovered after UV irradiation of gel samples. Within 30 min all examined gels fully degrade to fluid phase. We would therefore expect total recovery of DNA at the same range as for irradiated gel samples with doxorubicin (above 90%, as the material loss during size-exclusion chromatography are already taken into account). The value of 42% was initially a little puzzling. However, after careful inspection of the used aliquots and the remaining glass vials we have noticed subtle flakes of precipitation that were most likely some form of DNA aggregation (no such material was found in analogous DOX/DKP gels). This insoluble material was obviously retained on the top of SEC columns during filtration so it could not be properly re-solubilized.

In summary, we have recovered around half of the total oligonucleotide material in soluble form after size-exclusion purification (Figure 52, top right). The outcome of these release experiments was also summarized on the **Fig. 5** of the manuscript. The total material recovery in a soluble form is compared on the bottom of the Figure 52 for both types of the guest molecules in darkness and upon irradiation.

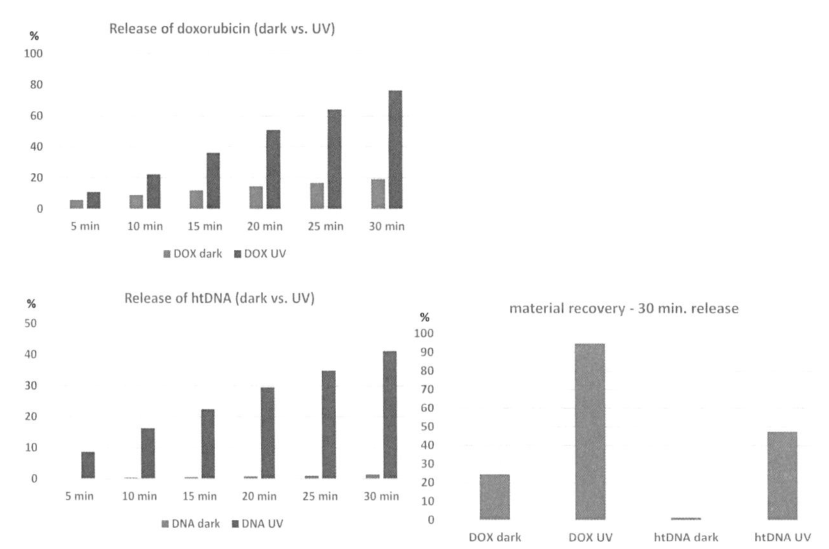

Figure 52. Guest release experiments. Release of doxorubicin (top left) upon UV irradiation (red) or in darkness by diffusion (blue); release of htDNA (bottom left) upon UV irradiation (red) or in darkness by diffusion (blue); total recovery of encapsulated material (bottom right) after 30 min of the release experiments – in darkness ("dark") or under UV-light irradiation ("UV") – from left to right: doxorubicin recovered in absence of light, doxorubicin recovered from UV-irradiated gels, soluble htDNA recovered in absence of light, soluble htDNA recovered from UV-irradiated samples.

5.2.3.11 Influence of NaCl concentration on rigidity of the hydrogels

Three gels with constant concentration of **1** (15 g/L) and variable concentrations of NaCl in water were characterized by rheology. Additionally, the results for a gel sample composed of 30 g/L of **1** in 2 M aqueous NaCl solution are demonstrated. Based on these results, we can see that the rigidity of gels consistently increases with the NaCl content, and the increase is the fastest in the range between 0 mM and 200 mM NaCl.

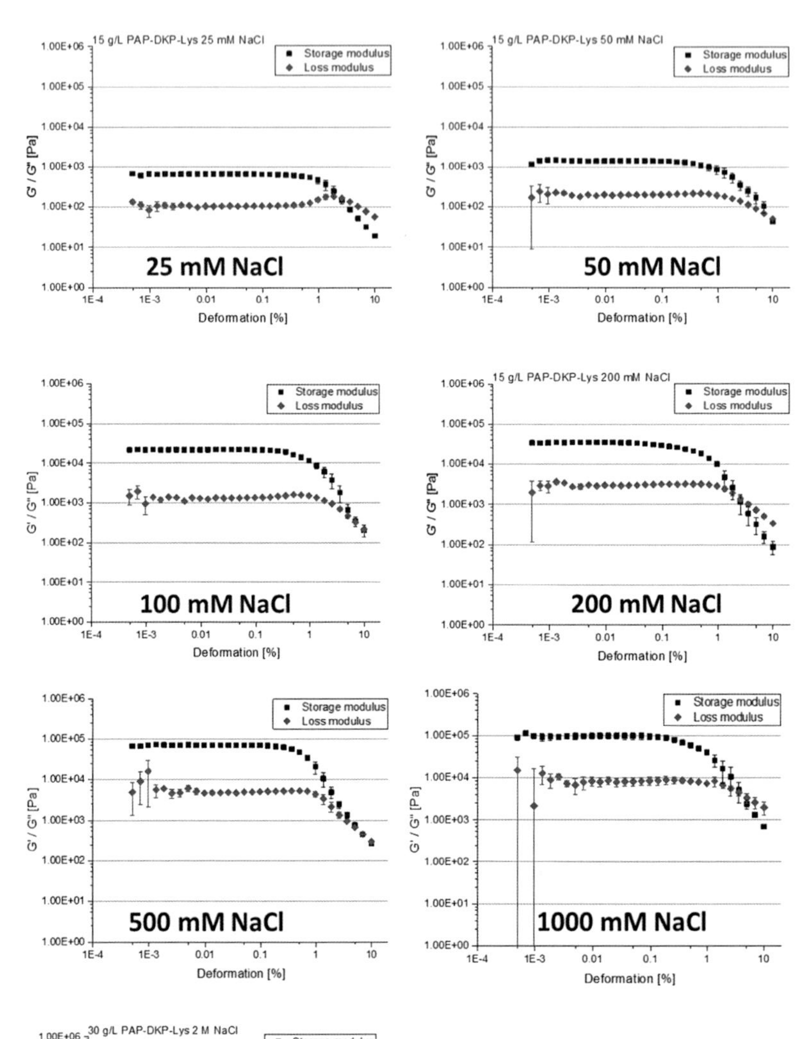

Figure 53. Strain sweep experiments: hydrogels composed of 15 g/L 1 and various amounts of NaCl (25 mM – 1000 mM) in water.

Bottom – the analogous experiment for 30 g/L of 1 and 2 M NaCl.

5.3 F_2-PAP-DKP-Lys

The following chapters about the properties of F_2-PAP-DKP-Lys **2** are quoted verbatim from the supporting information.[298b]

5.3.1 Synthesis of F_2-PAP-DKP-Lys

Boc-F_2-PAP-OH, (*S,E*)-2-((*tert*-butoxycarbonyl)amino)-3-(4-((2,6-difluorophenyl)diazenyl)phenyl)propanoic acid 12

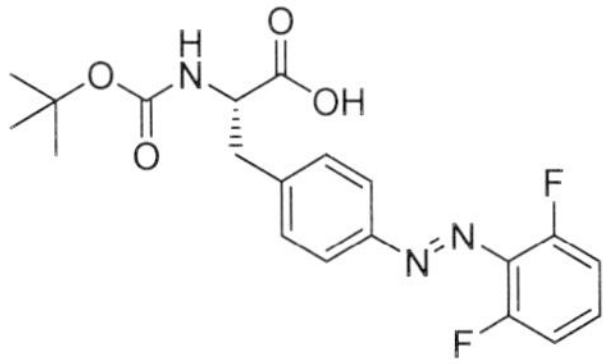

2,6-difluoronitrosobenzene **11** was prepared *in situ* from 2,6-difluoroaniline **35** (2.60 g, 20.1 mmol, 1.00 eq.) dissolved in CH_2Cl_2 (80 mL) and Oxone® (24.6 g, 40.0 mmol, 2.00 eq.) dissolved in water (240 mL). Both solutions were mixed and stirred vigorously for 4 h at rt. The mixture, initially colourless, became dark-green upon stirring. Afterwards stirring was switched off, the organic phase was separated and washed with 150 mL of water in a separation funnel. The organic phase was then dried with anhydrous sodium sulphate, filtered and evaporated to dryness at 40 °C yielding pale-brown solid with tendency for sublimation upon longer evaporation or exposure to high vacuum. The product was directly used as a crude for the following reaction.[2] 2,6-difluoronitrosobenzene **11** (crude, 20.1 mmol, 1.33 eq.) and (*S,E*)-2-((*tert*-butoxycarbonyl)amino)-3-(4-aminophenyl)propanoic acid **8** (abbr. Boc-Phe(4-NH_2)-OH) (4.20 g, 15.0 mmol, 1.00 eq.) were dissolved in glacial AcOH (80 mL) in a round-bottom flask and stirred overnight at rt. The solvent was then evaporated under reduced pressure and the residual oil crude was co-evaporated three times with PhMe (100 mL). This way, most of the unreacted 2,6-difluoronitrosobenzene was removed from the crude and was found in the distilled PhMe. The non-volatile crude was purified by silica gel column chromatography. The crude was dissolved in CH_2Cl_2 and applied on the column that was pre-equilibrated with CH_2Cl_2. The column was eluted with pure CH_2Cl_2 until the initial yellow fractions (containing mainly the residual nitrosobenzene) were collected and the flow-through liquid became colourless. The pure product was then eluted with 2% MeOH in CH_2Cl_2 and 10% MeOH in CH_2Cl_2 (R_f = 0.4 in CH_2Cl_2: MeOH 10:1). After evaporation of the solvents and drying *in vacuo*, 4.26 g Boc-F_2-PAP-OH **12** (10.5 mmol, 70% yield) was obtained as red solid.

^{1}H NMR (400 MHz, DMSO): δ = 12.71 (s, 1H), 7.80 (d, J = 8.3 Hz, 2H), 7.51 (dd, J = 13.3, 8.3 Hz, 3H), 7.31 (t, J = 8.8 Hz, 2H), 7.22 (d, J = 8.5 Hz, 1H), 4.18 (td, J = 10.3, 4.6 Hz, 1H), 3.24 – 2.82 (m, 2H), 1.31 (s, 9H) ppm. **^{13}C NMR (101 MHz, DMSO):** δ = 173.4, 156.0 (d, J = 4.4 Hz), 155.5, 153.5 (d, J = 4.3 Hz), 151.3, 143.4, 131.8 – 131.3 (m), 130.7 – 130.3 (m), 130.4, 122.5, 113.2 – 112.8 (m), 78.1, 54.9, 36.4, 28.1 ppm. **^{19}F NMR (376 MHz, acetic acid-*d*$_4$):** δ = −122.52 ppm. **HRMS (ESI+):** *m/z* calcd for $[C_{20}H_{21}O_4N_3F_2]^+$: 405.1500 Da $[M+H]^+$, found: 405.1498 Da (Δ = 0.5 ppm). **IR (ATR):** ṽ = 3346 (w), 2973 (vw), 1722 (w), 1683 (m), 1613 (w), 1588 (w), 1522 (m), 1479 (w), 1460 (w), 1444 (w), 1393 (w), 1368 (w), 1340 (w), 1239 (m), 1156 (m), 1103 (w), 1052 (w), 1023 (m), 946 (vw), 892 (w), 844 (w), 786 (m), 743 (w), 637 (w), 596 (w), 573 (w), 525 (w), 511 (w), 483 (w) cm^{-1}. **elemental analysis calcd for $C_{20}H_{21}F_2N_3O_4$ (%):** C: 59.25, H: 5.22, F: 9.37, N: 10.37, O: 15.79, found: C: 59.04, H: 4.98, N: 10.34.

Boc-F$_2$-PAP-Lys(Boc)-OMe, methyl *N*^{6}-(*tert*-butoxycarbonyl)-*N*^{2}-((*S*)-2-((*tert*-butoxycarbonyl)amino)-3-(4-((*E*)-(2,6-difluorophenyl)diazenyl)phenyl)propanoyl)-*L*-lysinate 14

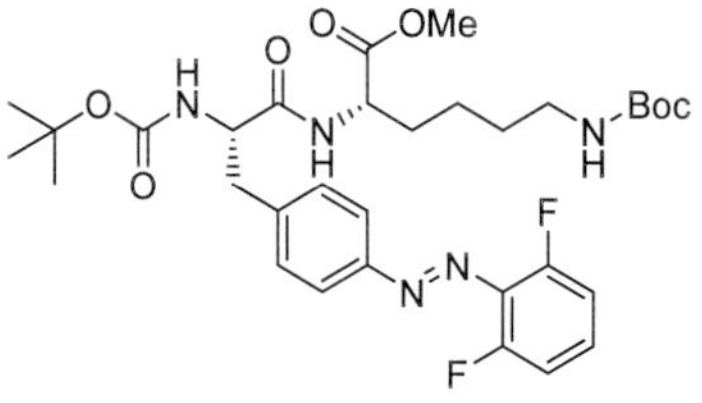

F$_2$-PAP(Boc)-OH **12** (3.28 g, 8.42 mmol, 1.00 eq.), HBTU (3.19 g, 8.42 mmol, 1.00 eq.) and DIPEA (3.5 mL, 21 mmol, 2.50 eq.) were dissolved in anhydrous DMF (25 mL) and stirred for 15 min at rt under argon. The same amount of DIPEA (3.5 mL, 21.0 mmol, 2.50 eq.) was added together with solid *ω*-*N*-Boc-lysine methyl ester hydrochloride (2.50 g, 8.42 mmol, 1.00 eq.). The reaction mixture was stirred further at rt under argon and the reaction progress was followed by TLC. After 2 h, full conversion of the starting material **12** was observed. The reaction mixture was quenched with sat. aqueous NH_4Cl solution (200 mL) and extracted once with EtOAc (200 mL). The organic layer was washed with sat. aqueous NH_4Cl solution (3×200 mL), brine (1×200 mL), dried with anhydrous Na_2SO_4 and the solvent was evaporated under reduced pressure to dryness. The crude was purified by silica gel column chromatography. Initial elution with cH:EtOAc 3:1 washed out colourless non-polar impurities. Then the column was eluted with cH:EtOAc 2:1. Evaporation of combined orange-red fractions and drying under reduced pressure (oil pump) resulted in 4.65 g Boc-F_2-PAP-Lys(Boc)-OMe **14** (7.18 mmol, 85% yield) as red solid.

TLC: $R_f = 0.45$ (cH:EtOAc 1:1). **^{1}H NMR (400 MHz, acetic acid-*d*$_4$):** δ = 7.92 (d, *J* = 7.9 Hz, 2H), 7.50 (d, *J* = 8.1 Hz, 2H), 7.49 – 7.35 (m, 1H), 7.15 (t, *J* = 8.8 Hz, 2H), 4.82 – 4.50 (m, 2H), 3.75 (d, *J* = 19.2 Hz, 3H), 3.41 – 2.98 (m, 4H), 1.98 – 1.84 (m, 1H), 1.74 (dt, *J* = 13.6, 7.2 Hz, 1H), 1.51 (s, 6H), 1.47 (s, 6H), 1.42 (s, 9H) ppm. **^{13}C NMR (101 MHz, acetic acid-*d*$_4$):** δ = 174.9, 174.3, 158.9, 158.7 (d, *J* = 4.4 Hz), 158.2, 156.2 (d, *J* = 4.3 Hz), 154.0, 143.3, 133.3 – 132.7 (m), 132.3, 132.1, 124.6, 114.2 (d, *J* = 22.5 Hz), 82.2, 81.3, 57.4, 54.2, 53.9, 41.7, 39.6, 32.9, 30.8, 29.4, 29.3, 24.2 ppm. **^{19}F NMR (376 MHz, acetic acid-*d*$_4$):** δ = −122.88 ppm. **HRMS (ESI+):** *m/z* calcd for $[C_{32}H_{43}F_2N_5O_7]^+$: 648.3203 Da $[M+H]^+$, found: 648.3205 Da (Δ = 0.3 ppm). **IR (ATR):** ṽ = 3332 (vw), 2931 (vw), 1743 (vw), 1683 (w), 1656 (w), 1614 (vw), 1516 (w), 1470 (w), 1390 (vw), 1365 (w), 1242 (w), 1163 (w), 1047 (vw), 1012 (w), 851 (vw), 781 (w), 715 (vw), 636 (vw), 529 (vw), 432 (vw) cm^{-1}. **elemental analysis calcd for $C_{21}H_{21}F_4N_3O_4$ (%):** C: 55.39, H: 4.65, F: 16.69, N: 9.23, O: 14.05, found: C: 55.17, H: 4.47, N: 9.07.

H-F_2-PAP-Lys-OMe, methyl *N*^{2}-((*S*)-2-amino-3-(4-((*E*)-(2,6-difluorophenyl)diazenyl)phenyl]propanoyl)-*L*-lysinate 14d

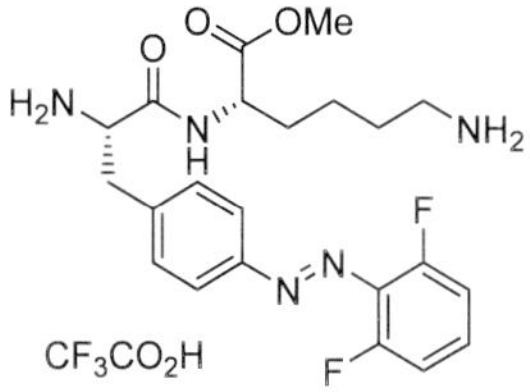

Boc-F_2-PAP-Lys(Boc)-OMe **14** (4.35 g, 6.72 mmol, 1.00 eq.) was dissolved in CH_2Cl_2 (80 mL) yielding red-orange solution. Trifluoroacetic acid (TFA, 80 mL) was added at rt and the colour of the mixture changed to deep red. The mixture was stirred for 1 h at rt. Then PhMe (100 mL) was added and the solvent was removed under reduced pressure. The crude product was obtained quantitatively with 4.54 g H-F_2-PAP-Lys-OMe **14d** (6.72 mmol, quant.) as a TFA salt and can be used directly after co-evaporation in the next reaction. For analysis the crude can be purified by silica gel column chromatography 2% TFA, 13% MeOH in CH_2Cl_2 ($R_f = 0.25$ in 2% TFA, 13% MeOH in CH_2Cl_2).

^{1}H NMR (400 MHz, acetic acid-*d*$_4$): δ = 8.07 (d, *J* = 8.4 Hz, 2H), 7.69 (d, *J* = 8.4 Hz, 2H), 7.61 (tt, *J* = 8.4, 5.9 Hz, 1H), 7.31 (t, *J* = 8.7 Hz, 2H), 4.88 (t,

J = 7.0 Hz, 1H), 4.77 (dd, J = 8.9, 4.8 Hz, 1H), 3.90 (s, 3H), 3.57 (dd, J = 7.0, 3.1 Hz, 2H), 3.26 (t, J = 7.5 Hz, 2H), 2.18 – 2.00 (m, 1H), 1.92 (td, J = 8.9, 8.4, 5.1 Hz, 3H), 1.64 (t, J = 7.7 Hz, 2H) ppm. **^{13}C NMR (101 MHz, acetic acid-*d*$_4$):** δ = 173.6, 170.7, 164.0, 163.6, 158.7 (d, J = 4.3 Hz), 156.1 (d, J = 4.3 Hz), 154.5, 140.0, 132.9 (t, J = 10.5 Hz), 132.7 (d, J = 10.7 Hz), 132.5, 124.9, 114.4 (d, J = 5.0 Hz), 114.2 (d, J = 4.8 Hz), 56.3, 54.3, 54.0, 41.4, 38.5, 32.5, 28.1, 23.8 ppm. **^{19}F NMR (376 MHz, acetic acid-*d*$_4$):** δ = −123.29 ppm. **HRMS (ESI+):** *m/z* calcd for $[C_{22}H_{28}F_2N_5O_3]^+$: 448.2155 Da $[M+H]^+$, found: 448.2154 Da (Δ = 0.2 ppm). **IR (ATR):** $\tilde{\nu}$ = 3314 (vw), 2943 (w), 1666 (m), 1611 (w), 1556 (w), 1470 (w), 1441 (w), 1352 (w), 1295 (vw), 1201 (m), 1182 (m), 1126 (m), 1061 (w), 1026 (m), 982 (w), 897 (w), 863 (vw), 837 (w), 800 (w), 788 (m), 722 (m), 625 (w), 598 (vw), 584 (w), 559 (vw), 527 (w), 488 (vw), 410 (vw) cm^{-1}. **elemental analysis calcd for $C_{26}H_{33}F_8N_5O_9$ (%):** C: 43.89, H: 4.67, F: 21.36, N: 9.84, O: 20.24, found: C: 43.82, H: 4.32, N: 9.84.

F$_2$-PAP-DKP-Lys, (3*S*,6*S*)-3-(4-aminobutyl)-6-(4-((*E*)-(2,6-difluorophenyl)diazenyl)benzyl)piperazine-2,5-dione 2

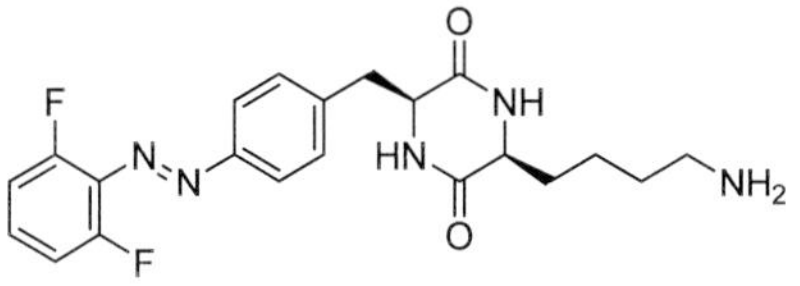

The crude H-F$_2$-PAP-Lys-OMe **14d** from the previous experiment (4.54 g, 6.72 mmol, 1.00 eq.) was dissolved in 2-butanol (130 mL). It was mixed with glacial AcOH (1.2 mL, 21.0 mmol, 3.13 mmol), *N*-methylmorpholine (800 µL, 7.28 mmol, 1.08 mmol) and *N,N*-diisopropyl-*N*-ethylamine (DIPEA, 1.6 mL, 9.19 mmol, 1.37 mmol). The resulting mixture was refluxed for 2 h in an oil bath with an external heating of 120 °C, then cooled down. Next, solvents were evaporated to dryness on the rotavap and the crude was recrystallized from acetonitrile (MeCN, 600 mL). The gel-like precipitate was filtered off, dried on the filter and *in vacuo* using high-vacuum oil pump overnight resulting in 2.99 g F$_2$-PAP-DKP-Lys **2** (5.65 mmol, 84% yield) as red solid. Fractional recrystallization over three steps in MeCN is necessary to obtain analytic purity and the given yield. Purity was determined by analytical HPLC with standard setting over 20 min run with a gradient of 5-95% MeCN in diH_2O, with 0.1% TFA.

TLC: R_f = 0.28 (5% TFA, 15% MeOH in CH_2Cl_2). **^{1}H NMR (400 MHz, acetic acid-*d*$_4$):** δ = 8.05 (d, J = 8.1 Hz, 2H), 7.60 (d, J = 8.4 Hz, 3H), 7.27 (t, J = 8.9 Hz, 2H), 4.77 (t, J = 4.8 Hz, 1H), 4.16 (dd, J = 8.5, 4.3 Hz, 1H), 3.47 (ddd, J = 79.9,

13.9, 4.8 Hz, 2H), 3.09 (t, J = 7.4 Hz, 2H), 1.64 (dq, J = 16.7, 10.4, 8.8 Hz, 3H), 1.31 (dq, J = 15.3, 8.0 Hz, 2H), 0.95 (tq, J = 14.1, 8.9, 7.0 Hz, 1H) ppm. **^{13}C NMR (101 MHz, acetic acid-d_4):** δ = 171.7, 170.7, 158.6 (d, J = 4.4 Hz), 156.1 (d, J = 4.3 Hz), 154.3, 141.4, 133.2, 132.9 (t, J = 10.4 Hz), 132.6 (t, J = 10.4 Hz), 124.6, 114.4 – 114.0 (m), 57.3, 55.6, 41.1, 40.6, 34.8, 27.7, 22.6 ppm. **^{19}F NMR (376 MHz, acetic acid-d_4):** δ = −123.07 ppm. **HRMS (ESI+)** *m/z* calcd for $[C_{21}H_{24}F_2N_5O_2]^+$: 416.1893 Da $[M+H]^+$, found: 416.1893 Da (Δ = 0.2 ppm); **IR (ATR):** ṽ = 2954 (w), 1661 (m), 1614 (m), 1458 (w), 1336 (w), 1242 (w), 1200 (m), 1176 (m), 1128 (m), 1025 (w), 832 (w), 784 (w), 720 (w), 599 (vw), 581 (vw), 560 (vw), 510 (vw), 485 (vw), 434 (w) cm^{-1}.

CiproFAzo, (*S,E*)-7-(4-(2-amino-3-(4-((2,6-difluorophenyl)diazenyl)phenyl)propanoyl)piperazin-1-yl)-1-cyclopropyl-6-fluoro-4-oxo-1,4-dihydroquinoline-3-carboxylic acid 15

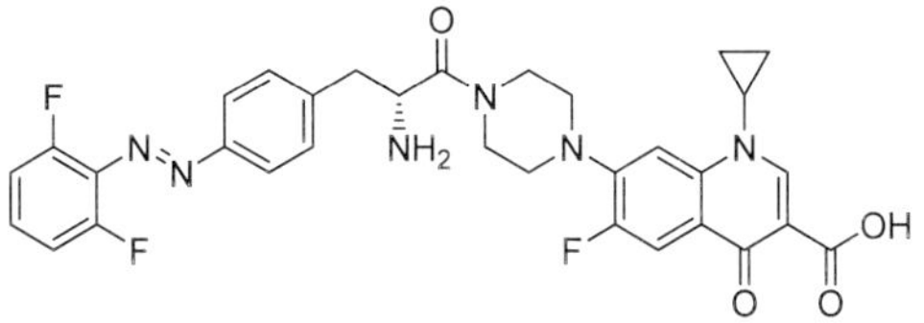

F_2-PAP(Boc)-OH **12** (0.500 g, 1.23 mmol, 1.00 eq.), HBTU (0.468 g, 1.23 mmol, 1.00 eq.) and DIPEA (0.43 mL, 2.47 mmol, 2.00 eq.) were dissolved in anhydrous DMF (6.17 mL). The reaction was stirred for 10 min at rt under argon. Then a solution of ciprofloxacin (0.407 g, 1.23 mmol, 1.00 eq.) and DIPEA (0.159 mg, 1.23 mmol, 1.00 eq.) dissolved in anhydrous DMF (6.17 mL) was added to the reaction mixture. The reaction was stirred at rt under argon for further 2 h and quenched with aqueous ammonium chloride (12 mL), extracted with CH_2Cl_2 (3×100 mL), washed with aqueous ammonium chloride (3×10 mL), brine (10 mL), dried with anhydrous sodium sulphate and evaporated under reduced pressure. The crude was obtained in 0.885 g Boc-CiproFAzo (1.23 mmol, quant.) as an orange solid (R_f = 0.13-0.30 in 1% TFA in EtOAc) and used without further purification in the next step. The crude (873 mg, 1.22 mmol, 1.00 eq.) was dissolved in CH_2Cl_2 (12 mL), Trifluoroacetic acid (TFA, 12 mL) and 1 vol% Triisopropylsilane (0.24 mL) at rt. The mixture was stirred for 1 h at rt, evaporated under reduced pressure, dissolved in a minimal amount of MeOH (5 mL) and precipitated by adding the solution into stirred diethyl ether (50 mL). After filtration and high-vacuum drying, 0.746 g of the crude was obtained. The product was purified by HPLC with the following settings: C_{18} column, detection at 330 nm, flow rate

15 mL/min, rt, injection at 5% of the mixture of ACN:MeOH 1:1 in diH_2O with 0.1% TFA, elution conditions: 30 min gradient from 20-80% ACN:MeOH 1:1 in diH_2O with 0.1% TFA, retention time: 22 min. After lyophilisation 679 mg CiproFAzo **15** (0.927 mmol, 76%) was obtained as an orange solid.

^{1}H NMR (400 MHz, acetic acid-*d*$_4$): δ = 8.78 (s, 1H), 8.01 (d, J = 8.0 Hz, 2H), 7.94 (d, J = 13.1 Hz, 1H), 7.69 (d, J = 8.0 Hz, 2H), 7.47 (ddd, J = 20.5, 11.4, 4.8 Hz, 2H), 7.13 (t, J = 9.3 Hz, 2H), 5.26 (dd, J = 10.1, 5.4 Hz, 1H), 4.23 (d, J = 14.3 Hz, 1H), 4.08 – 3.24 (m, 9H), 3.15 (t, J = 8.7 Hz, 1H), 1.55 (s, 2H), 1.36 (s, 2H) ppm. **^{13}C NMR (101 MHz, acetic acid-*d*$_4$):** δ = 178.3 (d, J = 2.5 Hz), 170.4, 169.0, 163.4 (q, J = 37.0 Hz), 158.5 (d, J = 4.2 Hz), 156.4, 156.0 (d, J = 4.2 Hz), 154.4, 153.9, 149.9, 146.8 (d, J = 10.5 Hz), 140.8, 140.0, 133.3 (t, J = 10.3 Hz), 132.9, 132.0 (t, J = 9.6 Hz), 125.2, 121.6 – 120.9 (m), 117.9 (d, J = 289.0 Hz), 114.4 (dd, J = 21.0, 2.7 Hz), 113.4 (d, J = 23.7 Hz), 108.6, 107.5, 52.4, 50.3 (dd, J = 15.5, 3.8 Hz), 47.2, 44.0, 38.9, 37.4, 9.3 (d, J = 7.7 Hz) ppm. **^{19}F NMR (376 MHz, acetic acid-*d*$_4$):** δ = −122.00, −122.40 ppm. **HRMS (ESI+):** *m/z* calcd for $[C_{32}H_{31}N_5O_6F]^+$: 619.2242 Da $[M+H]^+$, found: 619.2242 Da (Δ = 0.1 ppm). **IR (ATR):** $\tilde{\nu}$ = 2913 (vw), 1664 (w), 1626 (w), 1452 (m), 1385 (w), 1334 (w), 1265 (w), 1243 (w), 1200 (w), 1178 (w), 1131 (w), 1022 (w), 944 (vw), 884 (vw), 832 (w), 786 (w), 749 (vw), 720 (w), 665 (vw), 619 (vw), 579 (vw), 563 (vw), 508 (w), 396 (vw) cm^{-1}.

N-(4-carboxybutanoyl)-ciprofloxacin, 7-(4-(4-carboxybutanoyl)piperazin-1-yl)-1-cyclopropyl-6-fluoro-4-oxo-1,4-dihydroquinoline-3-carboxylic acid 69

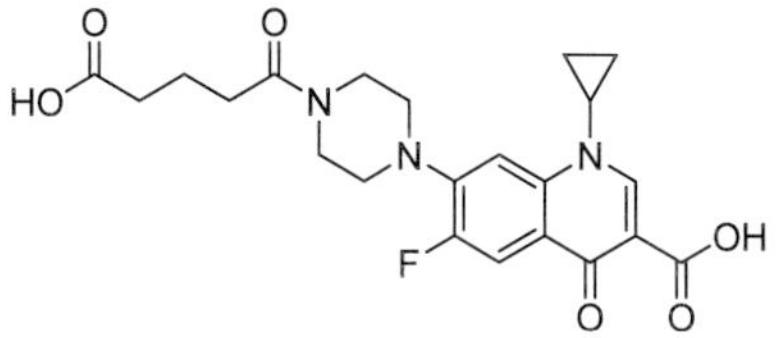

A mixture of ciprofloxacin (500 mg, 1.51 mmol, 1.00 eq.), trimethylamine (458 mg, 4.53 mmol, 3.00 eq.) and glutaric anhydride (172 mg, 1.51 mmol, 1.00 eq.) in CH_2Cl_2 (30 mL) was allowed to stir 3 d at rt.[303] The product was precipitated by adding the reaction mixture into diluted aqueous HCl (30 mL, 1 mol/L). The precipitate was washed with diH_2O (10 mL), Et_2O (10 mL) and CH_2Cl_2 (10 mL) and dried under airstream and high vacuum. The 544 mg N-(4-carboxybutanoyl)-ciprofloxacin **69** (1.22 mol, 81%) were used without further purification in the next reaction, but can be purified by recrystallization in AcOH.

^{1}H NMR (300 MHz, D_2O): δ = 8.35 (s, 1H), 7.57 (d, *J* = 13.4 Hz, 1H), 7.23 (d, *J* = 7.3 Hz, 1H), 3.71 (d, *J* = 5.6 Hz, 4H), 3.45 – 3.28 (m, 1H), 3.27 – 2.95 (m, 4H), 2.47 (t, *J* = 7.8 Hz, 2H), 2.23 (t, *J* = 7.4 Hz, 2H), 1.82 (t, *J* = 7.7 Hz, 2H), 1.24 (d, *J* = 6.8 Hz, 2H), 0.92 (s, 2H) ppm. **^{13}C NMR (101 MHz, D_2O):** δ = 182.5, 175.2 (d, *J* = 2.4 Hz), 174.4, 172.3, 154.1, 151.6, 146.9, 143.5 (d, *J* = 10.7 Hz), 138.3, 121.9 (d, *J* = 7.1 Hz), 116.4, 111.4 (d, *J* = 22.8 Hz), 106.1, 49.7 (d, *J* = 38.4 Hz), 45.5, 41.5, 36.8, 34.7, 32.4, 22.1, 7.4 ppm. **^{19}F NMR (376 MHz, acetic acid-*d*$_4$):** δ = −121.76 ppm. **HRMS (ESI+):** *m/z* calcd for $[C_{22}H_{25}N_3O_6F]^+$: 446.1722 Da $[M+H]^+$, found: 446.1721 Da (Δ = 0.3 ppm). **IR (ATR):** ṽ = 3042 (vw), 1715 (w), 1624 (w), 1492 (w), 1447 (w), 1383 (w), 1339 (w), 1295 (w), 1269 (w), 1244 (w), 1205 (w), 1145 (w), 1030 (w), 980 (w), 932 (w), 885 (w), 832 (w), 801 (w), 775 (w), 745 (w), 702 (w), 666 (vw), 604 (w), 550 (vw), 535 (vw), 481 (w), 504 (w) cm^{-1}.

CiproDKP, 1-cyclopropyl-7-(4-(5-((4-((2*S*,5*S*)-5-(4-((*E*)-(2,6-difluorophenyl)diazenyl)benzyl)-3,6-dioxopiperazin-2-yl)butyl)amino)-5-oxopentanoyl)piperazin-1-yl)-6-fluoro-4-oxo-1,4-dihydroquinoline-3-carboxylic acid 16

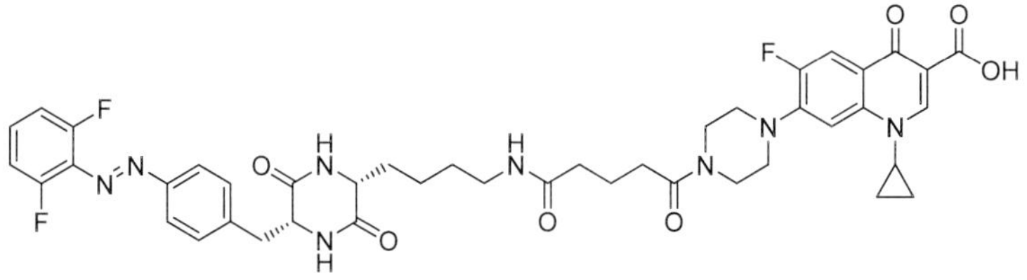

N-(4-carboxybutanoyl)-ciprofloxacin **69** (300 mg, 0.673 mmol, 1.00 eq.), HBTU (255 mg, 0.673 mmol, 1.00 eq.) and DIPEA (122 mg, 0.943 mmol, 1.40 eq.,) was dissolved in anhydrous DMF (4.7 mL). The reaction was stirred for 10 min at rt under argon. Then a solution of F_2-PAP-DKP-Lys **2** (280 mg, 0.673 mmol, 1.00 eq.) and DIPEA (87 mg, 0.673 mmol, 1.00 eq.) dissolved in anhydrous DMF (3.37 mL) was added to the reaction mixture. The reaction was stirred at rt for 2 h and quenched with aqueous ammonium chloride (12 mL). The resulting orange precipitate was filtered, washed with water (2×10 mL) and diethyl ether (2×10 mL). The product was purified by HPLC with following settings: C_{18} column, detection at 330 nm, flow rate 15 mL/min, rt, injection at 5% of the mixture of ACN:MeOH 1:1 in diH_2O with 0.1% TFA, elution conditions: 30 min gradient from 20-80% ACN:MeOH 1:1 in diH_2O with 0.1% TFA, retention time: 28.5 min.

After lyophilisation 192 mg of the CiproDKP **16** (0.276 mmol, 41%) was obtained as an orange solid.

^{1}H NMR (400 MHz, acetic acid-d_4): δ = 9.01 (s, 1H), 8.16 (d, *J* = 13.1 Hz, 1H), 8.02 (d, *J* = 8.0 Hz, 2H), 7.71 (d, *J* = 5.9 Hz, 1H), 7.57 (d, *J* = 8.1 Hz, 3H), 7.25 (t, *J* = 8.9 Hz, 2H), 4.75 (t, *J* = 4.7 Hz, 1H), 4.11 (dd, *J* = 8.2, 4.3 Hz, 1H), 3.98 (d, *J* = 24.2 Hz, 5H), 3.67 – 3.48 (m, 5H), 3.33 (dd, *J* = 13.8, 4.8 Hz, 1H), 3.28 – 3.09 (m, 2H), 2.66 (t, *J* = 7.6 Hz, 2H), 2.46 (t, *J* = 7.1 Hz, 2H), 2.07 (t, *J* = 7.4 Hz, 2H), 1.56 (d, *J* = 6.5 Hz, 3H), 1.41 (s, 4H), 1.17 (dt, *J* = 15.9, 8.3 Hz, 2H), 0.89 (dd, *J* = 13.8, 6.4 Hz, 1H) ppm. **^{13}C NMR (101 MHz, acetic acid-d_4):** δ = 179.4, 176.3, 174.9, 172.7, 171.8, 170.6, 158.6 (d, *J* = 4.3 Hz), 156.7, 156.1 (d, *J* = 4.2 Hz), 154.3, 154.2, 150.3, 147.2 (d, *J* = 10.6 Hz), 141.5, 141.2, 133.2, 132.8, 132.6, 124.7, 121.4 (d, *J* = 7.9 Hz), 114.2 (d, *J* = 21.5 Hz), 113.7 (d, *J* = 23.8 Hz), 108.7, 107.7, 57.3, 55.9, 51.5, 50.9, 47.3, 43.3, 40.7 (d, *J* = 10.7 Hz), 37.5, 36.5, 35.1, 33.8, 29.9, 23.1 (d, *J* = 10.4 Hz), 9.3 ppm. **^{19}F NMR (376 MHz, acetic acid-d_4):** δ = −121.78, 122.60 ppm. **HRMS (ESI+):** *m/z* calcd for $[C_{43}H_{46}N_8O_7F]^+$: 843.3442 Da $[M+H]^+$, found: 843.3441 Da (Δ = 0.1 ppm). **IR (ATR):** ṽ = 3043 (vw), 2930 (vw), 1724 (vw), 1663 (w), 1626 (w), 1505 (w), 1455 (w), 1387 (w), 1334 (w), 1302 (vw), 1242 (w), 1209 (w), 1148 (w), 1022 (w), 950 (vw), 886 (vw), 831 (w), 806 (w), 784 (w), 747 (w), 704 (w), 621 (vw), 599 (vw), 581 (vw), 562 (vw), 510 (vw), 485 (vw), 438 (vw), 404 (vw) cm^{-1}. **elemental analysis calcd for $C_{45}H_{46}F_6N_8O_9$ (%):** C: 56.48, H: 4.85, F: 11.91, N: 11.71, O: 15.05, found: C: 56.09, H: 4.85, N: 11.67.

5.3.2 Photophysical properties of the gelator F_2-PAP-DKP-Lys 2

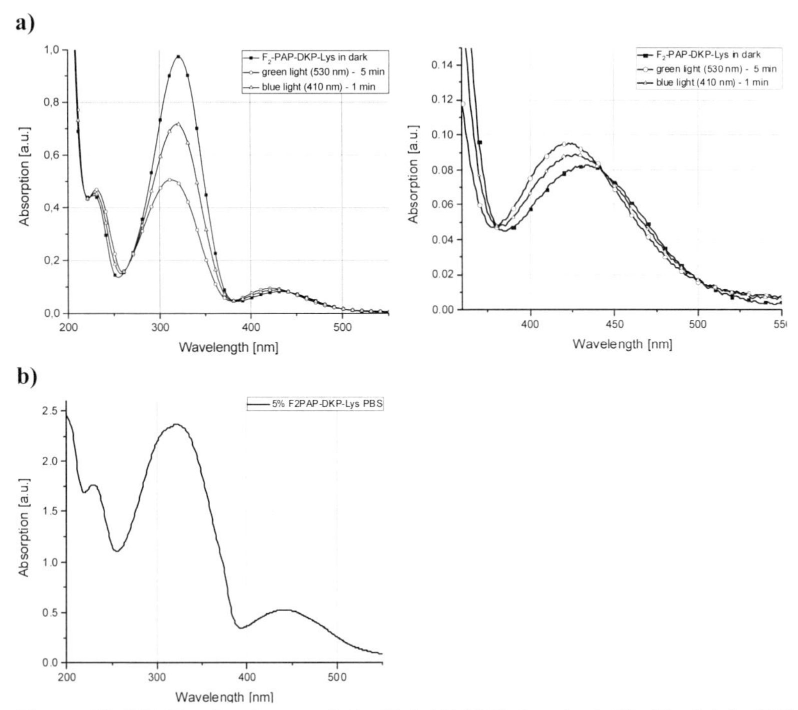

Figure 54. UV-Vis absorbance of the F_2-PAP-DKP-Lys 2. a) 72 μM of 1 in PBS buffer pH 7.4 in absence of light or upon irradiation with green (530 nm) or violet (410 nm) light – 3 W LED diodes (right side – magnification) [*spectrum measured in a quartz cuvette*]; b) 50 g/L (5%) of the gelator 1 in PBS buffer pH 7.4 (gel, > 95% *trans*-1) *(measurement of a thin layer between two quartz plates; the artefact at 318 nm occurs due to switching of the light sources).*

The 'dark state' refers to the *trans*-isomer obtained by thermal equilibration during the purification procedure (crystallization from MeCN) during synthesis of compound **2** (F_2-PAP-DKP-Lys) and its further storage in absence of light. Daylight irradiation results in decrease of the *trans*-isomer ratio in the mixture.

The samples were irradiated with various LED diodes in the range 365 – 530 nm, until the photostationary state (PSS) was achieved, followed by the absorption measurement at 315 nm (the λ_{max} of *trans*-**2**) .The relative depletion of the *trans*-isomer in the respective PSSs is plotted in comparison to the 'dark state' (Figure 55). This plot illustrates trends the change of the *trans*/*cis* ratio. However, it cannot directly serve for calculations of percentage of the photoisomers, as the summary absorption of both isomers is always observed.

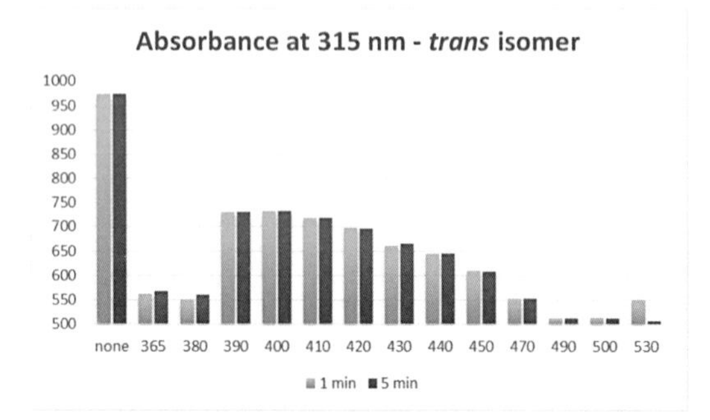

Figure 55. UV-Vis absorbance of the F_2-PAP-DKP-Lys 2 (72 μM in PBS buffer pH 7.4) at 315 nm upon sample irradiation with different wavelengths in the range of 365-530 nm.

For the HPLC-based quantification of the *cis*- and *trans*-isomer ratio of **2**, the isosbestic point at λ=224 nm was chosen, because at this point the molar absorptivity of both photoisomers is equal and a medium absorption 0.4450 ± 0.0035 au was determined for the 72 μM solution of F_2-PAP-DKP-Lys **2** in PBS. Due to different polarity of the azobenzene photoisomers, both isomers of the compound **2** exhibit different retention times and can be independently quantified.

A solution containing 72 μM F_2-PAP-DKP-Lys in PBS was irradiated for 5 min with 530 nm (3 W green LED), or with 410 nm (3 W violet LED). Then 50 μL of each solution, as well as 50 μL of the non-irradiated solution were injected in the HPLC under following conditions: run of 20 min, gradient 5% H_2O to 95% ACN with 0.1% TFA (8 nm slit, wavelength 224 nm). The sample irradiated with 530 nm contained 58% *cis* and 42% *trans*-isomer, and the sample irradiated with 410 nm – 35% *cis* and 65% *trans*-isomer, whereas the non-irradiated sample contained over 99% *trans*-**2**.

An analogous experiment was performed using DMSO as a solvent instead of the aqueous buffer. 25 mM solution of F_2-PAP-DKP-Lys in DMSO was irradiated by 10 W LED 523 nm or 3 W 410 nm for 30 min. Directly after irradiation, the solutions were analysed by HPLC with the following settings: C_{18} column, injection 1 μL, 4 nm slit, wavelength 224 nm, 20 min gradient from 5-95% ACN in diH_2O with 0.1% TFA. The results are summarized in Table 11.

Table 11. Quantification of *cis:trans* ratio by HPLC of 2.

Sample	Peak min	Area, 224 nm au^2	Ratio %	Isomer
not irradiated	10.3	87.9	0.9	cis
	11.2	9776.2	99.1	trans
523 nm, 30 min	10.3	6024.3	65.7	cis
	11.2	3142.4	34.3	trans
410 nm, 30 min	10.4	1983.1	20.4	cis
	11.2	7742.4	79.6	trans

Overall, the photostationary states of the compound **2** is to some extent depend on the solvent. However, neither in DMSO nor in aqueous buffer more than 70% of the *cis*-isomer can be achieved. Although this ratio is sufficient to achieve the desired macroscopic effects, in particular to dissipate hydrogel samples (*vide infra*), for other experiments (e.g.to measure the lifetime of *cis*-**2**) the isomers had to be separated by preparative HPLC. Due to the extraordinarily long lifetime of the *cis*-**2** at rt (Figure 60), both photoisomers can be then separately stored in darkness for several days without significant interconversion.

5.3.2.1 CD spectra of F_2-PAP-DKP-Lys

The spectra of the CD spectra of *E*- and *Z*-F2-PAP-DKP-Lys were measured with a CD spectrometer Jasco J-810. The sample was prepared according to the gel preparation by heating up the sample and the final concentration was 290 μM F2-PAP-DKP-Lys in 13.7 mM NaCl or 137 mM NaF. The *Z*-F2-PAP-DKP-Lys was isolated by preparative HPLC prior to the measurement. The distance of the quartz cuvette was 1 mm and the inside of the CD spectrometer was flushed with an inert argon atmosphere.

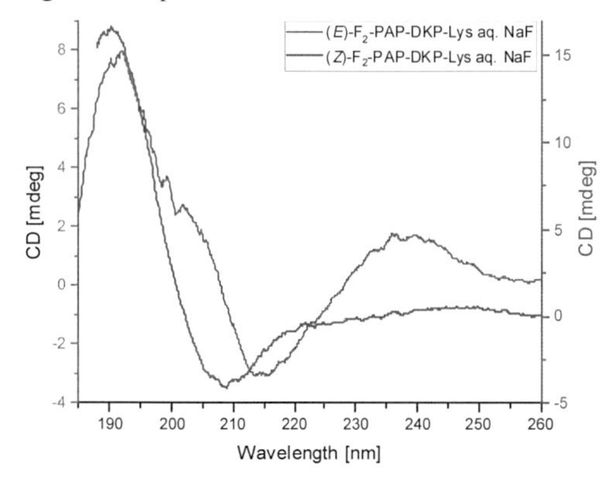

Figure 56. CD spectra of *E* and *Z*-F_2-PAP-DKP-Lys in 137 mM NaF.

5.3.2.2 Isomerization of F_3-PAP-OH, (*S,E*)-2-((*tert*-butoxycarbonyl)amino)-3-(4-((2,4,6-trifluorophenyl)diazenyl)phenyl)propanoic acid 13

The F_3-PAP-OH was prepared by Dr. Pianowski according the procedure of F_2-PAP-OH and characterised by Johannes Karcher in his master thesis to evaluate the properties F_2-PAP-OH and compare it to the similar derivative F_3-PAP-OH.[243a]

(*E*)- F_3-PAP-OH – ^{1}H NMR (300 MHz, ACN-d_3): δ =7.79 (m_c, 2 H, A_r-C_q-N), 7.43 (m_c, 2 H, A_r-C_q), 7.02 (m_c, 2 H, A_r-CF), 5.57 (bs, 1 H, NH), 4.41 (bs, 1 H, CH), 3.22 (m_c, 1 H, CH_2), 3.00 (m_c, 1 H, CH_2), 1.35 (s, 9 H, *tert*-Butyl) ppm, -CO_2H not visible.

(*Z*)- F_3-PAP-OH – ^{1}H NMR (300 MHz, ACN-d_3): δ =7.21 (m_c, 2 H), 6.90 (m_c, 2 H), 6.81 (m_c, 2 H), (5.49 (bs, 1 H, NH), 4.30 (bs, 1 H, CH), 3.01 (m_c, 1 H, CH_2), 2.87 (m_c, 1 H, CH_2), 1.34 (s, 9 H, *tert*-Butyl) ppm, -CO_2H not visible.

Table 12. *E/Z* ratio of F_3-PAP-OH in ACN.

Procedure	*E* / *Z* ratio	Peaks for calculation
Dark state	89% / 11%	7.79 / 7.21
30 min 460 nm	58% / 42%	7.79 / 7.21
30 min 365 nm	55% / 45%	7.79 / 7.21
30 min 523 nm	33% / 67%	7.79 / 7.21

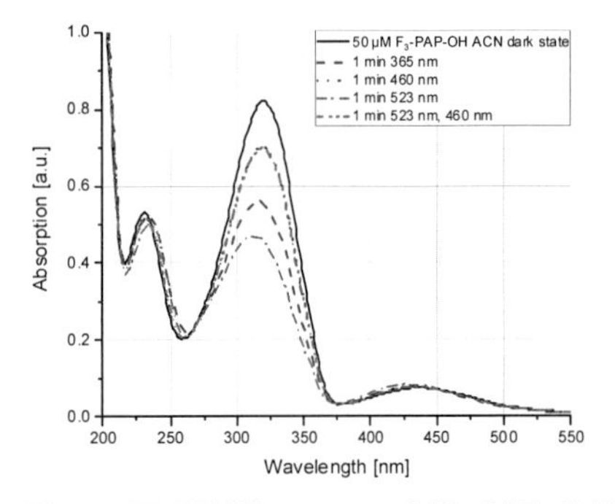

Figure 57. UV-Vis spectra of 50 µM F_3-PAP-DKP-Lys in ACN.

5.3.2.3 Isomerization of F_2-PAP-OH, (*S,E*)-2-((*tert*-butoxycarbonyl)amino)-3-(4-((2,6-difluorophenyl)diazenyl)phenyl)propanoic acid 12

(*E*)-F_2-PAP-OH – ^{1}H NMR (300 MHz, ACN-*d*$_3$): δ =7.80 (m_c, 2 H, A_r-C_q-N), 7.43 (m_c, 3 H, A_r-C_q, A_r-CH-CF), 7.17 (m_c, 2 H, A_r-CF), 5.58 (bs, 1 H, NH), 4.41 (bs, 1 H, CH), 3.23 (m_c, 1 H, CH_2), 3.01 (m_c, 1 H, CH_2), 1.35 (s, 9 H, *tert*-Butyl) ppm, -CO_2H not visible.

(*Z*)-F_2-PAP-OH – ^{1}H NMR (300 MHz, ACN-*d*$_3$): δ =7.17 (m_c, 5 H), 6.92 (m_c, 3 H), (5.45 (bs, 1 H, NH), 4.29 (bs, 1 H, CH), 3.01 (m_c, 1 H, CH_2), 2.85 (m_c, 1 H, CH_2), 1.35 (s, 9 H, *tert*-Butyl) ppm, -CO_2H not visible.

Table 13. *E/Z* ratio of F_2-PAP-OH in ACN

Procedure	*E* / *Z* ratio	Peaks for calculation
Dark state	93% / 7%	7.80 / 6.92
30 min 460 nm	43% / 57%	7.80 / 6.92
30 min 365 nm	46% / 54%	7.80 / 6.92
30 min 523 nm	17% / 83%	7.80 / 6.92

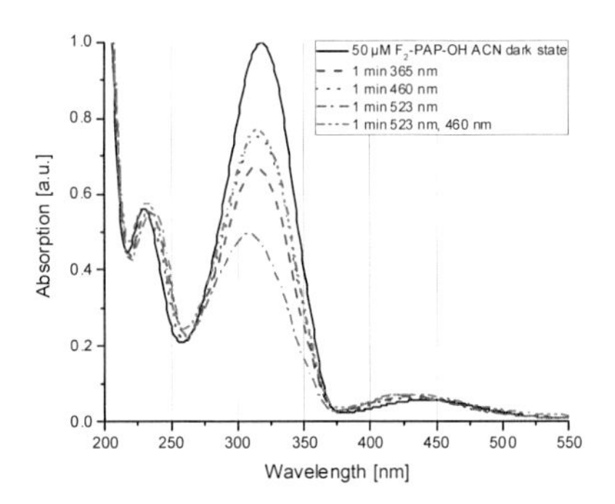

Figure 58. UV-Vis spectra of 50 µM F_2-PAP-OH in ACN.

5.3.2.4 Stability of F_2-PAP-DKP-Lys against glutathione

For prospective applications, the stability of gelator **1** under physiological conditions had to be investigated as well. Reduction of azobenzene derivatives with thiol groups to arylhydrazines in biological systems is considered the most serious limitation for their *in vivo* application as photoswitches. Therefore, we investigated the stability of **1** in the reducing environment. A solution of 5 mM F_2-PAP-DKP-Lys **1** and 5 mM reduced glutathione was prepared in PBS and incubated in darkness at rt. After 6 d there was no significant degradation of the F_2-PAP-DKP-Lys by HPLC under aforementioned conditions (injection 5 μL) (chapter 5.1.7). Then the experiment was discontinued.

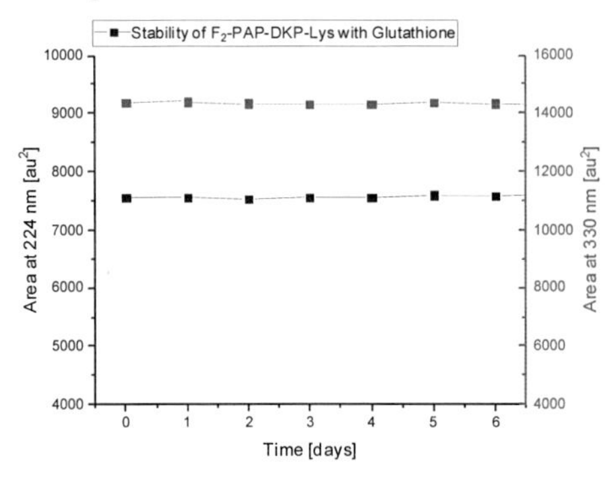

Figure S3. Stability of F_2-PAP-DKP-Lys against glutathione.

5.3.2.5 Lifetime of the *Z*-F_2-PAP-DKP-Lys

The *cis*-isomers of azobenzenes are thermally less stable than the respective *trans*-forms and the *cis*-isomer undergoes thermal back-isomerization to the *trans*-isomer. However, the rate of that equilibration depends on the substitution pattern on both aromatic rings of the azobenzene system and can span from microseconds to years. For the *ortho*-fluorinated azobenzenes, lifetime of the *cis*-isomer is enhanced in comparison to the non-substituted azobenzene. Here we wanted to compare our results with the existing literature data.[2,4] For the lifetime measurement, the *cis*-F_2-PAP-DKP-Lys (*cis*-**2**) was purified by preparative HPLC (same conditions like **CiproDKP**) and its thermal decay to the *trans*-isomer was measured by analytical HPLC (by using the same conditions like for quantification of the *cis:trans* ratio). The half-life of *cis*-F_2-PAP-DKP-Lys (*cis*-**2**) was determined with 36 ± 1 min at 100 °C in PBS (Figure 59). It comes in agreement with the observation that, upon green light dissipation of hydrogels composed of *trans*-**2** to the respective fluid (sol), the fluid (containing now mostly the *cis*-F_2-PAP-DKP-Lys, or else *cis*-**2**) can solidify again to the hydrogel after incubation of the sample for 2 h at 100 °C and cooling it to the rt in darkness. Also, during the synthesis of **2** (*vide supra*), the reaction mixture that is warmed up to 120 °C for 90 min provides the product **2** entirely as the *trans*-isomer. The lifetime of *cis*-**2** was also measured at other temperatures. The half-life at 60 °C is 46 ± 1 h and at rt (22 °C): 109 ± 1 d (Figure 60), which correlates with the long stability (at the rt) of the fluid (sol) obtained after irradiation of the hydrogel with green light. The half-life of *cis*-F_2-PAP-DKP-Lys at rt (22 °C) measured in acetonitrile (MeCN) is 60 ± 1 d (Figure 60). This decrease of thermal stability in less polar solvents was previously described by Bleger *et al.* for other fluoro-azobenzenes.[30a]

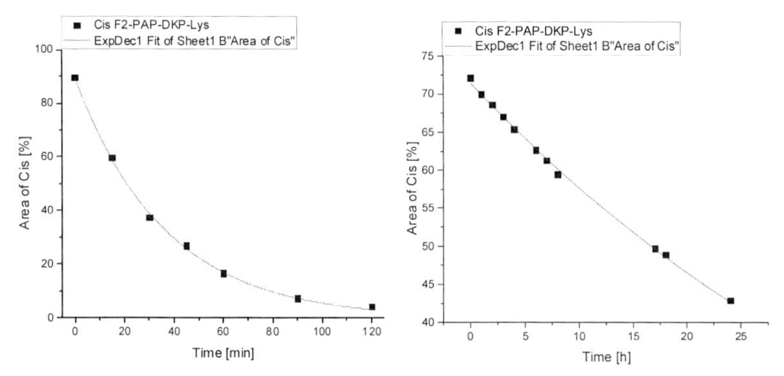

Figure 59. 36 ± 1 min half-life of *cis*-F_2-PAP-DKP-Lys at 100 °C (left), 46 ± 1 h half-life of *cis*-F_2-PAP-DKP-Lys at 60 °C in PBS (right).

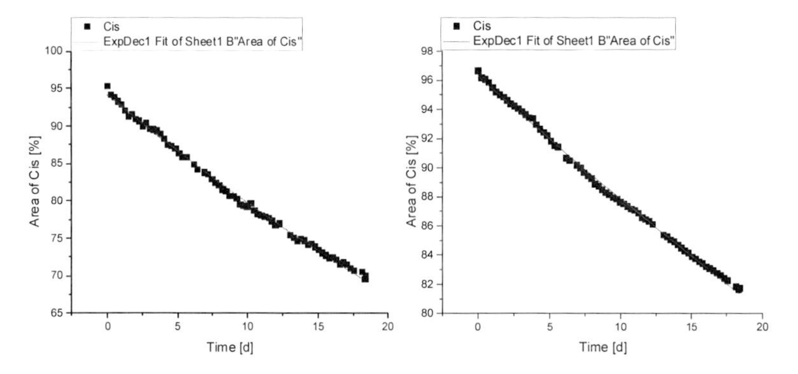

Figure 60. 60 ± 1 d half-life of *cis*-F_2-PAP-DKP-Lys at rt (22 °C) in ACN (left); 109 ± 1 d half-life of *cis*-F_2-PAP-DKP-Lys at rt (22 °C) in PBS (right).

5.3.3 Gelation properties of F_2-PAP-DKP-Lys

5.3.3.1 Measuring the gel-to-sol transition temperature of hydrogels 2

The gel-to-sol transition temperature is characteristic for the particular gel composition and can be used to estimate its relative stability in comparison to the other gel samples. The value of that transition, referred to as "the melting temperature" of a gel, was measured according to the following protocol: in a 1.5 mL glass vial (crimp top, 12×32 mm), the photochromic gelator **2** was added to PBS buffer pH 7.4 (500 µL) (mass of the gelator and its approximate final percentage in the gel are listed in Table 14). The crimped vial was warmed up vertical in a water bath at 80 °C for 5 min. The suspension was dissolved in the crimped vial after heating it up to the boiling point (< 1 min) with a heat gun. The hot fluid turned to an orange solution and this fluid gelated upon cooling to rt. (typically within 15 min). Before all measurements, the hydrogel (unless indicated otherwise) was kept overnight in darkness at rt in order to allow equilibration of the components, which in turn minimized statistical deviation of the measured behaviour.

To measure the melting temperature, a sample of the hydrogel prepared above was swimming horizontally on the surface of a slowly stirred (60 rpm) water bath at 25 °C. The bath was then warmed up with the heating rate of ca. 2 °C/min. The hydrogel starts melting slowly before the resulting sol (fluid) will abruptly flow down at the given gel-to-sol transition temperature. The measurement was repeated 5 times and the average transition temperature was reported as "the melting temperature" T_m (or gel-to-sol transition temperature). Gels prepared from **2** in PBS buffer pH 7.4 were stable and homogenous in the range between c.a. 40 g/L and 70 g/L (4% – 7% of **2**), also at ambient daylight. At the concentration of < 30 g/L (3%) gelation was slow (overnight), the resulting gel had low mechanical stability (almost immediately turned into liquid upon slight shaking of the vial) and its melting point was difficult to reproduce (Table 14).

Table 14. Melting temperatures of hydrogels comprising the gelator 1 in PBS buffer pH 7.4.

Composition of the gel x mg of **1** + 500 µL PBS	Approx. concentration	T_m °C	Gelation time at rt.
10	"2 %"	-	no gelation
15	"3 %"	67 ± 23	16 h
20	"4 %"	82 ± 8	2 h-16 h
25	**"5 %"**	**91 ± 3**	**< 1 h**
30	"6 %"	94 ± 6	< 1 h
35	"7 %"	91 ± 9	< 1 h

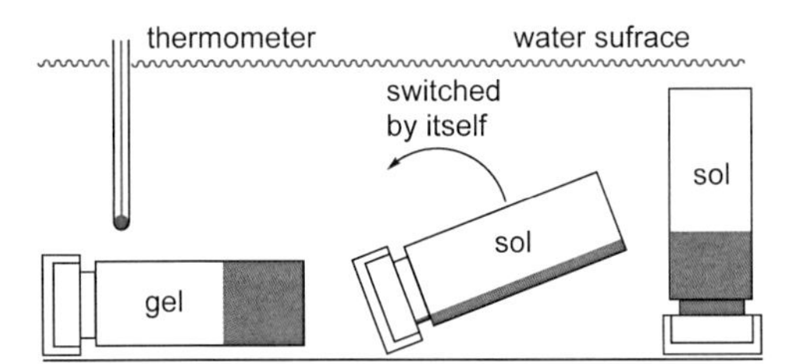

Figure 61. Schematic of measurement of the gel-to-sol transition temperature.

5.3.3.2 Maximum loading of ciprofloxacin in the hydrogel based on F_2-PAP-DKP-Lys

Regarding the prospective medical applications of our hydrogel materials for light-induced drug release we have investigated the upper limits of loading the bioactive guest without compromising physical properties of the gels. We examined the behaviour of the hydrogels comprised of 25.0 mg of F_2-PAP-DKP-Lys **2** in 500 µL of PBS buffer pH 7.4 with the increasing concentrations of the antibiotic ciprofloxacin. The gel could be doped with up to 10.0 mg of ciprofloxacin hydrochloride without influencing significantly the speed of gelation or the stability. Ciprofloxacin hydrochloride in the amount between 15.0 mg and 25.0 mg caused a prolonged gelation process and the gel needed up to 1 week to form. Addition of 50.0 mg ciprofloxacin hydrochloride (double the gelator mass) resulted in no gelation – the material remained transparent fluid upon prolonged storage in darkness at the rt. Overall, the supramolecular hydrogels demonstrate good loading capacity for encapsulated drugs. In the investigated system, the loaded cargo can reach up to 40% of the dry mass of the gelator and c.a. 2 wt% of the assembled gel.

Considering the ratio between molecular masses of ciprofloxacin (331.35 g/mol) and the gelator **2** (415.44 g/mol, respectively), the molar proportion guest/gelator in the optimally loaded (40% dry gelator mass) gel is 1:2. Thus, 1 mL of a loaded gel would contain 20 mg (60 µmol) of the antibiotic and 50 mg (120 µmol) of **2**. To achieve the therapeutically relevant concentration of ciprofloxacin (MIC = 24.1 nM) in the given area, the concentration of **2** would reach c.a. 50 nM. This is still four orders of magnitude below the measured value of toxicity for the compound **2** against mammalian cells (EC_{50} > 500 µM, Figure 87). Therefore, we believe that for high loading (> 1% of the dry gelator mass) of the gel with cargo molecules that possess activity in the nanomolar range (like ciprofloxacin or many anticancer agents), toxicity of the gelator **2** molecules co-released with the cargo will be negligible in comparison to the achieved therapeutic effect. And thus, the materials based on **2** seem suitable for applications as drug-releasing implants or micro-/nanogels.

5.3.3.3 Light-induced gel-to-sol transitions

First of all, we have to clearly indicate that the speed of light-induced gel-to-sol transition in our material depends on the factors like: concentration of the gelator, the presence of other components (like NaCl) in the gelation medium, the preparation protocol of the hydrogel, the ratio between height and diameter of the glass vial (and consequently, the area of light-absorbing surface of the gel), as well as the light intensity (including the respective distance of the light source from the irradiated sample). All the results reported here were measured in multiplicates and reproducible within the same experimental setup.

Samples of the hydrogels (3-7 wt% of **2** in PBS buffer pH 7.4, prepared accordingly to Table 14) in 1.5-mL-vials (crimp top, 12×32 mm) have been irradiated with green light (523 nm, 10 W LED diodes) at rt (22 °C). The vials were inverted every 5 min. After 30 min irradiation, the samples of 3% and 4% gels turned into homogenous fluids without need for any mechanical stimulation. The gel samples at the concentrations of 5% and 6% after 35 min irradiation became so unstable, that they dissipated to fluids upon inversion. The gel sample at the concentration of 7% has been irradiated for 120 min without an effect and it remained a stable gel. The hydrogel composition of 25.0 mg **2** in 500 μL PBS buffer pH 7.4 (Table 14 – the entry highlighted in blue) was selected as a reference for all further experiments due to its high mechanical stability, quick gelation upon cooling, and its reproducibly high melting temperature.

Figure S6. Transparent hydrogel 2 (5 wt%) (left) turns into sol after irradiation with green light (right).

Without inversion of the vial or any other mechanical stimulation, the light-induced gel-to-sol transition of the hydrogel is completed within 3 h. These are the optimal conditions for a stable and reproducible dissipation of our material with green light, which we applied to study the light-induced release of several bioactive compounds to the adjacent solution, as described below.

5.3.3.4 Rheology

The rheological characterization of hydrogels formed by the compound **2** in PBS buffer (pH 7.4) was performed on samples generated by cooling the warm solution (formed from 400 mg of **2** and 8.0 mL of buffer) directly on the rheometer plate from 95 °C to rt. Strain sweep experiments were performed at 10 rad/s to determine the linear viscoelastic regime and the mechanical strength of the hydrogel. Frequency experiments were performed at low strain within the linear viscoelastic region (LVR) of the sample. For regeneration experiments, the samples were exposed to a deformation of 100% for 30 sec to destroy the supramolecular network, afterwards the regeneration of G' was measured at low strain within the LVR.

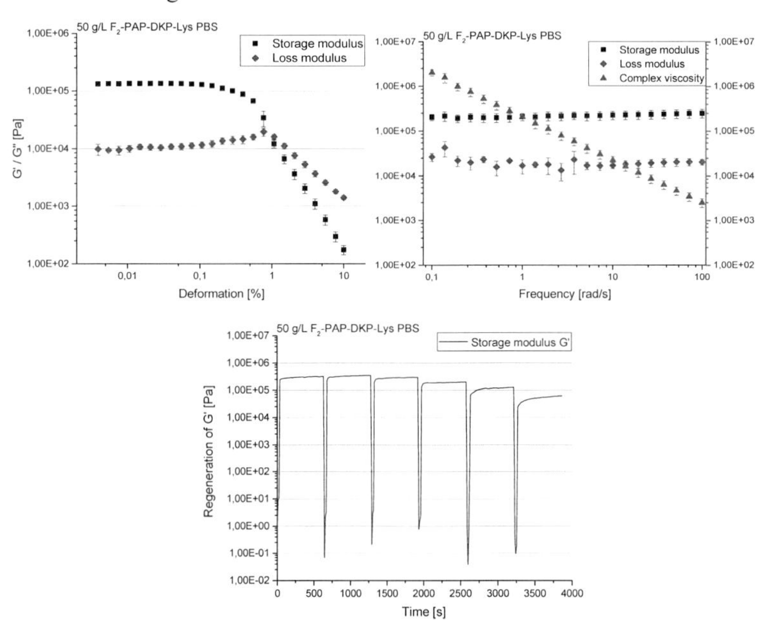

Figure 62. Rheological properties of the hydrogel (400 mg of 2 and 8 mL of PBS buffer pH 7.4). Strain sweep experiment (top left); frequency sweep experiment (top right); regeneration of G' after shearing the gel for 30 sec at 100% deformation (bottom).

5.3.3.5 TEM and NMR characterization

Diluted solutions were prepared from 10 volumes of diH_2O and one volume of a hydrogel, composed of 50 mg F_2-PAP-DKP-Lys **2** and 8 mg of NaCl or 8 mg of NaI dissolved in 1.0 mL of PBS pH 7.4. Carbon-coated copper grids (400 mesh) were covered with the material by short exposure on spray generated from the aforementioned solutions with ultrasounds.

The grids were dried under atmospheric pressure and subsequently examined using a Zeiss 912 Omega transmission electron microscope. Under these conditions, the fine fibrous structure of our material was revealed. In some cases, the samples were treated also with a staining of aqueous 1% Ammonium heptamolybdate tetrahydrate. The grids floated on top of a drop of the $(NH_4)_6Mo_7O_{24}$ solution for 1 min, directly after the ultrasound treatment. The excess of the staining solution was removed by a paper towel.

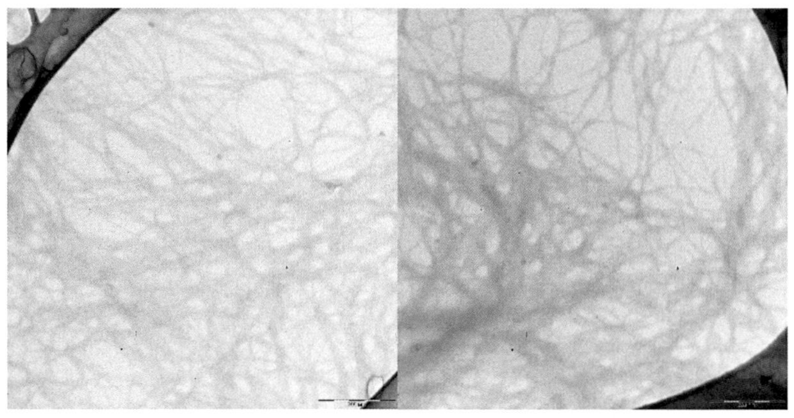

Figure 63. Fibres of F_2-PAP-DKP-Lys 1 without staining (scale 200 nm).

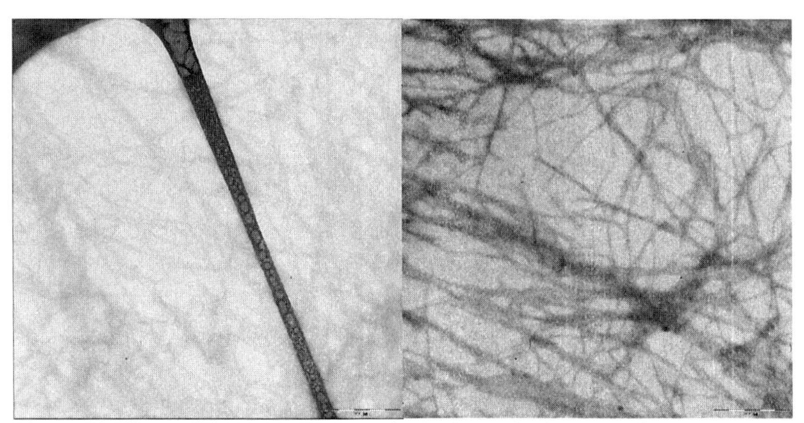

Figure 64. Fibres of F_2-PAP-DKP-Lys with NaI substitution (scale 200 nm).

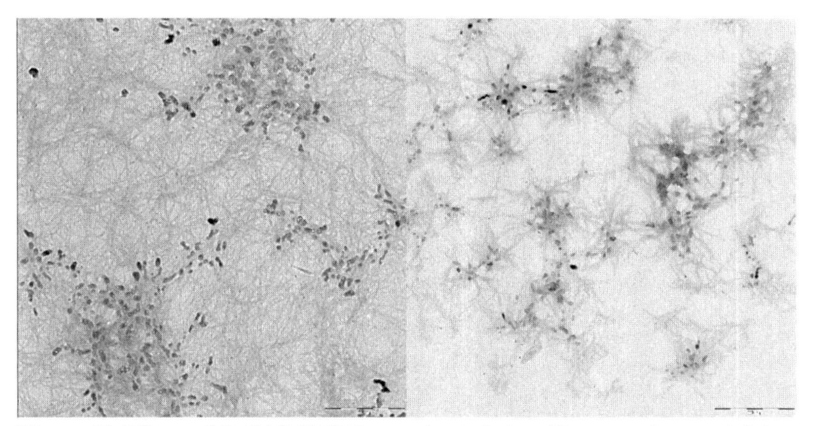

Figure 65. Fibres of F_2-PAP-DKP-Lys and artefacts with ammonium molybdate staining (scale 500 nm).

5.3.3.6 NMR investigations of the supramolecular hydrogel structure

Here we demonstrate the qualitative NMR analysis of the supramolecular structures present in our material.

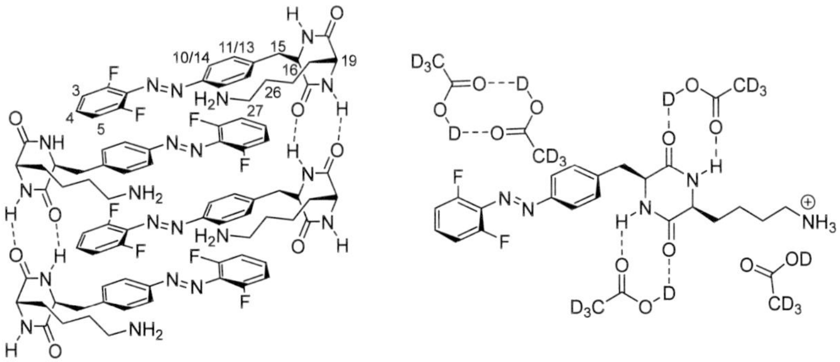

Figure 66. Hypothesis of supramolecular interactions of the *trans*-F_2-PAP-DKP-Lys 2 in aqueous solutions (left) and the reversible inhibition of these interactions in acetic acid.[197]

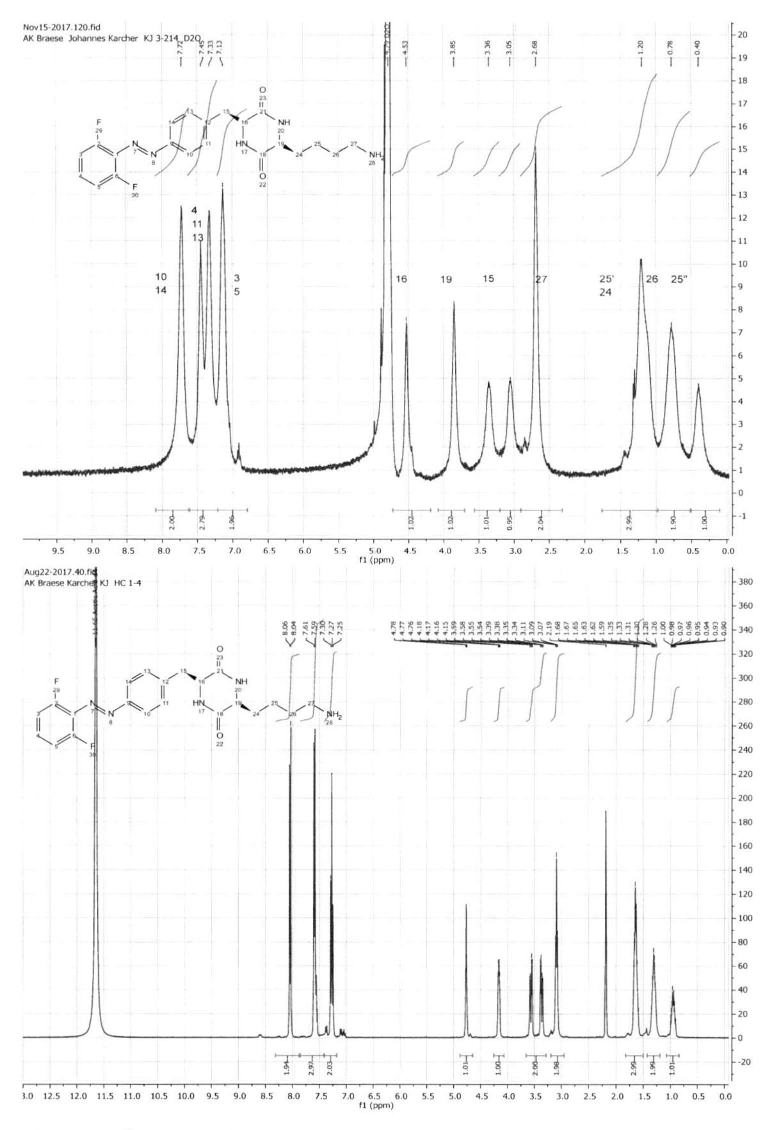

Figure 67. ^{1}H NMR of the hydrogel 50 g/L F_2-PAP-DKP-Lys 2, in D_2O (with 8 g/L NaCl, top), and in CD_3COOD (bottom).

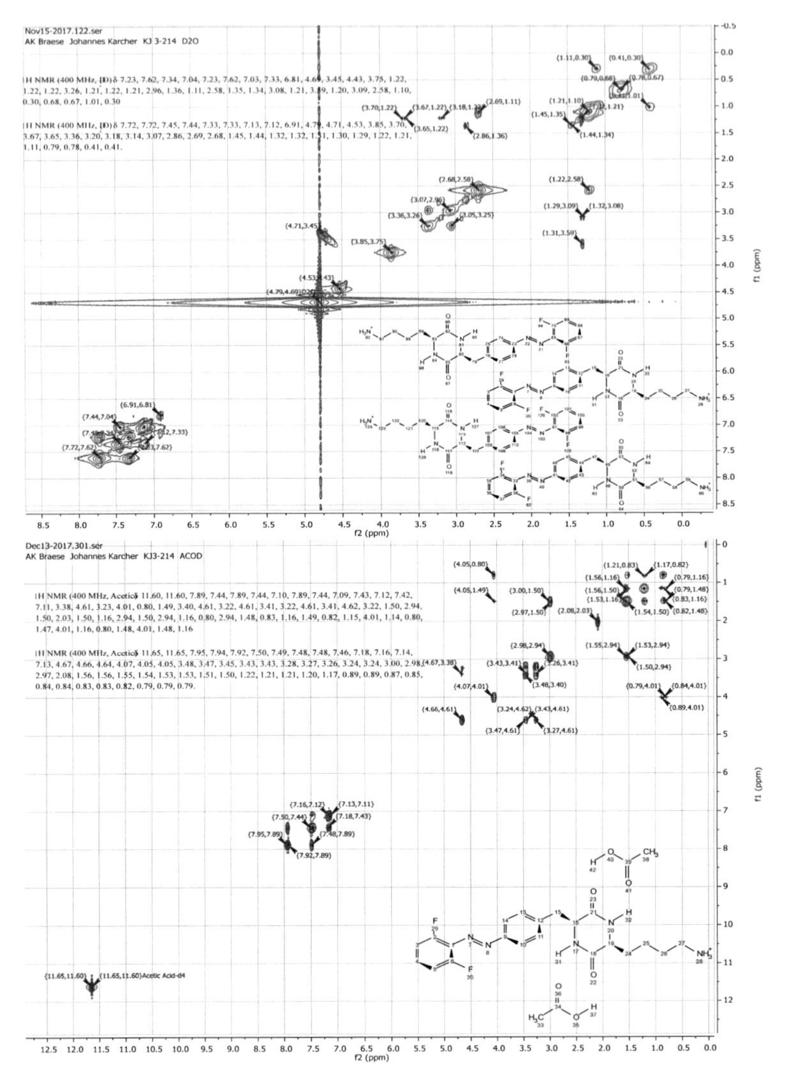

Figure 68. Comparison of COSY spectra of the hydrogel 50 g/L F2-PAP-DKP-Lys, 8 g/L NaCl in D_2O (top) and the COSY of 50 g/L F_2-PAP-DKP-Lys in acetic acid-d_4 (bottom).

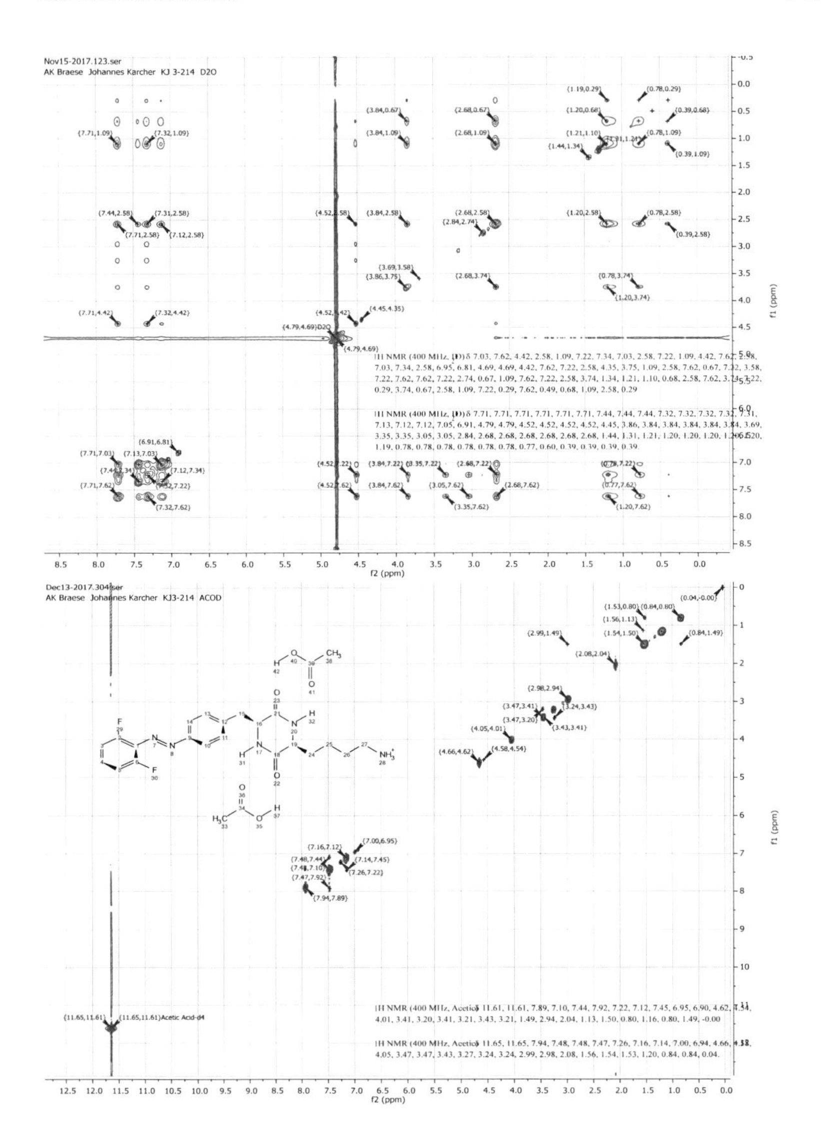

Figure 69. Comparison of NOESY spectra of the hydrogel 50 g/L F2-PAP-DKP-Lys, 8 g/L NaCl in D_2O (top) and the NOESY of 50 g/L F_2-PAP-DKP-Lys in acetic acid-*d_4* (bottom).

5.3.4 Guest release from hydrogels using green light

In order to study the guest release, the hydrogel based on F_2-PAP-DKP-Lys **2** was investigated in releasing various types of encapsulated guest molecules by means of diffusion (in darkness) or dissipation of the inner gel structure upon irradiation with green light. The composition of 25.0 mg of the gelator **2** in 500 µL of PBS buffer pH 7.4 was chosen as the gel matrix for all experiments in this section. As described in the 5.3.3 of the supporting information (Table 14), it forms a stable gel in absence of light that is completely converted to sol upon 3 h of irradiation with green light (523 nm, 10 W LED diode) without need for mechanical stimuli.

5.3.4.1 Preparation of the hydrogel samples

In a 1.5 mL-vial (crimp top, 12×32 mm) was added the photochromic gelator **2** (25.0 mg, powder) and PBS buffer pH 7.4 (495 µL). To this opaque suspension was added a 100× stock solution (5 µL) of the particular cargo molecule dissolved in PBS buffer or in DMSO. The crimped vial was warmed up vertically in a water bath at 80 °C for 5 min. The suspension was then homogenized inside the vial after heating it up to the boiling point (< 1 min) with a heat gun. The hot mixture turned into an orange solution (homogenous), which was gelated after 15 min at rt. Before the release experiment, the hydrogel was kept overnight in dark at rt.

The measurements of the release rate of ciprofloxacin hydrochloride, as well as of CiproDKP **16** and CiproFAzo **15** were performed in triplicates and the average values were taken for the final conclusions and result plots. The release processes of actinomycin, diclofenac, naproxen, and cytochrome c were carried out as single experiments, to provide the comparison among various classes of bioactive cargo molecules. The given concentration of the drugs and protein were chosen to obtain the optimal accuracy for the HPLC detection range. Although releasing of suspensions from the photodissipating gel would be in principle possible, but only homogenous mixtures were investigated with guest molecules that are soluble under the conditions of experiments.

5.3.4.2 Quantification of the passive diffusion – cargo "leaking" from hydrogels in darkness

500 µL of PBS buffer pH 7.4 was slowly added on top of a gel sample (on the wall of the vial) and immediately removed with a micropipette, to wash away unbound or loosely bound guest molecules from the surface. Addition of fresh 500 µL of PBS buffer followed. The gel was incubated together with the buffer on the top in darkness. 500 µL of the liquid was collected after 30 min by gently turning the vial

sideways and pipetting off the liquid from the side wall of the vial. Then, fresh 500 µL of PBS buffer was added on the side wall of the vial, incubated in darkness and removed after 30 min in the same way as described above. That process was repeated for the total duration of 180 min by collecting 7 subsequent volume aliquots. After that time, all hydrogels remained visually unaffected. By measuring the remaining liquid volume after removal of the last 500 µL aliquot from the top of the gel we estimated that the total decay of the gel volume was lower than 15%.

5.3.4.3 Procedure of the light-induced release

To measure the release process upon green light irradiation, the procedure described above was repeated in the same way, but after initial washing of the gel surface the sample was placed in an irradiation chamber and illuminated with one 10 W LED (523 nm). Short breaks in irradiation (< 30 sec) were taken for the replacement of 500 µL aliquot with fresh 500 µL of PBS buffer every 30 min, but the summary irradiation time was 180 min. The irradiation time was sufficient to fully convert all the gel samples into liquid. All aliquots were weighted before the HPLC measurement to calculate the released amount of the substance. The concentrations of the aliquots were calculated by a previously measured calibration curve of the substances. In all aliquots no precipitate was observed.

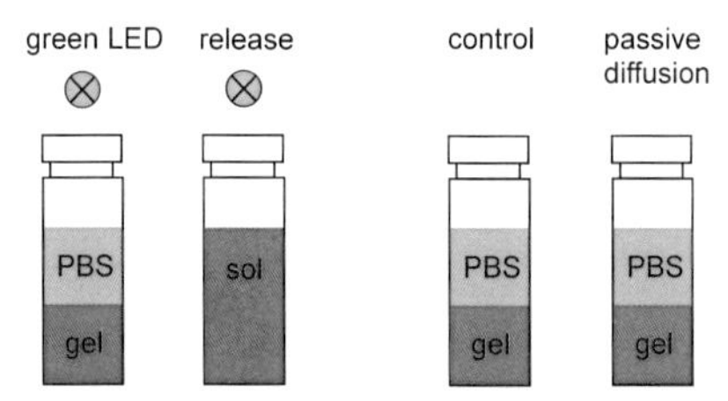

Figure 70. The light-triggered release of encapsulated molecules from hydrogel samples vs. passive diffusion of the cargo in the absence of light (schematic representation of the experiment).

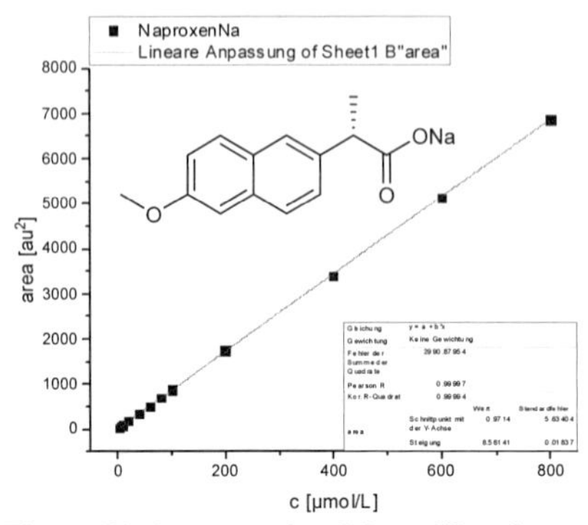

Figure 71. An example of the calibration curve of the naproxen sodium salt.

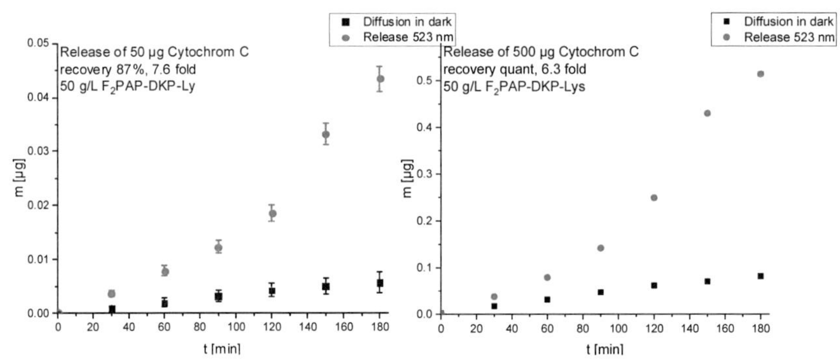

Figure 72. The release vs. retention efficiency at various cargo concentrations: comparison of releasing 50 µg cytochrome c (recovery 87%, 7.3-fold difference between the irradiated and non-irradiated sample) and 500 µg cytochrome c (quant. recovery, 6.3-fold difference).

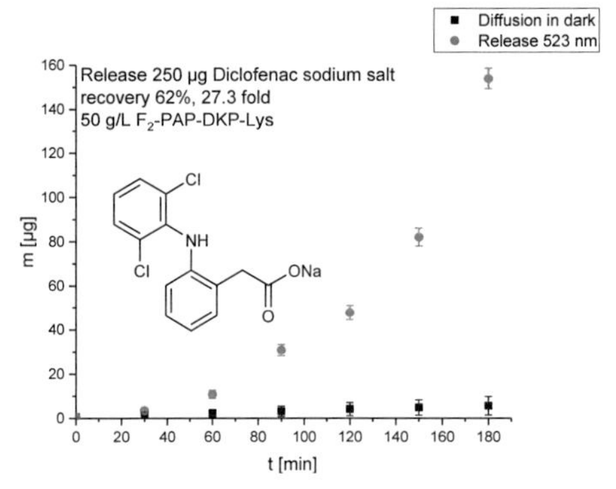

Figure 73. Release of 250 µg diclofenac sodium salt, recovery 62%, 27.3-fold difference.

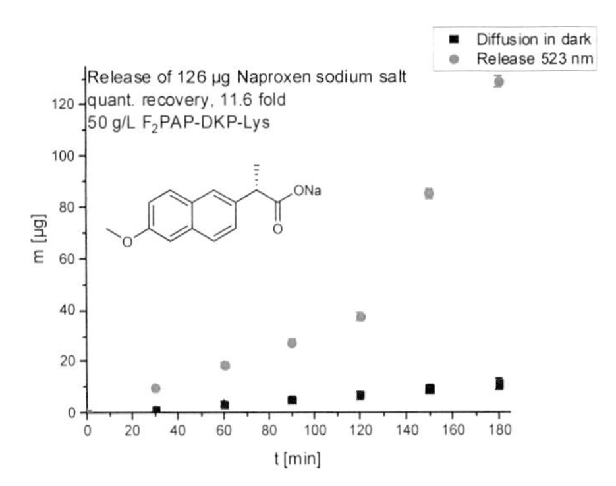

Figure 74. Release of 126 µg naproxen sodium salt, quant. recovery, 11.6-fold.

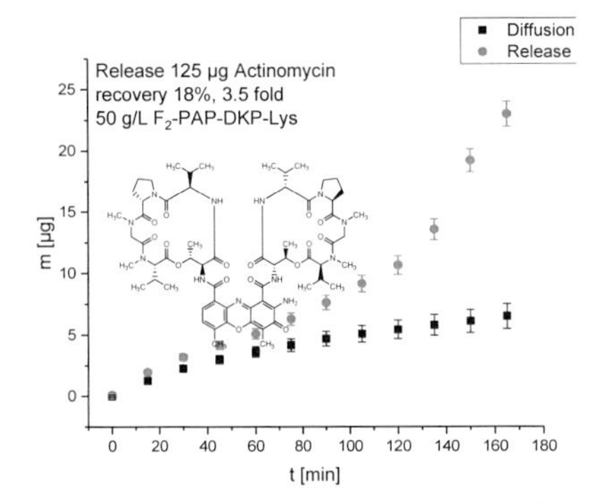

Figure 75. Release of 125 µg actinomycin, recovery 18%, 3.5-fold.

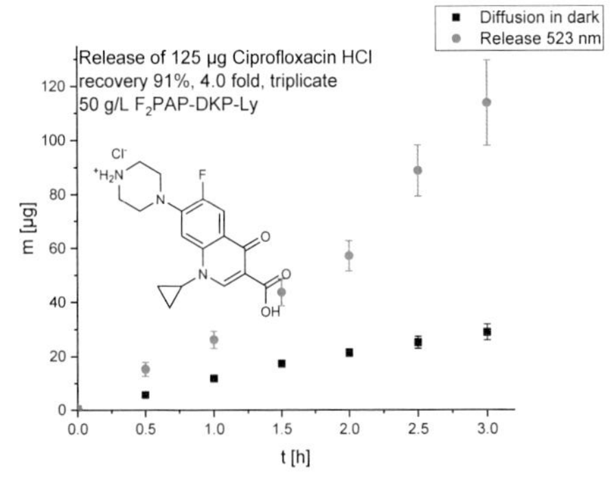

Figure 76. Release of 125 µg ciprofloxacin hydrochloride, recovery 91%, 4-fold (in triplicates).

Table 15. Physical parameters of the guest molecules encapsulated inside hydrogels in comparison to the relative rates of passive diffusion vs. light-induced release ("photomodulation of the release").

Cargo	Naproxen	Ciprofloxacin	Actinomycin	Diclofenac	Cytochrome C
Photomodulation of the release	11.6	4	4	27.3	6.3
MW of the cargo	230 Da	331 Da	1255 Da	296 Da	12384 Da
pK_A of the cargo	4.15	6.09	11.9	4.08	n.d. (pI = 10.0)
logP	3.18	0.28	1.6	4.51	n.d.

In most of the examined cases, the recovery of cargo molecules after gel irradiation is close to quantitative (> 90%) (see Figure 72 – Figure 76). In the cases with significantly smaller recovery (diclofenac Figure 73 and actinomycin Figure 75) we have observed degradation products, which are attributed to thermal decomposition upon gel loading at c.a. 100 °C. In the future, improved loading procedures will have to be developed.

By comparing the investigated cargo molecules (Table 15) it was assumed that the rate of passive diffusion in darkness (reflected by the calculated photomodulation, defined as release rate difference between the irradiated and non-irradiated sample) does not depend proportionally on the molecular weight of the guest, but rather on its combination with the relative cargo acidity. Relatively large basic guests, like actinomycin or cytochrome C, diffuse in darkness from the gel much quicker as the small acidic molecules (naproxen or diclofenac). Most disfavoured combination: low-MW neutral or basic molecules diffuse out of the gel volume relatively quickly, also without irradiation.

To address this issue and increase the scope of our drug release technology, an alternative encapsulation method was proposed for the cargo molecules with unfavourable physical parameters. By covalently coupling ciprofloxacin (Figure 76) to the photochromic fragment, or to the complete gelator **2** molecule, hybrid compounds were obtained with reduced basicity. Here the molecules **15** and **16** that, by virtue of favourable supramolecular interactions, probably intercalate directly into the hydrogel network structure, and their diffusion is significantly retarded (Figure 77). Thus, as demonstrated below (Figure 78**,** MIC 5.3.4.5, MTT 5.3.4.8), the primary bioactivity has been decreased by the covalent modification. However, this methodology can be optimised further by including, e.g. biodegradable linkers, which ultimately provide the cargo molecule in its active form. In contrast to this an alternative strategy for fluoroquinolones has been demonstrated in the literature by an irreversible activation with 380 nm, which cleaves a 7-aminocoumarin.[304]

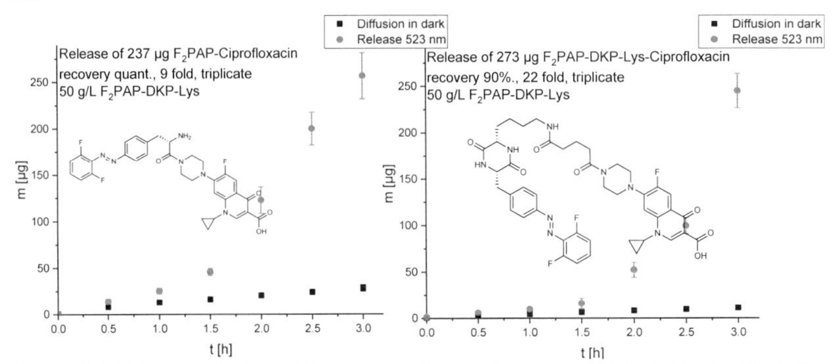

Figure 77. Release of 237 µg CiproFAzo 15, quant. recovery, 9-fold (in triplicates) (top); release of 237 µg CiproDKP 16, recovery 90%, 22-fold (in triplicates) (bottom).

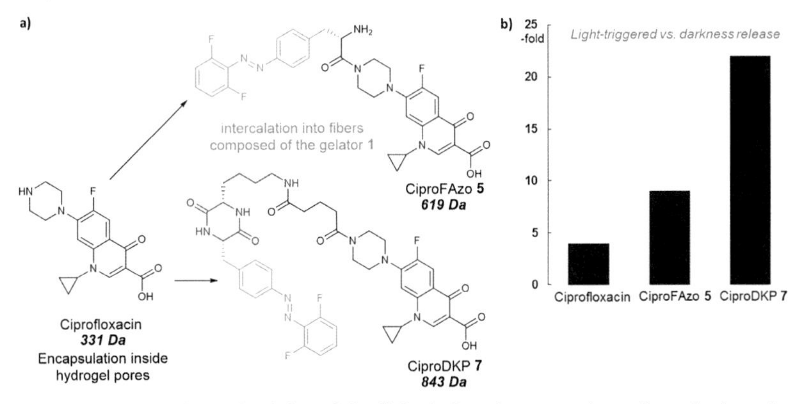

Figure 78. Improving selectivity of the light-induced cargo release from hydrogels. (a) Covalent hybrids of the antibiotic ciprofloxacin (black) with the azobenzene photoswitch (blue) and the 2,5-diketopiperazine moiety (red) – CiproFAzo 15 and CiproDKP 16, respectively. The latter molecule contains full structure of the gelator 2; (b) photomodulation of the release process of ciprofloxacin, CiproFAzo 15, and CiproDKP 16, encapsulated in the hydrogel composed of 2.

5.3.4.4 Photomodulation of the bacterial growth with green light

Antibacterial activity of the gelator molecule

For the initial test on antibacterial activity of the used gelators, we used *E. coli Bl21* bacterial cultures with an initial $OD_{600} = 0.005$ in 10 mL LB Müller broth. After addition of the investigated substance (the gelator **2,** or its non-fluorinated analogue), the bacterial cultures were incubated at 37 °C with shaking (180 rpm) in 50-mL Erlenmeyer flasks covered by aluminium foil. The OD_{600} was measured each hour. The results (Figure 79) indicate that no significant toxicity of either gelator: F_2-PAP-DKP-Lys **2** (final concentration = 0.5 g/L) and PAP-DKP-Lys[298a] against the bacteria has been detected under these conditions. There was also no significant difference, when 1% DMSO was present in the solution.

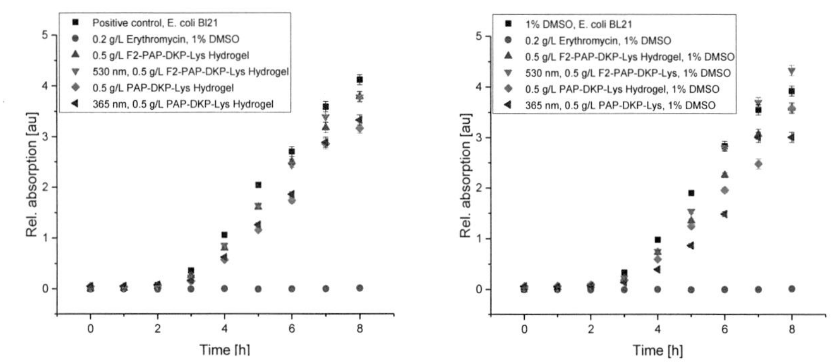

Figure 79. Initial tests of the antibacterial activity – growth of the *E. coli Bl21* bacterial cultures in presence and in absence of the gelator 2 and its non-fluorinated analogue PAP-DKP-Lys used in our previous studies. [5] At the concentration of 0.5 g/L (in darkness, or after irradiation) both hydrogels F_2-PAP-DKP-Lys 2 and PAP-DKP-Lys[298a] show no significant antibacterial activity in water nor in presence of 1% DMSO.

Based on these results, we assumed that all the antibacterial activity in following experiments can be attributed to the activity of respective antibiotic compounds and not the hydrogel matrix (it is unrelated to the presence of the gelator **2** in either of its photoisomeric forms).

5.3.4.5 Minimum inhibitory concentration (MIC)

The minimum inhibitory concentration (MIC) of ciprofloxacin, F_2-PAP-DKP-Lys **2**, CiproFAzo **5** and CiproDKP **7** was tested according to Wiegand *et al.*[305] However, the control strain of *E. coli ATCC 25922* did require a S2 safety protocol. Therefore, we had to switch to a common S1-biosafety-class strain: *E. coli K12* *wt.* After preparation of the McFarland 0.5 $BaSO_4$ turbidity standard (corresponding to OD_{600} = 0.095 au), an initial OD_{600} = 0.005 was used for every bacterial culture. The Mueller-Hinton growth medium was prepared without sodium chloride in diH_2O with 2.0 g/L beef extract, 17.5 g/L casein hydrolysate and 1.5 g/L starch. The MHB medium was adjusted with 25 mg/L $CaCl_2 \cdot 2H_2O$ and 12.5 mg/L $mgCl_2 \cdot 6H_2O$. The total volume of 200 μl was used in each well of the 96-well plate. All MIC values were performed in multiplicates with n =6 or higher. The stock solution of ciprofloxacin hydrochloride was prepared in diH_2O, and the gelator F_2-PAP-DKP-Lys **2** in PBS. Concentrated stock solutions of **CiproFAzo 15** and **CiproDKP 16** were prepared in DMSO, so that the final concentration of 0.1% DMSO is not exceeded in any of the wells. All photoswitchable azobenzenes were irradiated with 523 nm for 30 min at rt, which has previously yielded (Table 11) 58% of the *cis*-**2**-isomer of the gelator molecule **2**. The data was fitted by a logistic function for comparison:

$$y = \frac{A_1 - A_2}{1 + (x/x_0)^p} + A_2$$

Hereby the maximal non-toxic concentration of the hydrogel was calculated. The data of ciprofloxacin was fitted by the same function for a comparison, but not for the calculation of the MIC value.

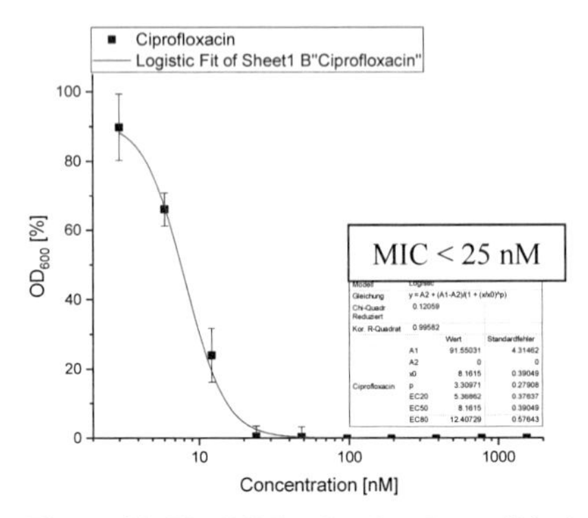

Figure 80. The MIC value for the antibiotic ciprofloxacin, against *E. coli K12 wt.*

The literature MIC value of 24.1 nM for the antibiotic ciprofloxacin (Silvia *et al.*)[202] was in agreement with our experiment (Figure 80).

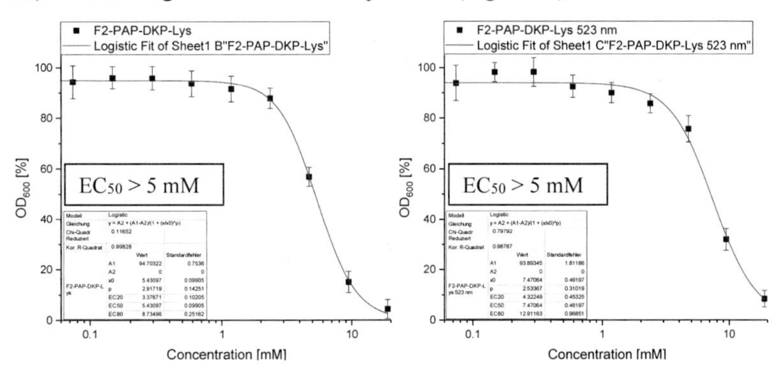

Figure 81. MIC of F_2-PAP-DKP-Lys 1, not irradiated (left) and irradiated (right), against *E. coli K12 wt.*

The EC_{50} value for *trans* F_2-PAP-DKP-Lys **2** was determined as > 5 mM (the numerical value of 5.4 ± 0.1 mM). No significant toxicity was observed at the concentration of **2** as high as 1 g/L, which is respectively a 50-fold dilution of the hydrogel applied for our investigation. There was also no significant difference upon irradiation with 523 nm light (predominant *cis*-isomer).

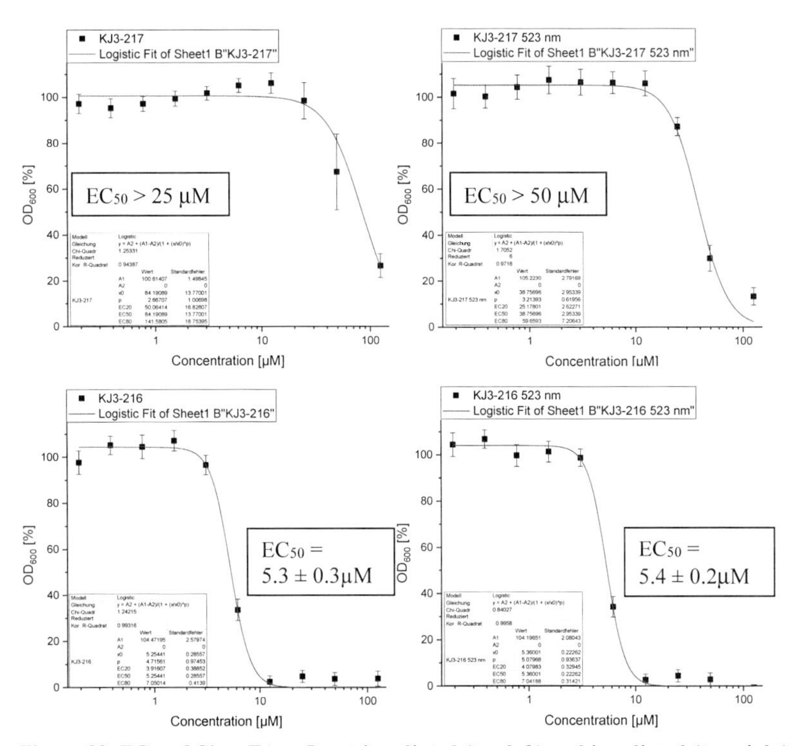

Figure 82. EC_{50} of CiproFAzo 5, not irradiated (top left) and irradiated (top right); EC_{50} of CiproDKP 7, not irradiated (bottom left) and irradiated (bottom right), against *E. coli K12 wt.*

Both derivatives CiproFAzo **5** and CiproDKP **7** did show significantly reduced activity against *E. coli K12 wt* in comparison with ciprofloxacin. The activity differences between the irradiated (mostly *cis*) and non-irradiated (entirely *trans*) samples were negligible. This result means that the molecules **5** and **7** alone, at least in the examined system, are not suitable as potential antibacterial photopharmacology agents, in contrary to previously described smaller ciprofloxacin-photoswitch conjugates.[42]

Hypothetically the increased distance between the ciprofloxacin pharmacophore and the photoswitch minimizes the activity difference of the *cis* and *trans* derivatives. This type of functionalization has significantly (up to 8-fold) decreased the passive diffusion of the cargo molecule from hydrogel pores, in comparison with unmodified ciprofloxacin (Figure 77 vs. Figure 76). Covalent binding of the gelator fragments most likely enhanced interactions with the supramolecular fibres of the hydrogel. The decreased diffusion indicates that the cargo molecule has been incorporated (intercalated) into the fibre structure, and not only absorbed into the pores between the neighbouring fibres. Therefore, it is a valuable prototype that indicates the way of cargo design for future combination of the optimal release and bioactivity.

5.3.4.6 *In vitro* release of ciprofloxacin from the irradiated and non-irradiated hydrogels

Before the experiments on photomodulation of the antibacterial activity using our photosensitive hydrogel material, the light-dependent release of ciprofloxacin from the gels was tested and optimised by *in vitro* experiments without bacteria.

In agreement with our previous observations, the hydrogel maintained mechanical stability in the absence of light. The gel sample in a glass vial was incubated for 24 h at rt. inside of a 50-mL Erlenmeyer flask containing 10 mL diH_2O shaken at 120 rpm, and the concentration of ciprofloxacin was determined every hour (by HPLC). After the 24 h experiment, the non-irradiated hydrogel still remained intact without observable degradation, whereas all the encapsulated ciprofloxacin diffused out of the hydrogel and was released in the surrounding aqueous solution.

The identical gel sample incubated for 3 h under irradiation (523 nm LED diode, 10 W) was added into a 50-mL Erlenmeyer filled with 10 mL diH_2O. The sol dissipated almost entirely. Over 85% of the loaded ciprofloxacin, released from the irradiated sample, was instantly available in the surrounding solution. Upon further shaking, the recovery of ciprofloxacin became quantitative within 24 h. For the experiment of the light-dependent ciprofloxacin release inside bacterial cultures, the protocol was modified towards a direct irradiated bacterial culture that contained samples of a hydrogel loaded with the antibiotic.

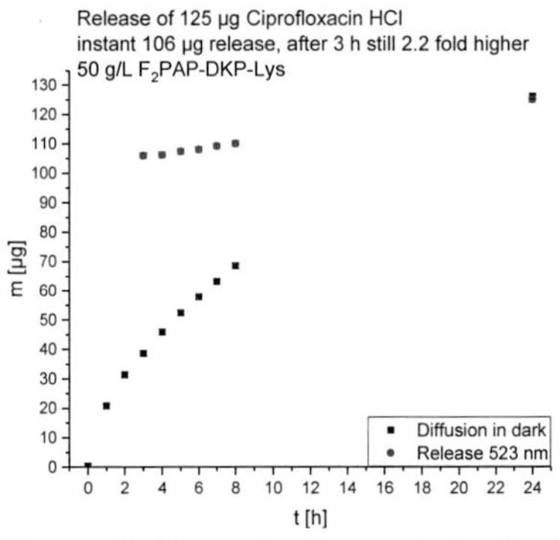

Figure 83. Example of an optimisation for the *in vivo* release.

5.3.4.7 Green light-induced modulation of the bacterial growth

Light-induced release of ciprofloxacin from our hydrogel material was performed directly inside bacterial cultures (*E. coli K12 wt*, pre-incubated to the OD_{600} = 0.005 in 20 mL Mueller-Hinton growth medium in 50-mL Erlenmeyer flasks). The bacterial growth in presence of the ciprofloxacin-loaded hydrogel has been followed in cultures irradiated with green light and, as a control, in identical cultures incubated in darkness. In contrast to the MIC test, which was performed under sterile conditions in a clean bench, these experiments were carried out on a regular lab bench with support of gas-flame sterilisation. All buffers, glassware and pipet tips were autoclaved prior to use. It was not necessary to change the preparation of the hydrogel to ensure a sterile sample, because boiling up the aqueous mixture with the antibiotic kept the sample sterile. There was observed no contamination any experiments including the sterility control.

The hydrogel samples were prepared in 0.7 mL crimp vials (40×7 mm) with 495 µL PBS with additional NaCl (to the final concentration of 9 g/L) and 25.0 mg of the gelator **2** (F_2-PAP-DKP-Lys). The hydrogel without antibiotic can be prepared in advance and stored without any degradation at rt in darkness. One day before the experiment, 5 µL of the 100× stock solution of ciprofloxacin hydrochloride (98.8 µM in diH_2O) was added and shortly boiled up analogue to the hydrogel preparation (5.3.3). The hydrogel with the antibiotics was stored 1-3 d in the fridge in darkness. The hydrogels for the release experiments (where the irradiation was intended) were prepared in transparent vials. Hydrogels for the control experiments (in darkness) were prepared in brown vials to avoid accidental irradiation or interactions with the daylight.

Following experiments and controls were made together and repeated three times:

1. Sterility control: 20 mL MH medium, without bacteria or antibiotics
2. Growth control: *E. coli K12 wt* (OD_{600} = 0.005) in 20 mL MH broth
3. Growth hydrogel: *E. coli K12 wt* (OD_{600} = 0.005) in 20 mL MH broth and 500 µL hydrogel (without antibiotic)
4. Growth sol: *E. coli K12 wt* (OD_{600} = 0.005) in 20 mL MH broth and 500 µL hydrogel (without antibiotic), which was irradiated by green light
5. MIC control: final concentration of 24.1 nM ciprofloxacin in 20 mL MH broth with *E. coli K12 wt* (OD_{600} = 0.005) (without the gelator **1**)
6. Diffusion of ciprofloxacin: *E. coli K12 wt* (OD_{600} = 0.005) in 20 mL MH broth and a sample of 500 µL hydrogel containing ciprofloxacin at the concentration of 988 nM (40x the MIC value)

7. Release of ciprofloxacin: *E. coli K12 wt* (OD_{600} = 0.005) in 20 mL MH broth and a sample of 500 μL hydrogel containing ciprofloxacin at the concentration of 988 nM (40x the MIC value), which was irradiated by green light

First 19 mL of MH broth were added into the 50-mL Erlenmeyer, which were covered on top with aluminium foil. Then 1 mL of diluted overnight culture *E. coli K12 wt* with an OD_{600} = 0.100 was added. All hydrogels in the crimped vials were dipped inside ethanol, wiped dry, opened by a glass cutter (final length 24 ± 1 mm) and put inside Erlenmeyer flasks containing the bacterial cultures. The experiments performed in absence of light (the "diffusion experiments") were wrapped with aluminium foil. All Erlenmeyer were put inside of the same water bath (height ca. 2 cm) and incubated at 22 °C.

The drug release experiments were irradiated for 3 h inside the water bath with two green 10 W LED diodes (523 nm) from a short distance (< 5 cm). After that time, irradiated vials still contained a significant amount of the gel. This might be due to the other experimental setup and the less favourable irradiation geometry that allows less light per a surface unit in comparison with the previous experiments. However, it was estimate that over 50% of the desired antibiotic concentration was released in the medium by that time. At that moment, the irradiation was ceased, and the experiment was continued in darkness (all the flasks were wrapped up with aluminium foil). The flasks were then placed into an incubator at 37 °C with 120 rpm and the cultures were incubated for the following 6 h. Every hour, the incubation was interrupted for ca. 5 min to collect 500 μL-aliquots of the medium to measure the OD_{600}. The results have been presented in the main part (Figure 30).

Figure 84. Picture of the setup for the light-induced release of ciprofloxacin inside bacterial cultures.

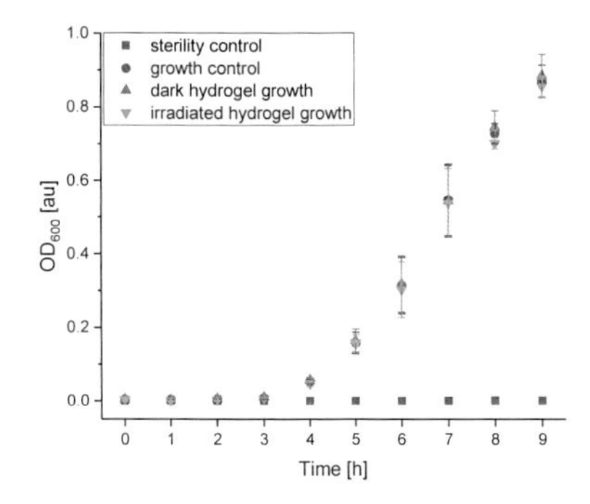

Figure 85. Control experiments 1.-4. for the inhibition of bacterial growth – medium without and with bacteria, medium + bacteria + gel (irradiated or in darkness) in absence of antibiotic.

5.3.4.8 Cytotoxicity against mammalian cells

Ciprofloxacin inhibits prokaryotic DNA gyrase and type II topoisomerase. As these forms of enzymes do not occur in mammalian cells, toxicity of ciprofloxacin against, e.g. HeLa cells, is several orders of magnitude lower than against the bacteria. Although several azobenzene derivatives were reported as toxic, mainly due to anaerobic reduction to the respective hydrazine, the fluorinated azobenzenes used in our experiments are, due to the electron-poor aromatic ring, not susceptible to reduction under physiological conditions (as we have tested by prolonged incubation with glutathione). Moreover, numerous 2,5-diketopiperazine derivatives are physiologically active[173], and some of them were reported cytotoxic (plinabulin).[306] For all these reasons, the toxicity was a carefully investigated for the photochromic materials. To assess the influence on mammalian cells, all the newly synthesized final products, the substances **2**, CiproFAzo, and CiproDKP, were investigated with a cell metabolic activity MTT assay. This rapid colorimetric assay measures the reduction of the tetrazolium dye MTT (3-(4,5-dimethylthiazol-2-yl)-2,5-diphenyl-tetrazolium bromide) to the formazan by oxidoreductases inside the cells. Cytotoxic compounds inhibit the enzymatic cell activity and prevent the formazan production. This method was adapted according to the literature.[307]

5.3.4.8.1 Procedure of the MTT assay

The HeLa cells were cultured in DMEM, which was supplemented with 10% FCS and 1% Penicillin/Streptomycin and incubated at 37 °C in humidified atmosphere with 5% CO_2. The cell numbers were counted with an improved Neubauer chamber and a brightfield microscope. 100 µL of a culture containing 5×10^4 HeLa cells/mL were added into each well of a 96-well plate and incubated for 24 h. All outer wells were not used and filled instead with 200 µL DMEM medium to ensure an equal growth of the cells (n=6). The irradiated (30 min, 10 W, 523 nm) and not irradiated stock solutions of F_2-PAP-DKP-Lys, CiproFAzo and CiproDKP were prepared at the concentration of 1.00 mol/L in DMSO. The stock solution of ciprofloxacin hydrochloride (100 mM) was prepared in diH_2O. In all wells the medium was exchanged with 100 µL DMEM medium with 1 µM – 2000 µM of the substance and a final concentration of 0.2% DMSO was used including the growth control. To achieve homogeneity, the mixtures containing > 500 µM of CiproFAzo and CiproDKP were prepared by ultrasound treatment at 37 °C, before exchanging the medium. The plates were incubated for 3 d in the darkness under the aforementioned conditions. For the negative control, Triton-X was added with a final concentration of 1% 15 min before the MTT addition. Then 20 µL of a fresh prepared and sterile-filtered solution of 2.73 mg/mL MTT in DMEM was added into each well and incubated for 4 h at 37 °C. Afterwards 100 µL commercial stop solution from Promega was added and the plates were incubated at 4 °C under humid conditions overnight to prevent evaporation. The formazan conversion was determined by an ELISA reader at 570 nm at rt and additionally at 700 nm to control the complete solubilisation. The absorption of doxorubicin was measured at 595 nm, because of the overlapping UV-Vis spectra of formazan at 570 nm. The data was fitted by the following logistic function with good correlation for mammalian cells:

$$y = \frac{A_1 - A_2}{1 + (x/x_0)^p} + A_2$$

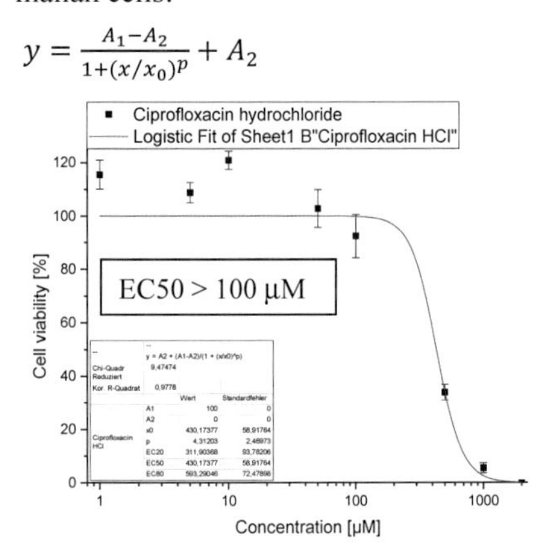

Figure 86. Logistic fit of the ciprofloxacin in the MTT assay, against human HeLa cell lines.

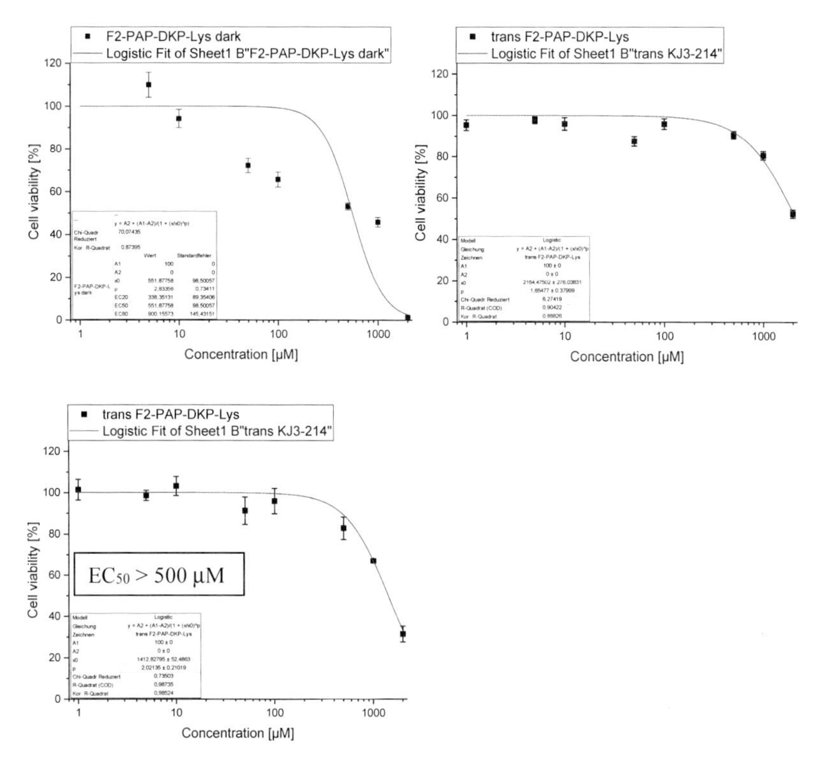

Figure 87. Cytotoxicity of the gelator *trans*-2 (F2-PAP-DKP-Lys, non-irradiated), against human HeLa cell lines (triplicates). $EC_{50} > 500$ µM.

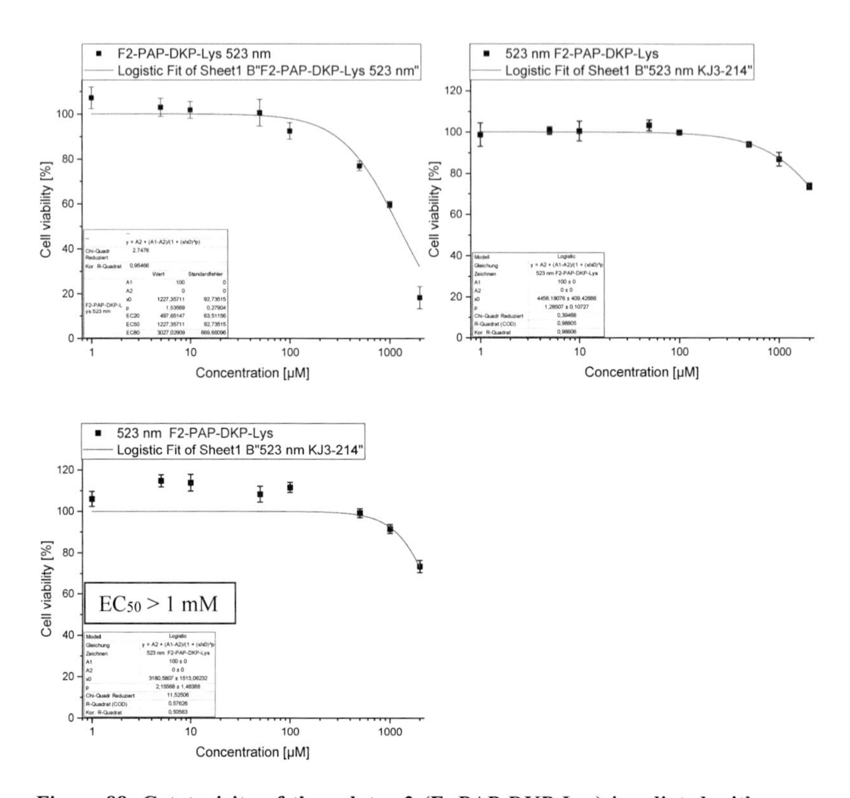

Figure 88. Cytotoxicity of the gelator 2 (F_2-PAP-DKP-Lys) irradiated with green light (523 nm), against human HeLa cell lines (triplicates). EC_{50} > 1 mM.

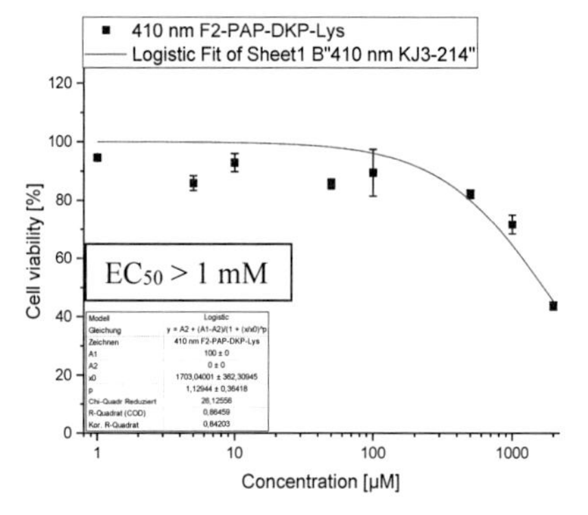

Figure 89. Cytotoxicity of the gelator 2 (F_2-PAP-DKP-Lys) irradiated with green light (523 nm) and with violet light (410 nm), against human HeLa cell lines. EC_{50} > 1 mM.

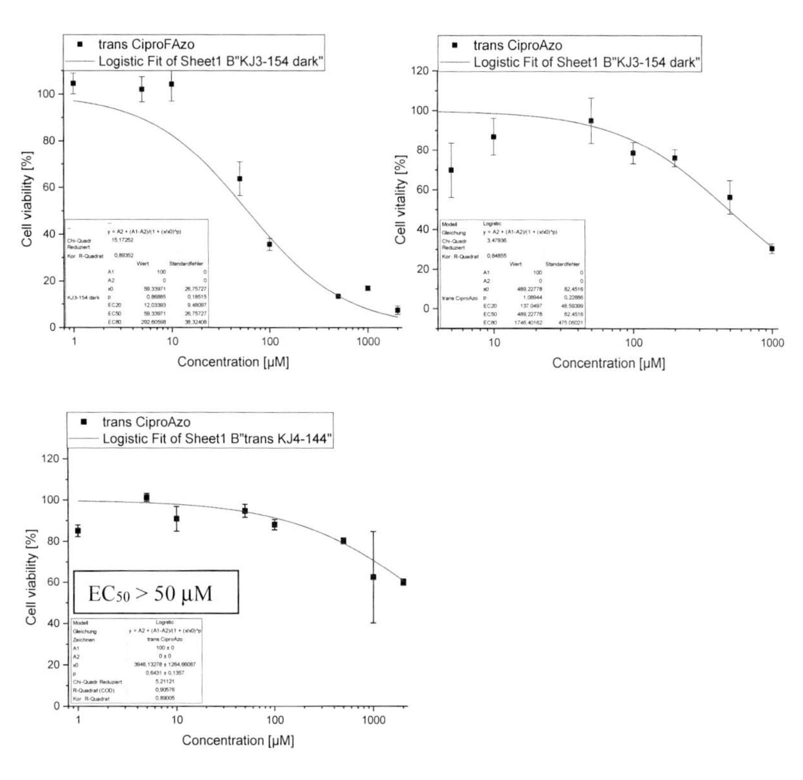

Figure 90. Cytotoxicity of the CiproFAzo *trans*-15 (non-irradiated) against human HeLa cell lines (triplicates). EC_{50} > 50 µM.

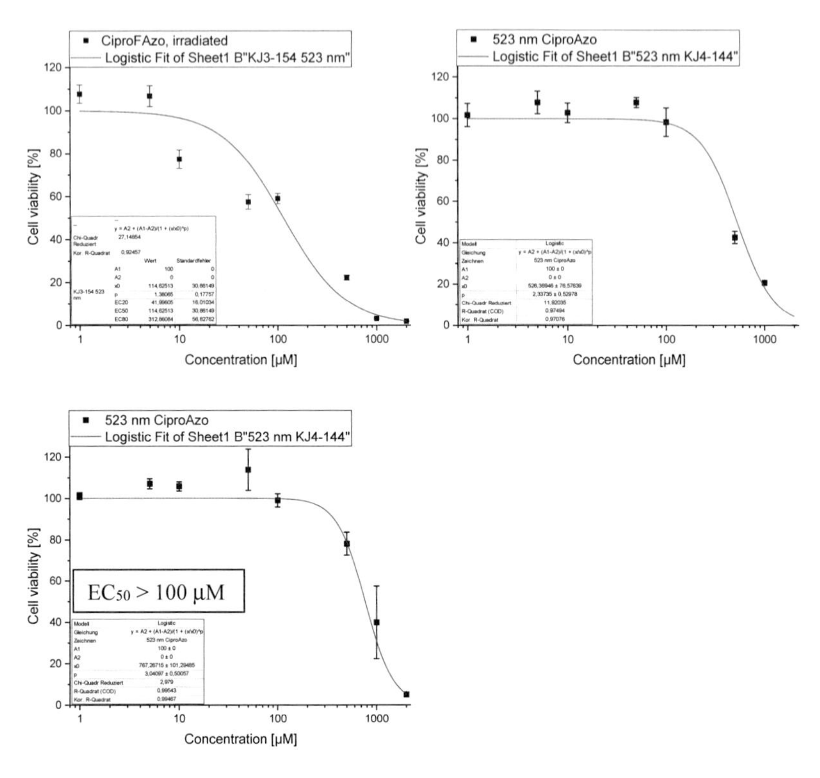

Figure 91. Cytotoxicity of the CiproFAzo 15 irradiated with green light (523 nm), against human HeLa cell lines (triplicates). EC_{50} > 100 µM.

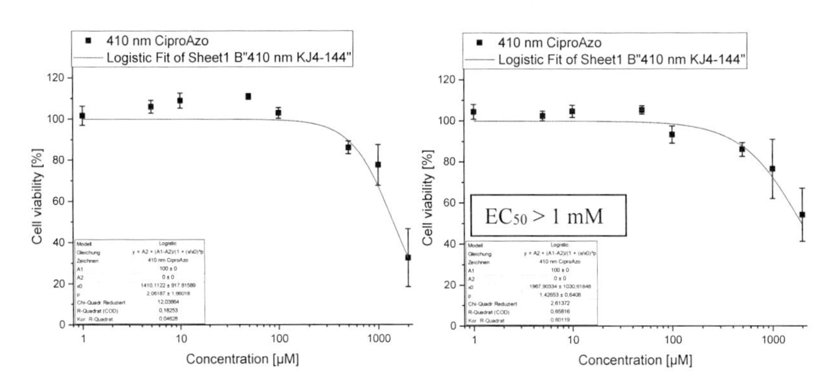

Figure 92. Cytotoxicity of the CiproFAzo 15 irradiated with green light (523 nm) and with violet light (410 nm), against human HeLa cell lines. EC_{50} > 1 mM.

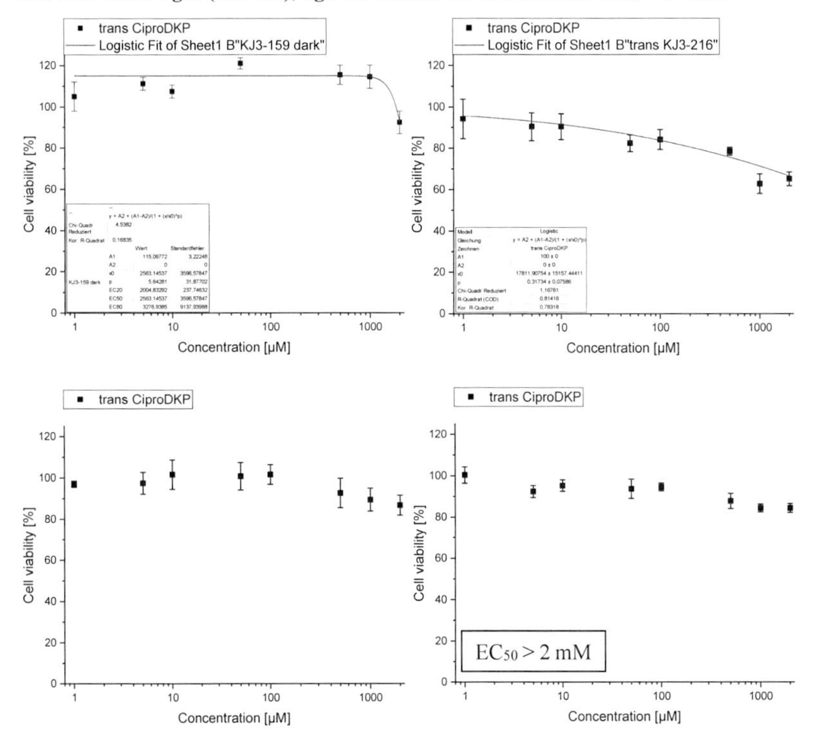

Figure 93. Cytotoxicity of the CiproDKP *trans*-16 (non-irradiated) against human HeLa cell lines (quadruplicates). EC_{50} > 2 mM.

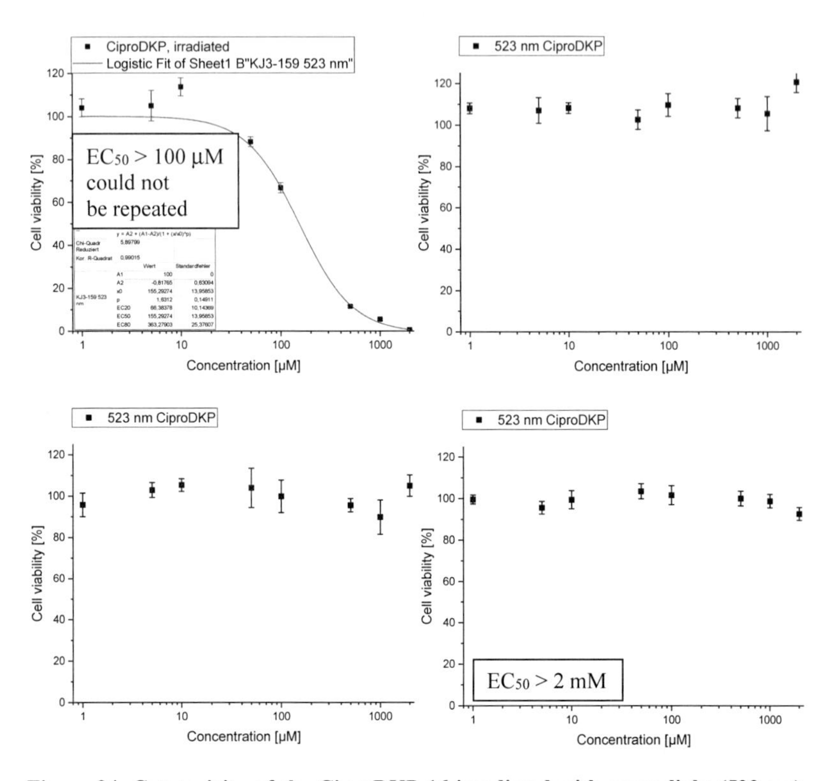

Figure 94. Cytotoxicity of the CiproDKP 16 irradiated with green light (523 nm), against human HeLa cell lines (quadruplicates). EC_{50} > 2 mM. In the first assay, an apparent EC_{50} value of > 100 µM (c.a. 155 µM approximated by logistic fit) has been obtained. However, this result could not be repeated and has to be considered an experimental artefact.

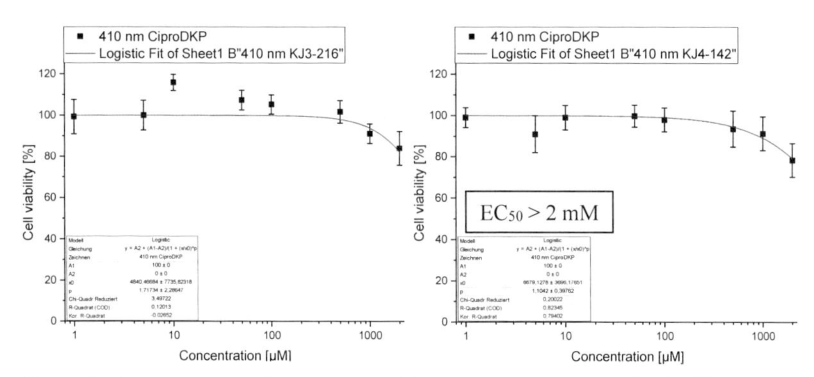

Figure 95. Cytotoxicity of the CiproDKP 16 irradiated with green light (523 nm) and with violet light (410 nm), against human HeLa cell lines (duplicates). EC_{50} > 2 mM.

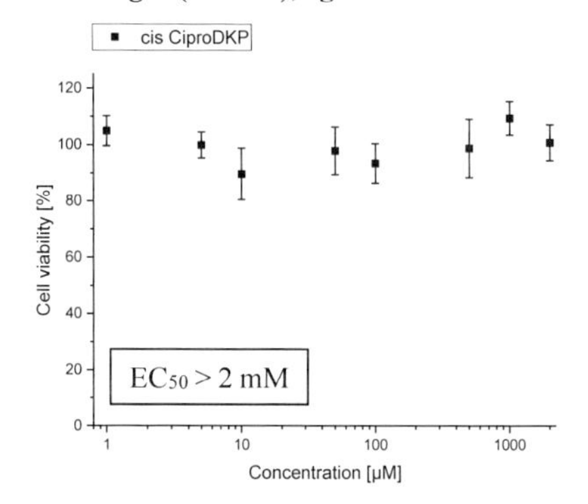

Figure 96. Cytotoxicity of the CiproDKP 7 irradiated with green light (523 nm) and purified with HPLC to obtain pure *cis*-16, against human HeLa cell lines. EC_{50} > 2 mM.

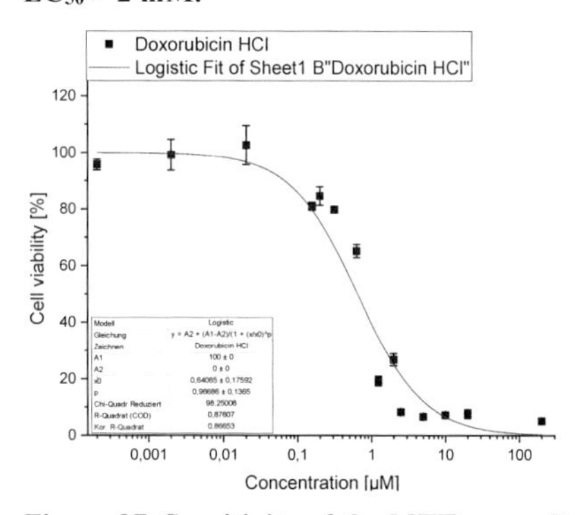

Figure 97. Sensitivity of the MTT assay. Determination of cytotoxicity for the commonly used cytotoxic anticancer drug doxorubicin. The measured EC_{50} range (0.5-1.0 µM) is in agreement with literature data (cf. 0.29-0.44 µM[207]).

5.4 F_4-PAP-DKP-Lys

5.4.1 Synthesis of F_4-PAP-DKP-Lys

Boc-Ala(I)-OMe, (*R*)-2-((*tert*-butoxycarbonyl)amino)-3-iodopropanoate 18

The methods of Boc-Ala(I)-OMe **18** are cited from the bachelor thesis of Christian Hald, which was supervised by Johannes Karcher.[308] This procedure and the procedures of Richard F. W. Jackson and Manuel Perez-Gonzalez[309] were reproduced by undergraduate students under the supervision of Johannes Karcher to provide constant replenishment.

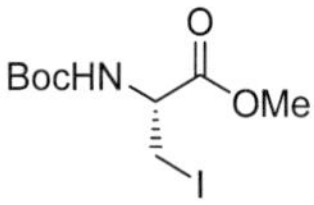

Triphenylphosphine (16.3 g, 62.3 mmol, 1.30 eq.) was dissolved in dry CH_2CL_2 (200 mL) under argon. Imidazole (4.24 g, 62.3 mmol, 1.30 eq.) was added and the mixture was stirred until complete dissolution.[310] The solution was cooled to 0 °C and the iodine (15.8 g, 62.3 mmol, 1.30 eq.) was added in four batches to maintain temperature and darkness. The slurry solution was stirred at rt for 10 min and cooled to 0 °C again. A solution of methyl (*tert*-butoxycarbonyl)-*L*-serinate (Boc-Ser(OH)-OMe, 10.5 g, 47.9 mmol, 1.00 eq.) in dry CH_2CL_2 (50 mL) was added dropwise over 1 h and in darkness. Then the reaction mixture was stirred at rt and the reaction progress was followed by TLC. After 2 h, full conversion of the starting material Boc-Ser(OH)-OMe was observed. The reaction mixture was filtered through a short column with silica gel and cH:EtOAc (1:1) as an eluent. The solvent was evaporated under reduced pressure yielding a brown oil. The crude was purified by silica gel column chromatography. Initial elution with pure *n*-hexane washed out non-polar impurities. Then the column was eluted with 5% diethyl ether in *n*-hexane and 10% diethyl ether in *n*-hexane. Evaporation of combined fractions and drying under reduced pressure resulted in 11.6 g Boc-Ala(I)-OMe **18** (35.2 mmol, 73% yield) as a colourless oil, which solidified as a white solid in the freezer at −20 °C.

TLC: R_f = 0.55 (*n*-hexane:EtOAc 4:1). **^{1}H NMR (300 MHz, $CDCl_3$):** 5.35 (d, *J* = 6.0 Hz, 1H), 4.61 – 4.43 (m, 1H), 3.78 (s, 3H), 3.55 (s, 2H), 1.44 (s, 9H) ppm. **^{13}C NMR (300 MHz, $CDCl_3$):** 170.1, 154.9, 80.6, 53.8, 53.1, 28.4, 7.9 ppm. **HRMS (ESI+):** *m/z* calcd for $[C_9H_{16}INO_4]^+$: 329.0119 Da $[M+H]^+$, found: 329.0119 Da (Δ = 0.1 ppm). **IR (ATR):** ṽ = 3347 (w), 2981 (w), 1731 (s), 1688 (s), 1520 (s), 1438 (m), 1392 (w), 1367 (m), 1312 (m), 1272 (m), 1248 (m), 1220 (m), 1156 (s), 1058 (m), 1008 (m), 978 (m), 905 (w), 862 (m), 790 (m), 775 (m), 760 (w), 741 (w), 645 (m), 584 (m), 479 (w), 435 (w) cm^{-1}.

Boc-Phe(3,5-F_2-4-NH_2)-OH, methyl (*S*)-3-(4-amino-3,5-difluorophenyl)-2-((*tert*-butoxycarbonyl)amino)propanoate 19

The synthesis starting from Boc-Phe(3,5-diF_2-4-NH_2)-OH **19** to Boc-F_4-PAP-OMe **20** was initially synthesized by Dr. Pianowski, Johannes Karcher and Christian Hald.[308] The methods to Boc-F_4-PAP-OH are cited from the master thesis of Susanne Kirchner[209], which included further optimisations and were reproduced by Johannes Karcher.

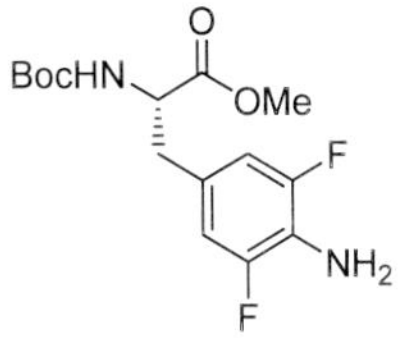

Dry DMF (30 mL) was added to zinc dust (10 Å) (2.98 g, 45.6 mmol, 3.00 eq.) under argon followed by iodine (0.289 g, 2.28 mmol, 0.150 eq.) and the mixture was stirred until the solution turned clear again. Methyl (*R*)-2-((*tert*-butoxycarbonyl)amino)-3-iodopropanoate **18** (Boc-Ala(I)-OMe, 5.00 g, 15.2 mmol, 1.00 eq.) was added followed by iodine (0.289 g, 2.28 mmol, 0.150 eq.) and the solution was stirred for 15 min until it had cooled down to rt again. $Pd_2(dba)_3$ (0.348 g, 0.380 mmol, 0.0250 eq.), SPhos (0.312 g, 0.760 mmol, 0.0500 eq.) and 4-bromo-2,6-difluoroaniline **17** (4.11 g, 19.8 mmol, 1.30 eq.) were added and the reaction mixture was stirred under argon for 3 d at rt.[208] The crude product was filtrated over Celite®, concentrated *in vacuo* and purified by silica gel column chromatography (cH:EtOAc 5:1 with 1% triethylamine, $R_f = 0.21$) to yield 3.01 g Boc-Phe(3,5-diF_2-4-NH_2)-OH) **19** (11.5 mmol, 75%) as a slightly brown solid. To achieve good and repeatable yields, it was critical to dry all substrates as well as the zinc dust under high vacuum overnight prior to the reaction.

^{1}H NMR (400 MHz, $CDCl_3$): $\delta = 6.59$ (d, $J = 8.5$ Hz, 2H), 5.01 (d, $J = 7.8$ Hz, 1H), 4.50 (dd, $J = 13.3, 5.7$ Hz, 1H), 3.72 (s, 3H), 3.59 (s, 2H), 2.95 (ddd, $J = 34.1, 14.0, 5.5$ Hz, 2H), 1.42 (s, 9H) ppm. **^{13}C NMR (101 MHz, $CDCl_3$):** $\delta = 172.2$ (s), 155.2 (s), 152.0 (dd, $J = 240.7, 8.4$ Hz), 125.5 – 125.0 (m), 122.9 (t, $J = 16.3$ Hz), 112.0 (dd, $J = 14.8, 7.2$ Hz), 80.2 (s), 54.5 (s), 52.5 (s), 37.5 (s), 28.4 (s) ppm. **^{19}F NMR (377 MHz, $CDCl_3$):** $\delta = -136.56$ (s) ppm. **HRMS (EI+):** *m/z* calcd for $C_{15}H_{20}F_2N_2O_4$: 330.1391 Da [M], found: 330.1392 Da ($\Delta = 0.3$ ppm). **IR (ATR):**

ṽ = 3456 (vw), 3360 (m), 2978 (vw), 2954 (vw), 1731 (m), 1688 (m), 1676 (m), 1591 (w), 1519 (m), 1438 (m), 1281 (m), 1248 (m), 1156 (m), 1051 (w), 1020 (m), 991 (w), 962 (m), 841 (m), 761 (w), 593 (m), 411 (vw) cm^{-1}.

Boc-F_4-PAP-OMe, methyl (*S*,*E*)-2-((*tert*-butoxycarbonyl)amino)-3-(4-((2,6-difluoro-phenyl)diazenyl)-3,5-difluorophenyl)propanoate 20

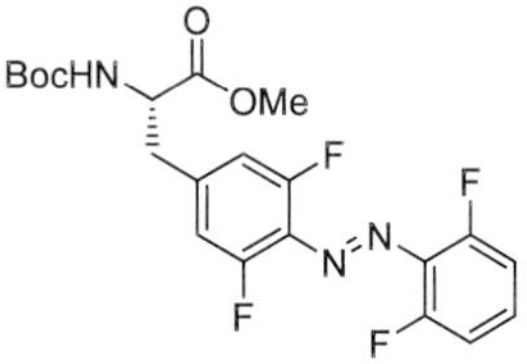

Oxone® (12.3 g, 40.0 mmol, 2.20 eq.) in diH_2O (120 mL) was added to 2,6-difluoroaniline **35** (2.35 g, 18.2 mmol, 1.00 eq.) dissolved in CH_2Cl_2 (72 mL). The mixture was stirred vigorously at rt for 3 h. The layers were separated, and the organic layer was washed with 1N HCl (50 mL), water (3×50 mL), dried with Na_2SO_4 and concentrated *in vacuo*. The crude product **11** was used for the synthesis of compound **20** without any further purification.[2] The crude product 1,3-difluoro-2-nitrosobenzene **11** (1.95 g, 13.6 mmol, 1.50 eq.) and Boc-Phe(3,5-diF_2-4-NH_2)-OH **19** (3.00 g, 9.08 mmol, 1.00 eq.) were dissolved in AcOH:PhMe:TFA (6:6:1) (130 mL) and the reaction mixture was stirred at rt for 3 d. H_2O (60 mL) was added and the crude product was extracted with EtOAc (3×50 mL). The combined organic layers were dried with Na_2SO_4 and concentrated *in vacuo*. The crude product was purified by silica gel column chromatography (cH:CH_2Cl_2 1:1, R_f (CH_2Cl_2) = 0.25) to yield 2.12 g Boc-F_4-PAP-OMe **20** (4.65 mmol, 51%) as an orange solid.

^{1}H NMR (400 MHz, $CDCl_3$): δ = 7.37 (tt, J = 8.3, 5.9 Hz, 1H), 7.06 (t, J = 8.7 Hz, 2H), 6.86 (d, J = 9.8 Hz, 2H), 5.10 (d, J = 7.5 Hz, 1H), 4.62 (d, J = 6.9 Hz, 1H), 3.77 (s, 3H), 3.14 (ddd, J = 53.2, 13.8, 5.8 Hz, 2H), 1.76 (s, 2H), 1.44 (s, 7H) ppm. **^{13}C NMR (101 MHz, $CDCl_3$):** δ = 171.6 (s), 157.0 (dd, J = 14.9, 4.4 Hz), 155.1 (s), 154.4 (dd, J = 15.7, 4.4 Hz), 141.7 (t, J = 9.9 Hz), 132.0 (t, J = 9.9 Hz), 131.5 (t, J = 10.5 Hz), 130.6 (t, J = 10.0 Hz), 113.7 (dd, J = 21.1, 2.7 Hz), 112.8 (dd, J = 21.1, 2.9 Hz), 80.6 (s), 54.1 (s), 52.8 (s), 38.3 (s), 28.4 (s) ppm. **^{19}F NMR (377 MHz, $CDCl_3$):** δ = −129.81 (s), −130.40 (s) ppm. **HRMS (FAB+, 3-NBA):** *m/z* calcd for $C_{21}H_{22}F_4N_3O_4$: 456.1546 Da [M+H], found: 456.1547 Da (Δ = 0.2 ppm). **IR (ATR):** ṽ = 3348 (w), 2980 (vw), 2954 (vw), 1733 (w), 1680 (m), 1626 (w), 1576 (w), 1519 (m), 1470 (m), 1437 (w),

1393 (w), 1349 (w), 1271 (w), 1242 (m), 1212 (m), 1158 (m), 1049 (w), 1024 (m), 845 (w), 786.4 (m), 743 (w), 605 (w), 513 (w), 471 (w), 435 (vw), 383 (vw) cm^{-1}. **elemental analysis calcd for $C_{21}H_{21}F_4N_3O_4$ (%):** C: 55.39, H: 4.65, F: 16.69, N: 9.23, O: 14.05, found: C: 55.17, H: 4.47, N: 9.07.

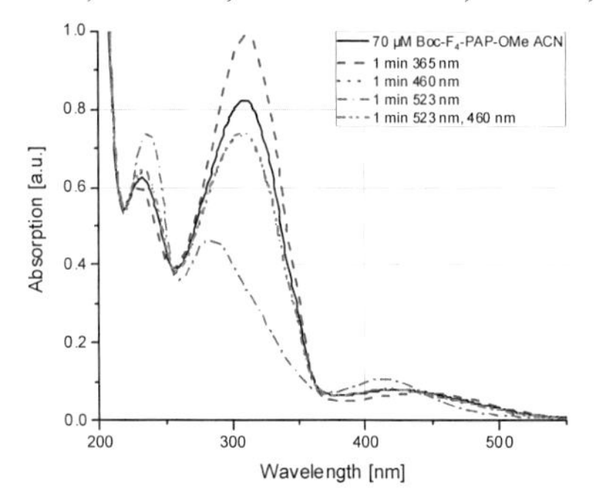

Figure 98. UV-Vis of 70 µM Boc-F_4-PAP-OMe in ACN, 1 min irradiation with 365, 460 and 523 nm.

Boc-F_4-PAP-OH, (*S*,*E*)-2-((*tert*-butoxycarbonyl)amino)-3-(4-((2,6-difluorophenyl)-diazenyl)-3,5-difluorophenyl)propanoic acid 21

Boc-F_4-PAP-OMe **20** (1.90 g, 4.17 mmol, 1.00 eq.) was dissolved in in MeCN (19 mL) and subsequently LiOH (1.90 g, 79.3 mmol, 19.0 eq.) in H_2O (19 mL) was added. The reaction mixture was stirred at rt for 15 min. The reaction was quenched by adding aqueous HCl (2 M, 114 mL). The mixture was extracted with EtOAc (2×100 mL), washed with brine (20 mL) and the solvent was removed under reduced pressure. The crude product was purified by silica gel column chromatography (CH_2Cl_2:MeOH 9:1, R_f = 0.55) to yield 1.84 g Boc-F_4-PAP-OH **21** (4.18 mmol, 93%) as an orange solid.

^{1}H NMR (400 MHz, $CDCl_3$): δ = 7.65 (br s, 1H), 7.35 (tt, J = 8.4, 5.9 Hz, 1H), 7.04 (t, J = 8.7 Hz, 2H), 6.94 (t, J = 9.5 Hz, 2H), 5.20 (d, J = 7.2 Hz, 1H), 4.53 (dd, J = 71.9, 5.2 Hz, 1H), 3.18 (ddd, J = 76.7, 13.7, 6.6 Hz, 2H), 1.42 (s, 9H) ppm. **^{13}C NMR (101 MHz, $CDCl_3$):** δ = 175.0 (s), 156.9 (dd, J = 14.8, 4.3 Hz), 155.5

(s), 154.3 (dd, $J = 15.5, 4.3$ Hz), 141.8 (s), 132.0 (d, $J = 9.6$ Hz), 131.5 (t, $J = 10.1$ Hz), 130.6 (t, $J = 9.8$ Hz), 113.8 (d, $J = 20.5$ Hz), 112.7 (dd, $J = 21.0$, 2.8 Hz), 81.0 (s), 54.1 (s), 37.9 (s), 28.3 (s) ppm. **^{19}F NMR (377 MHz, $CDCl_3$):** $\delta = -125.11$ (s), -125.64 (s) ppm. **HRMS (FAB+):** *m/z* calcd for $C_{20}H_{20}F_4N_3O_4$: 442.1390 Da [M+H], found: 442.1388 Da ($\Delta = 0.4$ ppm). **IR (ATR):** $\tilde{\nu} = 3340$ (vw), 2981 (vw), 2932 (vw), 1682 (w), 1626 (w), 1577 (w), 1519 (w), 1471 (w), 1446 (w), 1368 (w), 1242 (w), 1160 (w), 1062 (w), 1024 (m), 845 (w), 786 (w), 744 (w), 626 (w), 512 (vw), 472 (vw) cm^{-1}.

Boc-F$_4$-PAP-Lys(Boc)-OMe, methyl N^6-(*tert*-butoxycarbonyl)-N^2-((*S*)-2-((*tert*-butoxycarbonyl)amino)-3-(4-((*E*)-(2,6-difluorophenyl)diazenyl)-3,5-difluorophenyl)propanoyl)-*L*-lysinat 22

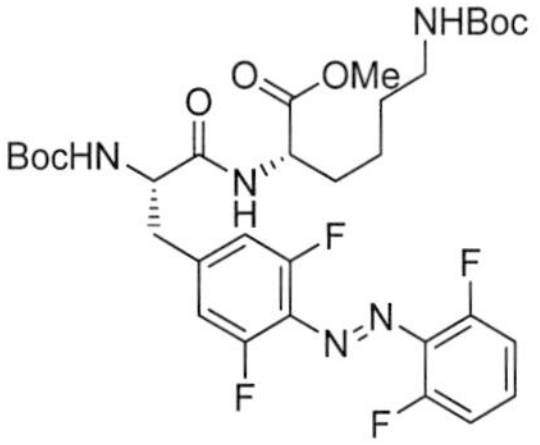

Boc-F$_4$-PAP-OH **21** (3.09 g, 6.99 mmol, 1.00 eq.), HBTU (2.65 g, 4.61 mmol, 1.00 eq.) and DIPEA (3.0 mL, 17.5 mmol, 2.50 eq.) were dissolved in anhydrous DMF (23.3 mL) and stirred for 10 min at rt under argon. The same amount of DIPEA (3.0 mL, 17.5 mmol, 2.50 eq.) was added together with *ω*-*N*-Boc-lysine methyl ester hydrochloride H-Lys(Boc)-OMe·HCl (2.10 g, 7.06 mmol, 1.01 eq.) dissolved in DMF (23.3 mL). The reaction mixture was stirred at rt and the reaction progress was followed by TLC. After 2 h, full conversion of the starting material was observed. The reaction mixture was quenched with sat. aqueous NH_4Cl solution (200 mL) and extracted once with EtOAc (200 mL). The organic layer was washed with sat. aqueous NH_4Cl solution (3×200 mL), brine (1×200 mL), dried with anhydrous Na_2SO_4 and the solvent was evaporated under reduced pressure. The crude mixture was purified by silica gel column chromatography. The column was started with 40% EtOAc in cH to wash out colourless non-polar impurities ($R_f = 0.13$). Then the product was eluted with EtOAc:cH 1:1 ($R_f = 0.38$). Evaporation of combined orange fractions and drying under high vacuum resulted in 4.22 g Boc-F$_4$-PAP-Lys(Boc)-OMe **22** (6.17 mmol, 88% yield) of the linear dipeptide as orange solid.

^{1}H NMR (500 MHz, DMSO): δ = 8.33 (d, J = 7.5 Hz, 1H), 7.62 (tt, J = 8.5, 6.1 Hz, 1H), 7.35 (t, J = 8.9 Hz, 2H), 7.28 (d, J = 11.1 Hz, 2H), 7.04 (d, J = 8.8 Hz, 1H), 6.77 (t, J = 5.8 Hz, 1H), 4.27 (dtd, J = 27.4, 9.3, 8.3, 4.9 Hz, 2H), 3.62 (s, 3H), 3.06 (dd, J = 13.5, 4.3 Hz, 1H), 2.90 (q, J = 6.6 Hz, 2H), 2.83 (dd, J = 13.5, 10.4 Hz, 1H), 1.80 – 1.67 (m, 1H), 1.62 (dtd, J = 13.6, 9.1, 5.1 Hz, 1H), 1.33 (d, J = 31.8 Hz, 22H) ppm. **^{13}C NMR (126 MHz, DMSO):** δ = 172.4, 171.2, 155.6 (d, J = 4.5 Hz), 155.6, 155.4 (d, J = 4.8 Hz), 155.2, 153.5 (d, J = 4.1 Hz), 153.3 (d, J = 4.9 Hz), 145.3 (t, J = 10.2 Hz), 132.9 (t, J = 10.6 Hz), 130.8 (t, J = 10.0 Hz), 129.0 (t, J = 9.7 Hz), 114.0 (dd, J = 19.9, 3.2 Hz), 113.2 (dd, J = 19.6, 3.7 Hz), 78.2, 77.3, 54.8, 52.0, 51.9, 37.4, 30.7, 29.1, 28.3, 28.0, 22.6 ppm. **^{19}F NMR (471 MHz, DMSO):** δ = −122.26 (d, J = 11.5 Hz), -122.32 (dd, J = 10.3, 6.3 Hz) ppm. **^{1}H NMR (400 MHz, acetic acid-*d*$_4$):** δ = 7.46 (ddd, J = 14.3, 8.5, 5.8 Hz, 1H), 7.13 (t, J = 9.0 Hz, 2H), 7.07 (d, J = 10.6 Hz, 2H), 4.73 – 4.62 (m, 1H), 4.57 (dd, J = 8.9, 5.0 Hz, 1H), 3.72 (s, 3H), 3.20 (dd, J = 13.8, 6.4 Hz, 1H), 3.14 – 2.96 (m, 3H), 1.87 (dq, J = 13.5, 7.2 Hz, 1H), 1.71 (dq, J = 14.4, 7.9 Hz, 1H), 1.51 – 1.35 (m, 22H) ppm. **^{13}C NMR (101 MHz, acetic acid-*d*$_4$):** δ = 174.5, 174.4, 159.0, 158.6 (d, J = 4.3 Hz), 158.5 (d, J = 4.4 Hz), 158.1, 156.0 (d, J = 4.0 Hz), 155.9 (d, J = 4.6 Hz), 144.6, 133.5 (t, J = 9.7 Hz), 133.4, 131.9, 115.7 – 115.3 (m), 114.4 (dd, J = 21.0, 2.9 Hz), 82.4, 81.4, 56.9, 54.3, 53.9, 41.7, 39.6, 32.8, 30.8, 29.4, 29.3, 24.3 ppm. **^{19}F NMR (376 MHz, acetic acid-*d*$_4$):** δ = −126.35, -126.92 ppm. **HRMS (FAB+):** *m/z* calcd for $C_{32}H_{42}O_7N_5F_4$: 684.3020 Da [M+H], found: 684.3023 Da (Δ = 0.4 ppm). **IR (ATR):** $\tilde{\nu}$ = 3323 (vw), 2936 (vw), 1747 (w), 1680 (w), 1654 (w), 1626 (w), 1577 (w), 1520 (w), 1472 (w), 1439 (w), 1366 (w), 1324 (vw), 1242 (w), 1162 (w), 1051)w), 1024 (w), 851 (w), 786 (w), 744 (w), 643 (w), 513 (vw), 472 (vw) cm^{-1}. **UV-Vis (MeCN):** λ_{max} = 314, 229, 454 nm.

H-F$_4$-PAP-Lys-OMe, methyl ((*S*)-2-amino-3-(4-((*E*)-(2,6-difluorophenyl)diazenyl)-3,5-difluorophenyl)propanoyl)-*L*-lysinate 22d

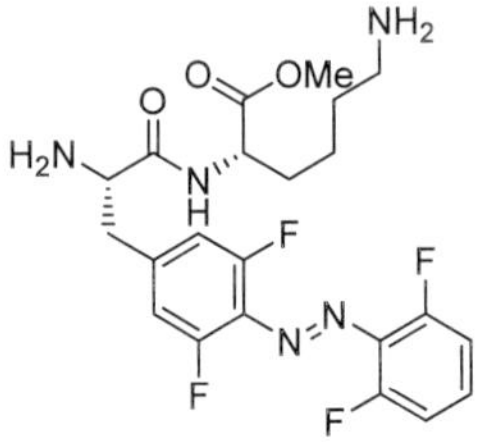

Boc-F$_4$-PAP-Lys(Boc)-OMe **22** (4.10 g, 6.17 mmol, 1.00 eq.) was dissolved in CH_2Cl_2 (61 mL). Trifluoroacetic acid (TFA, 61 mL) and 1 vol% Triisopropylsilane (1.2 mL, 6.17 mmol, 1.00 eq.) was added at rt and the reaction mixture was stirred for 1 h at rt. Then the reaction mixture was diluted with PhMe (100 mL) and evaporated under reduced pressure. After drying under high vacuum, the crude F$_4$-PAP-Lys-OMe **22d** was obtained as TFA salt (4.86 g). For the analytics, 50 mg of the product was dissolved in 2 mL DMSO and purified by preparative HPLC with the following settings: 10 mL/min, 40 min gradient 5-95% ACN in diH_2O with 0.1% TFA, detection at 330 nm, C_{18}-colum, retention 22.0 min.

^{1}H NMR (500 MHz, DMSO): δ = 9.03 (d, J = 7.4 Hz, 1H), 8.39 (s, 3H), 7.88 (s, 3H), 7.63 (tt, J = 8.5, 6.1 Hz, 1H), 7.36 (t, J = 8.9 Hz, 2H), 7.27 (d, J = 10.4 Hz, 2H), 4.32 (ddd, J = 9.0, 7.4, 5.2 Hz, 1H), 4.22 (s, 1H), 3.64 (s, 3H), 3.17 (ddd, J = 58.1, 14.0, 6.7 Hz, 2H), 2.77 (dq, J = 12.7, 5.6 Hz, 2H), 1.84 – 1.70 (m, 1H), 1.65 (dtd, J = 14.0, 9.3, 5.3 Hz, 1H), 1.55 (p, J = 7.6 Hz, 2H), 1.38 (td, J = 13.0, 9.5, 5.6 Hz, 2H) ppm. **^{13}C NMR (126 MHz, DMSO):** δ = 171.8, 167.9, 158.5 (q, J = 32.1 Hz), 155.6 (t, J = 4.3 Hz), 153.6 (t, J = 4.6 Hz), 141.3 (t, J = 10.3 Hz), 133.1 (t, J = 10.6 Hz), 130.8 (t, J = 10.1 Hz), 129.7 (t, J = 9.9 Hz), 117.0 (q, J = 298.2 Hz), 114.5 (dd, J = 20.4, 3.2 Hz), 113.3 (dd, J = 19.7, 3.6 Hz), 52.8, 52.1, 52.0, 38.5, 36.5, 30.4, 26.6, 22.2 ppm. **^{19}F NMR (471 MHz, DMSO):** δ = −121.86 (d, J = 11.2 Hz), -122.33 (dd, J = 9.9, 6.2 Hz) ppm. **^{1}H NMR (400 MHz, acetic acid-*d*$_4$):** δ = 7.50 (ddd, J = 14.3, 8.5, 5.8 Hz, 1H), 7.17 (t, J = 9.5 Hz, 4H), 4.73 (t, J = 6.8 Hz, 1H), 4.61 (dd, J = 9.0, 4.8 Hz, 1H), 3.77 (s, 3H), 3.39 (d, J = 6.9 Hz, 2H), 3.10 (t, J = 7.5 Hz, 2H), 1.92 (ddd, J = 13.8, 8.5, 5.0 Hz, 1H), 1.76 (tq, J = 14.1, 6.5 Hz, 3H), 1.50 (q, J = 8.2, 7.6 Hz, 2H), 1.09 (d, J = 4.4 Hz, 1H). **^{19}F NMR (376 MHz, acetic acid-*d*$_4$):** δ = −125.86, -126.89 ppm. **HRMS (FAB+):** *m/z* calcd for $C_{22}H_{26}O_3N_5F_4$: 484.1972 Da [M+H], found: 484.1971 Da (Δ = 0.2 ppm). **IR (ATR):** $\tilde{\nu}$ = 2954 (w), 1666 (m), 1627 (w), 1579 (w), 1472 (w), 1438 (w), 1243 (w), 1181 (m), 1131 (m), 1050 (w), 1023 (w), 837 (w), 798 (w), 743 (vw), 722 (m), 624 (vw), 597 (vw), 515 (vw), 481 (vw), 410 (vw) cm^{-1}. **UV-Vis (MeCN):** λ_{max} = 311, 229 nm.

F_4-PAP-DKP-Lys, (3*S*,6*S*)-3-(4-aminobutyl)-6-(4-((*E*)-(2,6-difluorophenyl)diazenyl)-3,5-difluorobenzyl)piperazine-2,5-dione

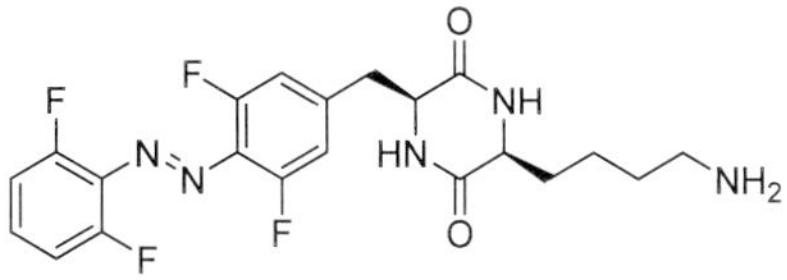

F F N N F HN NH O O NH2 F

The crude F_4-PAP-Lys-OMe **22d** (4.87 g, 6.17 mmol, 1.00 eq.) was dissolved in 2-butanol (260 mL). Subsequently glacial acetic acid (1302 µL, 22.8 mmol, 3.69 eq.), *N*-methylmorpholine (865 µL, 7.87 mmol, 1.28 eq.) and *N,N*-diisopropyl-*N*-ethylamine (DIPEA, 1562 µL, 8.97 mmol, 1.45 eq.) were added. The resulting mixture was refluxed for 2 h (oil bath, external heating 120 °C), then cooled down. Approximately 80% of the solvent was removed under reduced pressure. Cooling the residue down to rt resulted in orange gel-like precipitation. The precipitate was filtered off, washed with small amounts of cold ACN (3×20 mL) on the filter, dried on the filter and *in vacuo* using oil pump overnight resulting in 2.67 g F_4-PAP-DKP-Lys **3** (5.91 mmol, 96% yield) as orange solid. For the analytics the product was dissolved in DMSO and purified by preparative HPLC with the following settings: 10 mL/min, 40 min gradient 20-40% ACN in diH_2O with 0.1% TFA, detection at 330 nm, C_{18}-colum, retention 22.5 min. After lyophilisation, the pure *E*-F_4-PAP-DKP-Lys **3** was obtained as the TFA salt.

^{1}H NMR (400 MHz, acetic acid-d_4): δ = 7.49 (ddd, *J* = 14.3, 8.5, 5.9 Hz, 1H), 7.15 (t, *J* = 9.3 Hz, 2H), 7.09 (d, *J* = 10.4 Hz, 2H), 4.61 (t, *J* = 5.5 Hz, 1H), 4.14 (d, *J* = 6.4 Hz, 1H), 3.38 – 3.17 (m, 2H), 3.03 (t, *J* = 7.6 Hz, 2H), 1.82 – 1.53 (m, 3H), 1.46 – 1.17 (m, 3H) ppm (amine and amid not visible). **^{13}C NMR (101 MHz, acetic acid-d_4):** δ = 190.6, 189.3, 182.3 (d, *J* = 37.1 Hz), 177.4 (d, *J* = 4.0 Hz), 177.2 (d, *J* = 4.8 Hz), 174.8 (d, *J* = 4.0 Hz), 174.6 (d, *J* = 4.7 Hz), 161.6 (t, *J* = 9.6 Hz), 152.7 (t, *J* = 10.4 Hz), 152.3 – 151.6 (m), 151.4 – 150.9 (m), 136.7 (d, *J* = 290.6 Hz), 135.0 (d, *J* = 23.4 Hz), 133.2 (dd, *J* = 20.6, 3.3 Hz), 75.9, 74.6, 60.1, 59.6, 53.7, 46.8, 41.5 ppm. **^{19}F NMR (376 MHz, acetic acid-d_4):** δ = −126.28, -126.82 ppm. **^{1}H NMR (300 MHz, Deuterium Oxide, 137 mM NaCl):** δ = 7.63 – 7.41 (m, 1H), 7.17 (t, *J* = 9.8 Hz, 2H), 7.04 (d, *J* = 11.3 Hz, 2H), 4.56 (s, 1H), 4.01 (s, 1H), 3.35 (d, *J* = 14.1 Hz, 1H), 3.20 – 2.97 (m, 1H), 2.78 (t, *J* = 7.7 Hz, 2H), 1.55 – 1.20 (m, 3H), 1.11 – 0.72 (m, 3H) ppm. **HRMS (FAB+):**

m/z calcd for $C_{21}H_{22}O_2N_5F_4$: 452.1710 Da [M+H], found: 452.1708 Da (Δ = 0.4 ppm). **IR (ATR):** $\tilde{\nu}$ = 2952 (vw), 1664 (m), 1627 (w), 1578 (w), 1450 (w), 1332 (w), 1242 (vw), 1199 (w), 1174.4 (w), 1129 (w), 1052 (w), 1024 (w), 833 (w), 786 (w), 741 (w), 720 (w), 621 (vw), 480 (vw), 436 (vw), 382 (vw) cm^{-1}. **UV-Vis (MeCN):** λ_{max} = 312, 228, 456 nm.

5.4.1.1 Synthesis of plinabulin

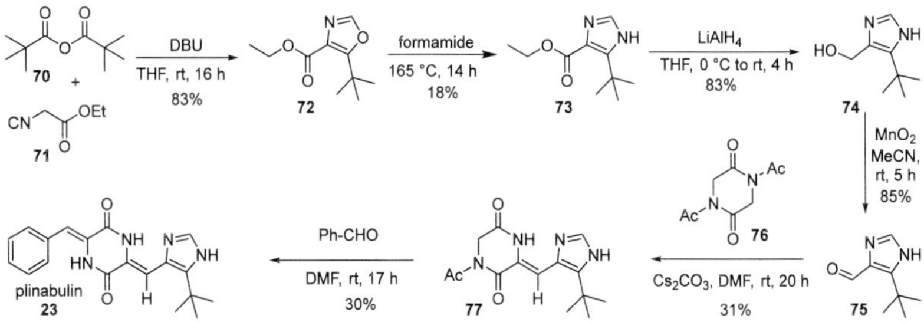

DBU
THF, rt, 16 h
83%
formamide
165 °C, 14 h
18%
$LiAlH_4$
THF, 0 °C to rt, 4 h
83%
MnO_2
MeCN,
rt, 5 h
85%
Cs_2CO_3, DMF, rt, 20 h
31%
Ph-CHO
DMF, rt, 17 h
30%

Scheme 14. Synthesis of a DKP-bearing antiproliferative agent plinabulin 23 as a suitable guest for selective light-triggered release from hydrogels.

The following synthesis was performed by the intern Anna-Lena Leistner under the supervision of Johannes Karcher and was used for the drug release.[311]

Ethyl 5-(*tert*-butyl)oxazole-4-carboxylate 72

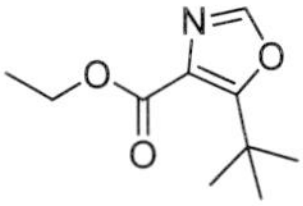

Pivalic anhydride **70** (24.5 mL, 22.5 g, 121 mmol, 1.10 eq.) and DBU (18.1 g, 121 mmol, 1.10 eq.) were added to a solution of ethyl isocyanoacetate **71** (12.4 g, 110 mmol, 1.00 eq.) in 100 mL of THF. The reaction mixture was stirred at rt overnight. The solvent was removed under reduced pressure and the residue was dissolved in 200 mL of EtOAc and washed with brine (150 mL), sat. aqueous Na_2CO_3 solution (75 mL), citric acid (10%, 100 mL) and dried over Na_2SO_4. The crude product was concentrated *in vacuo* and purified by silica gel column chromatography (cH:EtOAc 20:1 to 6:1, cH:EtOAc 8:1 R_f = 0.25) to yield 18.0 g ethyl 5-(*tert*-butyl)oxazole-4-carboxylate **72** (91.3 mmol, 83%) as a colourless oil.[312]

^{1}H NMR (400 MHz, $CDCl_3$): δ = 7.67 (s, 1H), 4.35 (q, *J* = 7.1 Hz, 2H), 1.42 (s, 9H), 1.37 (t, *J* = 7.1 Hz, 3H) ppm. **^{13}C NMR (101 MHz, $CDCl_3$):** δ = 166.0 (s), 162.2 (s), 147.3 (s), 126.0 (s), 61.2 (s), 33.4 (s), 28.2 (s), 14.3 (s) ppm. **HRMS**

(EI+): *m/z* calcd for $C_{10}H_{15}NO_3$: 197.1052 Da [M], found: 197.1050 Da (Δ = 0.1 ppm). **IR (ATR):** $\tilde{\nu}$ = 3119 (vw), 2975 (w), 2938 (w), 2909 (w), 2874 (vw), 1718 (m), 1572 (m), 1527 (m), 1463 (w), 1368 (m), 1334 (m), 1302 (m), 1254 (w), 1221 (w), 1180 (m), 1130 (m), 1097 (m), 1052 (s), 1023 (m), 942 (w), 844 (w), 794 (w), 709 (vw), 650 (m), 606 (vw), 542 (vw) cm^{-1}.

Ethyl 5-(*tert*-butyl)-1*H*-imidazole-4-carboxylate 73

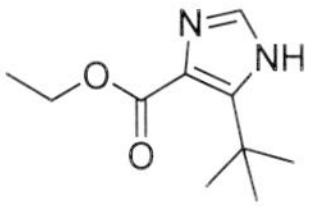

Ethyl 5-(*tert*-butyl)oxazole-4-carboxylate **72** (17.5 g, 88.7 mmol, 1.00 eq.) was dissolved in 106 mL of formamide (2.64 mol, 29.8 eq.) and the mixture was heated to 165 °C for 14 h. The mixture was cooled down, subsequently EtOAc (200 mL) and brine (100 mL) were added and the mixture was extracted with EtOAc (2×100 mL). The combined organic layers were washed with sat. aqueous solution of Na_2CO_3 (3×75 mL) and brine (3×75 mL), dried over Na_2SO_4 and concentrated *in vacuo*. The crude product was purified by silica gel column chromatography (MeOH (0.1% to 5%) in $CHCl_3$, R_f (2% MeOH in $CHCl_3$) = 0.25) to yield 3.19 g ethyl 5-(*tert*-butyl)-1*H*-imidazole-4-carboxylate **73** (16.3 mmol, 18%) as a yellow solid.[312]

^{1}H NMR (400 MHz, CDCl$_3$): δ = 7.52 (s, 1H), 4.36 (q, J = 7.1 Hz, 2H), 1.46 (s, 9H), 1.39 (t, J = 7.1 Hz, 3H) ppm. **^{13}C NMR (101 MHz, CDCl$_3$):** δ = 133.2 (s), 60.8 (s), 32.9 (s), 29.2 (s), 27.8 (s), 14.5 (s) ppm. **HRMS (EI+):** *m/z* calcd for $C_{10}H_{16}N_2O_2$: 196.1212 Da [M], found: 196.1211 Da (Δ = 0.5 ppm). **IR (ATR):** $\tilde{\nu}$ = 3119 (w), 2956 (w), 2907 (w), 2873 (w), 2841 (w), 2633 (vw), 1699, (m) 1559 (vw), 1498 (w), 1433 (m), 1365 (w), 1296 (m), 1224 (w), 1172 (w), 1147 (m), 1096 (w), 1058 (m), 1025 (m), 953 (m), 847 (w), 793 (w), 712 (vw), 662 (m), 562 (vw), 483 (vw), 400 (w) cm^{-1}.

(5-(*tert*-Butyl)-1*H*-imidazol-4-yl)methanol 74

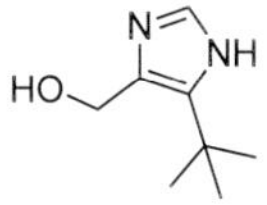

A solution of ethyl 5-(*tert*-butyl)imidazole-4-carboxylate **73** (2.56 g, 13.0 mmol, 1.00 eq.) in 80 mL of THF was cooled down to 0 °C and $LiAlH_4$ (1.05 g,

27.6 mmol, 2.00 eq.) was added in two portions with an interval of 1 h. The mixture was stirred for 20 h at rt, cooled down to 0 °C and 55 mL of EtOAc and 25 mL of H_2O were added. The mixture was stirred for 1 h, the precipitate was filtered off and the filtrate was dried over Na_2SO_4 and concentrated *in vacuo*. The residual solid was purified by silica gel column chromatography using (CH_2Cl_2:MeOH, 4:1, $R_f = 0.25$) to yield 1.75 g (5-(*tert*-butyl)-1*H*-imidazol-4-yl)methanol **74** (11.3 mmol, 83%) as a white solid.[313]

^{1}H NMR (400 MHz, DMSO): δ = 7.36 (s, 1H), 4.46 (s, 2H), 1.28 (s, 9H) ppm. **^{13}C NMR (101 MHz, DMSO):** δ = 131.8 (s), 56.0 (s), 31.9 (s), 30.7 (s), 30.6 (s), 27.4 (s) ppm. **HRMS (EI+):** *m/z* calcd for $C_8H_{14}N_2O$: 154.1106 Da [M], found: 154.1106 Da (Δ = 0.1 ppm). **IR (ATR):** $\tilde{\nu}$ = 3119 (w), 3065 (vw), 2959 (w), 2886 (w), 2838 (w), 2726 (w), 2695 (w), 2671 (w), 2635 (w), 1578 (w), 1492 (w), 1469 (w), 1390 (vw), 1356 (w), 1331 (vw), 1298 (w), 1263 (vw), 1230 (vw), 1189 (vw), 1163 (vw), 1093 (w), 1026 (w), 1012 (m), 953 (w), 823 (w), 765 (w), 715 (w), 649 (w), 593 (vw), 520 (w), 474 (vw), 407 (vw) cm^{-1}.

(5-(*tert*-Butyl)-1*H*-imidazol-4-carbaldehyde 75

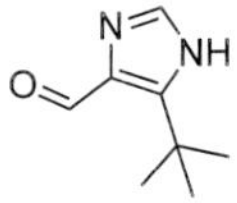

Activated MnO_2 (7.91 g, 87.5 mmol, 10.0 eq., c.a. 85%, < 10 µm) was added to a solution of (5-(*tert*-butyl)-1*H*-imidazol-4-yl)methanol **74** (1.35 g, 8.75 mmol, 1.00 eq.) in 80 mL of MeCN and the reaction mixture was stirred at rt for 4 h. Subsequently, the mixture was filtered through Celite® and the solvent was removed under reduced pressure to yield 1.13 g (5-(*tert*-butyl)-1*H*-imidazol-4-carbaldehyde **75** (7.42 mmol, 85%) as a pale-yellow solid.[313]

^{1}H NMR (400 MHz, $CDCl_3$): δ = 10.07 (s, 1H), 7.75 (s, 1H), 1.48 (s, 9H) ppm. **^{13}C NMR (101 MHz, $CDCl_3$):** δ = 183.8 (s), 155.3 (s), 130.1 (s), 128.7 (s), 35.2 (s), 31.6 (s) ppm. **HRMS (EI+):** *m/z* calcd for $C_8H_{12}N_2O$: 152.0950 Da [M], found: 152.0949 Da (Δ = 0.6 ppm). **IR (ATR):** $\tilde{\nu}$ = 3051 (w), 2963 (w), 2865 (w), 1669 (m), 1559 (w), 1506 (w), 1465, (w) 1435 (m), 1398 (w), 1377 (m), 1349 (w), 1295 (m), 1270 (w), 1204 (w), 1188 (w), 1056 (w), 1022 (w), 950 (w), 864 (m), 837 (m), 786 (m), 666 (m), 582 (w), 445 (vw), 417 (w) cm^{-1}.

(*Z*)-1-Acetyl-3-((5-(*tert*-butyl)-1*H*-imidazol-4-yl)methylene)piperazine-2,5-dione 77

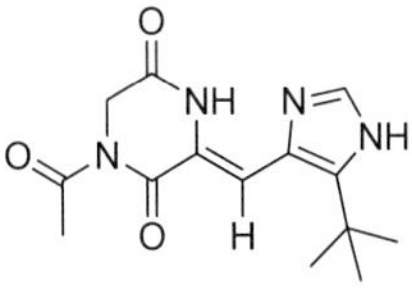

A mixture of (5-(*tert*-butyl)-1*H*-imidazol-4-carbaldehyde **75** (1.19 g, 7.79 mmol, 1.00 eq.), 1,4-diacetyl-piperazine-2,5-dione **76** (3.09 g, 15.6 mmol, 2.00 eq.) and Cs_2CO_3 (3.81 g, 11.7 mmol, 1.50 eq.) in dry DMF (10 mL) was stirred under argon at rt for 20 h. The reaction mixture was poured on ice water (20 mL) and the precipitate was filtered off and dried to yield 703 mg (*Z*)-1-acetyl-3-((5-(*tert*-butyl)-1*H*-imidazol-4-yl)methylene)piperazine-2,5-dione **77** (2.42 mmol, 31%) as a yellow solid.[313]

^{1}H NMR (400 MHz, $CDCl_3$): δ = 12.15 (br s, 1H), 9.26 (br s, 1H), 7.57 (s, 1H), 7.18 (s, 1H), 4.47 (s, 2H), 2.65 (s, 3H), 1.46 (s, 9H) ppm. **^{13}C NMR (101 MHz, $CDCl_3$):** δ = 173.0 (s), 162.6 (s), 160.7 (s), 141.0 (s), 132.5 (s), 131.7 (s), 124.0 (s), 109.0 (s), 77.4 (s), 46.6 (s), 32.0 (s), 30.9 (s), 27.5 (s) ppm. **HRMS (EI+):** *m/z* calcd for $C_{14}H_{18}N_4O_3$: 290.1379 Da [M], found: 290.1378 Da (Δ = 0.3 ppm). **IR (ATR):** $\tilde{\nu}$ = 2962 (w), 1671 (m), 1617 (w), 1506 (w), 1419 (m), 1366 (m), 1229 (m), 1190 (m), 1103 (w), 1051 (w), 1022 (w), 999 (w), 948 (w), 872 (w), 818 (w), 777 (w), 746 (w), 729 (w), 692 (w), 661 (w), 577 (w), 535 (w), 508 (vw), 471 (w), 437 (w) cm^{-1}.

Plinabulin, (3*Z*,6*Z*)-3-benzylidene-6-((5-(*tert*-butyl)-1*H*-imidazol-4-yl)methylene)piperazine-2,5-dione 23

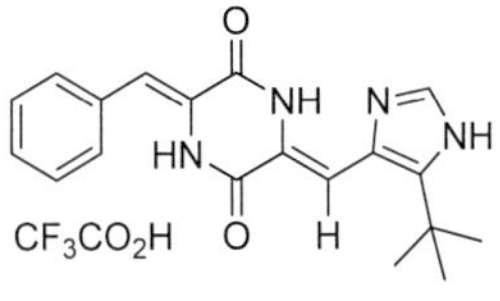

A mixture of (*Z*)-1-acetyl-3-((5-(*tert*-butyl)-1*H*-imidazol-4-yl)methylene)piperazine-2,5-dione **77** (200 mg, 0.689 mmol, 1.00 eq.), benzaldehyde (110 mg, 1.03 mmol, 1.50 eq.) and Cs_2CO_3 (336 mg, 1.03 mmol, 1.50 eq.) in 2.1 mL of dry DMF was stirred at rt under argon for 17 h. The reaction mixture was poured on ice water (20 mL) and the precipitate was filtered off. The crude product was purified by HPLC with 41 min gradient from 20-95% ACN in diH_2O with 0.1% TFA, retention = 21.7 min to yield 70 mg plinabulin **23** (0.208 mmol, 30%) as the TFA salt with a pale-yellow.[313]

^{1}H NMR (400 MHz, DMSO): δ = 12.59 (s, 1H), 12.03 (s, 1H), 10.13 (s, 1H), 8.05 (s, 1H), 7.53 (d, J = 7.2, 1.5 Hz, 2H), 7.42 (t, J = 8.4, 7.5, 6.9 Hz, 2H), 7.32 (tt, J = 7.3, 1.3 Hz, 1H), 6.81 (s, 1H), 6.76 (s, 1H), 1.38 (s, 9H) ppm. **^{13}C NMR (101 MHz, DMSO):** δ = 157.4 (s), 156.6 (s), 140.2 (s), 134.3 (s), 133.2 (s), 130.3 (s), 129.3 (s), 128.7 (s), 128.1 (s), 127.5 (s), 126.5 (s), 114.3 (s), 104.3 (s), 31.9 (s), 30.4 (s) ppm. **HRMS (FAB+):** *m/z* calcd for $C_{19}H_{20}N_4O_2$: 337.1665 Da [M], found: 337.1665 Da (Δ = 0.1 ppm). **IR (ATR):** $\tilde{\nu}$ = 2977 (vw), 1684 (w), 1644 (w), 1400 (w), 1353 (w), 1186 (w), 1144 (w), 953 (vw), 801 (w), 765 (w), 719 (w), 689 (w), 645 (vw), 520 (vw), 459 (vw), 442 (vw) cm^{-1}.

5.4.2 Photophysical properties of F$_4$-PAP-DKP-Lys

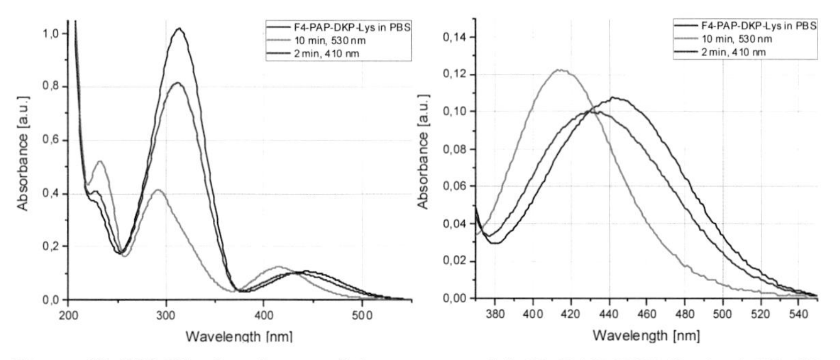

Figure 99. UV-Vis absorbance of the compound 3 (F$_4$-PAP-DKP-Lys). a) 55 µM of 3 in PBS buffer pH 7.4 in absence of light or upon irradiation with green (530 nm) or violet (410 nm) light – 3 W LED diodes (right side – magnification).

The 'dark state' refers to the *trans*-isomer obtained by thermal equilibration during the purification procedure (crystallization from 2-butanol/MeCN) during synthesis of the F$_4$-PAP-DKP-Lys and its further storage in absence of light. For UV-Vis measurements, the HPLC-purified F$_4$-PAP-DKP-Lys with > 99% *trans*-isomer content was used. Daylight irradiation results in decrease of the *trans*-isomer ratio in the mixture.

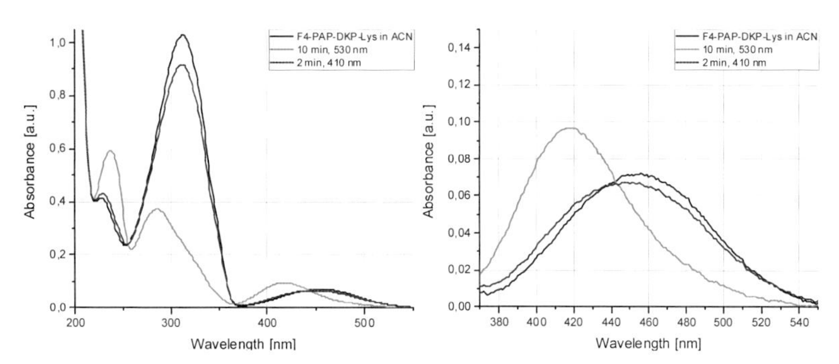

Figure 100. UV-Vis absorbance of the F_4-PAP-DKP-Lys. 55 µM gelator 1 in ACN pH 7.4 in absence of light or upon irradiation with green (530 nm) or violet (410 nm) light – 3 W LED diodes (right side – magnification).

For the HPLC-based quantification of the *cis*- and *trans*-isomer ratio of **3**, we have chosen λ = 254 nm as the isosbestic point with a medium absorption 0.1782 au ± 0.0012 au for the 55 µM solution of F_4-PAP-DKP-Lys **3** in PBS. Due to different polarity of the azobenzene photoisomers, both isomers of the compound **3** exhibit different retention times and can be independently quantified.

A 1.0 mM F_4-PAP-DKP-Lys in PBS was irradiated for 30 min with 530 nm (3 W green LED), or 10 min with 400 nm (3 W violet LED), or 10 min with 410 nm (3 W violet LED), or 30 min with 365 nm (3 W UV-LED). In case of 530 nm and 365 nm, the photostationary state (PSS) was not achieved within 10 min irradiation time. 25 µL of each solution, as well as 25 µL of the non-irradiated solution were injected in the HPLC under following conditions: run of 20 min, gradient 20% H_2O to 40% ACN with 0.1% TFA (4 nm slit, wavelength 254 nm). The sample irradiated with 530 nm contained 89% *cis* and 11% *trans*-isomer, and the sample irradiated with 410 nm – 23% *cis* and 77% *trans*-isomer, whereas the non-irradiated sample contained over 98% *trans*-**3**.

The same experiment was performed using acetonitrile (ACN) as a solvent instead of the aqueous PBS buffer. The solution with 1.0 mM F_4-PAP-DKP-Lys in ACN was irradiated the same time as the PBS samples, but after 10 min the PSS was achieved with all used light frequencies (530 nm (3 W green LED), 400 nm (3 W violet LED), 410 nm (3 W violet LED), or 365 nm (3 W UV-LED). The results are summarized in Table 16.

Table 16. Quantification of the *cis:trans* ratio in irradiated 1.0 mM samples of 3, by HPLC.

Wavelength nm	*Z* in PBS %	*Z* in ACN %
365	33	16
400	26	16
410	23	18
530	89	87
dark state	2	1

Overall, the values of photostationary states for the compound **3** to some extent depend on the solvent. Due to the extraordinarily long lifetime of the *cis*-**3** (Figure 101), both photoisomers can be separately stored in darkness at ambient temperature without significant interconversion.

5.4.2.1 Measurement of the lifetime of the *Z*-F_4-PAP-DKP-Lys

The *cis*-isomers of azobenzenes are thermally less stable than the respective *trans*-isomers and undergo spontaneous thermal equilibration. However, the rate of equilibration depends on the substitution pattern on both aromatic rings of the azobenzene system and can span from microseconds to years. For the *ortho*-fluorinated azobenzenes, lifetime of the *cis*-isomer is enhanced in comparison to the non-substituted azobenzene.

Our measurements of thermal isomerization parameters for the *cis*-**3** have been compared with the existing literature data.[2,4] For measurements, the F_4-PAP-DKP-Lys was purified by preparative HPLC, lyophilised, dissolved in ACN (1.0 mM) and irradiated for 10 min with 410 nm and its thermal decay to the *trans*-isomer was measured by analytical HPLC (by using the same conditions like for quantification of the *cis:trans* ratio). We determined the half-life of Z-F_4-PAP-DKP-Lys to be 70.9 ± 0.6 h at 60 ± 2 °C in ACN (Figure 101), which is comparable to values reported previously by Bleger *et al.* for the tetra-*ortho*-fluoroazobenzene and other fluoroazobenzenes.[30a]

This long half-life is in agreement with the following previous observation of the sol. Upon green light-induced dissipation of hydrogels of *trans*-**3** to the respective fluid sol, the fluid (containing mostly the Z-F_4-PAP-DKP-Lys) cannot solidify again to the hydrogel after incubation in darkness at rt for at least 7 d (hydrogel with 2% of **3**, n=5). During the last step of the synthesis of **3** (chapter 5.4.1), the reaction mixture that is warmed up to 120 °C for 2 h and under these conditions any *Z*-**3**, that accidentally formed, e.g. upon irradiation with daylight, will thermally back-isomerize, providing the product **3** almost exclusively in form of the *E*-isomer.

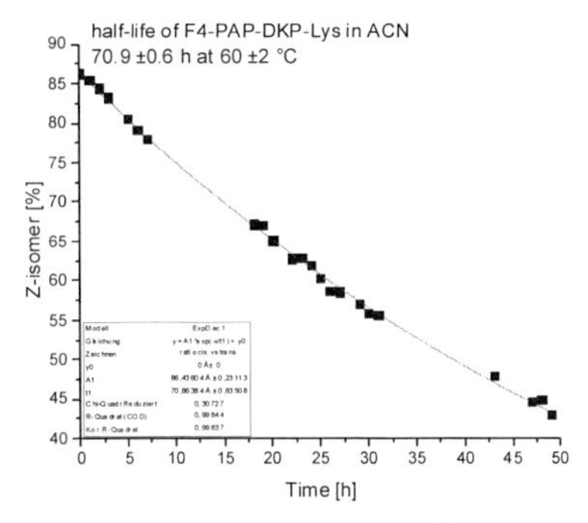

Figure 101. 70.9 ± 0.6 h half-life of *Z*-F_4-PAP-DKP-Lys at 60 ± 2 °C in acetonitrile.

5.4.2.2 Stability of F_4-PAP-DKP-Lys against glutathione

The stability of **3** was investigated in a reducing environment, emulating the intracellular redox potential values. A solution of 0.50 mM F_4-PAP-DKP-Lys (1.0 eq.) **3** and 5.0 mM reduced glutathione (10 eq.) was prepared in PBS and incubated in darkness at rt. After 3 d, 94% of **3** remained unaffected by HPLC (218 nm) (Figure 102). After 6 d 92% of the initial concentration of **3** was detected. The experiment was stopped after 6 d.

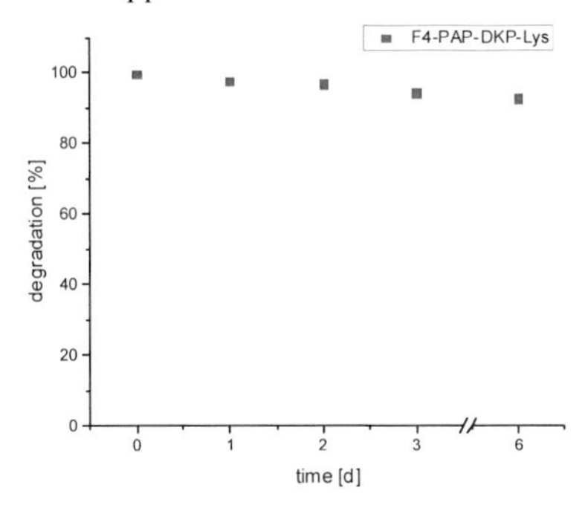

Figure 102. Stability of the gelator 3 (0.50 mM of F_4-PAP-DKP-Lys 3) against 5.0 mM reduced glutathione solution.

5.4.3 Gelation properties of F_4-PAP-DKP-Lys

5.4.3.1 Measuring the melting temperature of the hydrogels

The gel-to-sol transition temperature is characteristic for the particular gel composition and can be used to estimate its relative stability in comparison to the other gel samples. The value of that transition, referred to as "the melting temperature" of a gel, was measured according to the following protocol: in a 1.5 mL glass vial (crimp top, 12×32 mm), the photochromic gelator **3** was added to PBS buffer pH 7.4 (500 μL) (mass of the gelator and its approximate final percentage in the gel are listed in the Table 17). The crimped vial was warmed up vertical in a water bath at 80 °C for 5 min. The suspension was dissolved in the crimped vial after heating it up to the boiling point (< 1 min) with a heat gun. Prolonged boiling of the hydrogel under these conditions should be avoided, since decomposition of **3** can occur to unidentified products. The half-life of **3** of this thermal decomposition was determined with $\tau = 2.7\ h \pm 0.2\ h$ at 100 ± 2 °C in the PBS buffer.

The hot fluid turned to an orange solution and this fluid was gelated upon cooling to rt. (typically within 1 h). Before all measurement, the hydrogel was kept overnight in darkness at rt to allow equilibration of the components, which in turn minimized statistical deviation of the measured behaviour (unless indicated otherwise).

To measure the melting temperature, a sample of the hydrogel prepared above was swimming horizontally in the water of a slowly stirred (60 rpm) water bath at 25 °C. The bath was then warmed up with the heating rate of ca. 2 °C/min. The hydrogel starts melting slowly before the resulting sol (fluid) will abruptly flow down at the given gel-to-sol transition temperature. The measurement was done with 5 (n=5) identical samples and the average transition temperature was reported as "the melting temperature" T_m.

Gels prepared from **3** in PBS buffer pH 7.4 were stable and homogenous in the range between 15 g/L and 50 g/L (1.5% – 5% of **3**), also at ambient daylight. At the lower concentration of ≤ 12 g/L (1.2%) gelation was slow (several days), the resulting viscous gel had low mechanical stability (almost immediately turned into liquid upon slight shaking of the vial) and therefore the respective melting points have not been determined (Table 17).

Table 17. Gel-to-sol transition temperatures of hydrogels comprising the gelator 3 in PBS buffer pH 7.4.

Composition of the gel x mg of **1** + 500 μL PBS	Approx. concentration	T_m °C	Gelation time at rt
5	"1%"	-	n.d.
6.3	"1.2%"	-	n.d.
7.5	"1.5%"	46 ± 8	1 h-16 h
8.8	"1.7%"	**57 ± 3**	ca. 1 h
10	"2%"	82 ± 1	< 1 h

5.4.3.2 Light-induced gel-to-sol transitions

The speed of light-induced gel-to-sol transition of **3** depends on the factors like: concentration of the gelator, the presence of other components (like NaCl or TFA) in the gelation medium, the preparation protocol of the hydrogel, the ratio between height and diameter of the glass vial (and consequently, the area of light-absorbing surface of the gel), as well as the light intensity (including the respective distance of the light source from the irradiated sample). All the results reported here were measured in multiplicates and reproducible within the same experimental setup.

Samples of the hydrogels (1.5-2 wt% of **3** in PBS buffer pH 7.4, prepared according to Table 17 in 1.5-mL-vials (crimp top, 12×32 mm) have been irradiated with green light (530 nm, 3 W LED diodes) at rt (22 °C). The vials were inverted every 5 min. After 30 min irradiation, the samples of 1.5% and 1.7% gels turned into homogenous fluids without need for any mechanical stimulation upon inversion. The gel samples at the concentration of 2% after 30 min of irradiation became so unstable, that they dissipated to fluid upon slight shaking. The gel samples at the concentration of 3% and above have been irradiated for 60 min without an effect – they remained stable gels. The 2% hydrogel can reproducible (n=5) back-isomerized (*cis*-to-*trans*) from the sol to the hydrogel, after irradiation with 3 W LED diode (410 nm) for 60 min and incubation at rt overnight, without additional heating.

The gel composition 1.7% prepared from 8.8 mg of **3** in 500 μL PBS buffer pH 7.4 (Table 17 – the entry highlighted in blue) was selected as the most optimal formulation for the encapsulation and release experiments described below due to its mechanical stability, quick gelation upon cooling, and its efficient gel-to-sol transition without any mechanical stimulus.

Figure 103. Transparent hydrogel (1.7 wt%) (left) turns into sol after irradiation with green light (right).

5.4.3.3 Rheology of F_4-PAP-DKP-Lys

The rheological characterization of hydrogels formed by the compound **3** in PBS buffer pH 7.4 ("1.7%") was performed on samples generated by cooling the warm solution (formed from 140 mg of **3** and 8.00 mL of buffer) directly on the rheometer plate from 95 °C to rt. The hydrogel was covered with Parafilm and incubated overnight at rt in the dark. Strain sweep experiments were performed at 10 rad/s to determine the linear viscoelastic regime and the mechanical strength of the hydrogel at 20 °C. Frequency experiments were performed at low strain within the linear viscoelastic region (LVR) of the sample. For regeneration experiments, the samples were exposed to a deformation of 100% for 30 sec to destroy the supramolecular network, afterwards the regeneration of G' was measured at low strain within the LVR.

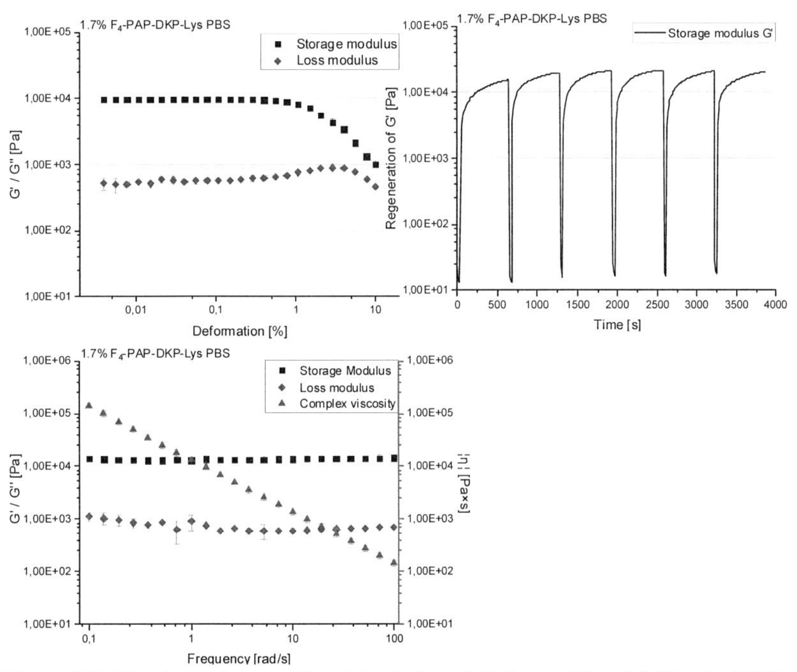

Figure 104. Rheological properties of the hydrogel (140 mg of 3 and 8.00 mL of PBS buffer pH 7.4). Strain sweep experiment (top left); frequency sweep experiment (bottom); regeneration of G' after shearing the gel for 30 sec at 100% deformation (top right).

5.4.3.4 TEM characterization

Diluted solutions were prepared one day in advance by the gel preparation method with a concentration of 5 mg/mL F_4-PAP-DKP-Lys **3** in diH_2O or 50 mM NaI. Carbon-coated copper grids (400 mesh) were covered with the material by dropping the prepared solutions on top of the grid, which was placed on top of a lint-free paper towel. The grids were dried under atmospheric pressure and subsequently examined using a Philips CM200 FEG transmission electron microscope, operated at 200 kV accelerating voltage. Under these conditions, the fine fibrous structure of our material was revealed.

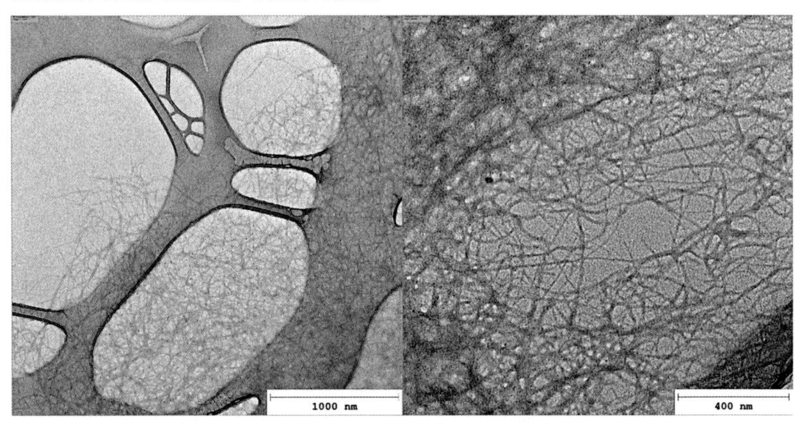

Figure 105. TEM of F_4-PAP-DKP-Lys 3 without staining.

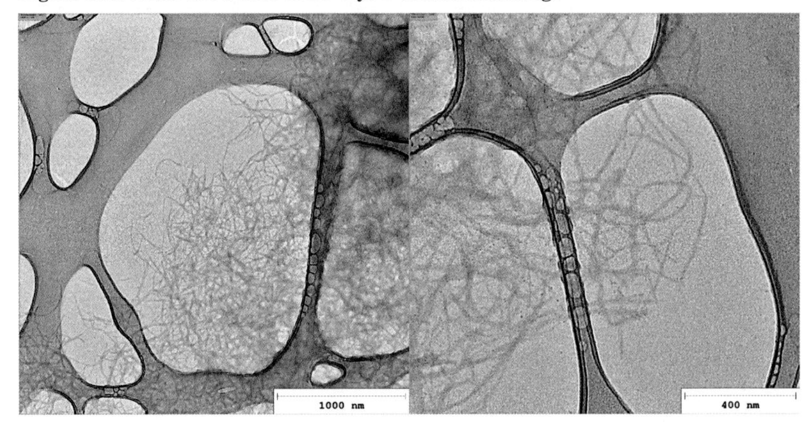

Figure 106. TEM of F_4-PAP-DKP-Lys 3 with NaI substitution.

5.4.3.5 Guest release from hydrogels using green light

The efficient guest release was investigated of hydrogels based on F_4-PAP-DKP-Lys **3** by releasing encapsulated molecules and comparing the diffusion (in darkness) or dissipation of the inner gel structure upon irradiation with green light. The hydrogel composition was chosen with 8.8 mg gelator **3** in 500 μL PBS buffer pH 7.4 as the gel matrix for all experiments in this section. As described in the chapter 5.4.3 in the experimental part (Table 17), this composition forms a stable hydrogel in absence of light, and this gel is completely converted to sol upon 30 min of irradiation with green light (530 nm, 3 W LED diode) without need for mechanical stimuli.

5.4.3.5.1 Preparation of the hydrogel samples

In a 1.5-mL glass vial (crimp top, 12×32 mm) was added the photochromic gelator **3** (8.8 mg, powder) and PBS buffer pH 7.4 (495 μL). To this opaque suspension was added a 100× stock solution (5 μL) of the particular cargo molecule dissolved in in DMSO (for plinabulin·TFA) or diH_2O (ciprofloxacin·HCl). The crimped vial was warmed up vertically in a water bath at 80 °C for 5 min. The suspension was then homogenized inside the vial after heating it up to the boiling point (< 1 min) with a heat gun. The hot mixture turned into an orange solution (homogenous), which was gelated after 1 h at rt. Before a release experiment, the hydrogels were kept overnight in dark at rt.

The measurements of the release rate of plinabulin were performed in triplicates and the average values were taken for the final conclusions and result plots. The release processes of ciprofloxacin hydrochloride were carried out as single experiments, to provide the comparison among previous F_2-PAP-DKP-Lys hydrogels.[298b] The given concentration of the drugs were chosen to obtain the optimal accuracy for the HPLC detection range.

5.4.3.5.2 Quantification of the passive diffusion – cargo “leaking” from hydrogels 3 in darkness

PBS buffer (500 μL) pH 7.4 was slowly added on top of a gel sample (on the wall of the vial) and immediately removed with micropipette, to wash away unbound or loosely bound guest molecules from the surface. Addition of fresh PBS (500 μL) buffer followed. The gel was incubated together with the buffer on the top in darkness. After 5 min 500 μL of the liquid was collected by gently turning the vial sideways and pipetting off the liquid from the side wall of the vial. Then, fresh PBS buffer (500 μL) was added on the side wall of the vial, incubated in darkness and removed after 5 min in the same way as described above. That process was repeated for the total duration of 30 min by collecting 7 subsequent volume aliquots. After that time, all gels remained visually unaffected. By measuring the remaining liquid volume after removal of the last 500 μL aliquot from the top of the gel we estimated that the total decay of the gel volume was lower than 15%.

5.4.3.5.3 Procedure of the light-induced release

To measure the release process upon green light irradiation, the procedure described above was repeated analogue, but after initial washing of the gel surface the sample was placed in an irradiation chamber and illuminated with one 3 W LED (530 nm, distance = 5 cm). Short breaks in irradiation (< 30 sec) were taken for the replacement of 500 μL aliquot with fresh 500 μL of PBS buffer every 5 min, but the summary irradiation time was 30 min. The irradiation time was sufficient to fully convert all the gel samples into sol. All aliquots were weighted before the HPLC measurement to calculate the released amount of the substance. The concentration of the aliquots was calculated by a previously measured calibration curve of the substances. In all aliquots no precipitate of ciprofloxacin was observed by the naked eye.

It was reported, that solubility of plinabulin in aqueous solutions is low with < 3 μmol/L.[213] This is one of the major obstacles for broad pharmaceutical applications of this compound. Therefore, stock solutions of plinabulin were prepared in DMSO and diluted 100-fold in aqueous solutions for the final formulations. For quantification of our release experiments we have determined the solubility of plinabulin (as the TFA salt, so in the form measured by the HPLC runs in our setup) in 1% DMSO in PBS to be 7.5 ± 1.7 μmol/L.

More concentrated solutions, prepared by 100-fold dilution of homogenous DMSO solutions in PBS, formed visible precipitation after c.a. 1 h. However, hydrogels (1.7% of F_4-PAP-DKP-Lys·TFA **3**) prepared with plinabulin at the concentrations up to 250 μmol/L were still transparent and no precipitation of the plinabulin was observed overnight (n=7). In combination with the unusually low leaking rate (below 1%) of plinabulin measured in darkness and its molecular structure, similar to the gelator **3**, it was hypothesized that these effects can be overall explained by plinabulin molecules being integrated into the supramolecular structure of the hydrogel. Some sort of association between the gelator **3** and plinabulin molecules in solution may also explain the fact that concentrations of plinabulin in PBS buffer (still in homogenous solution) measured upon certain light-induced release experiments have been up to 32.7 μmol/L, which is 4.4 times higher than the solubility of plinabulin·TFA) in 1% DMSO in PBS alone. Still, to prevent incorrect assignment of the plinabulin concentrations in the diffusion and light-induced release experiments, 50 µL of AcOH (ca 10 vol%) were added to all aliquots after the removal of the aliquots from the gel surface. Acetic acid was chosen to dissolve aggregates or precipitations and reduce the hypothetical supra-molecular interactions between plinabulin and molecules of the gelator **3** for the quantification purpose. It was not possible to reduce the concentration of plinabulin much below its solubility limits, because the detection by the HPLC would become inaccurate.

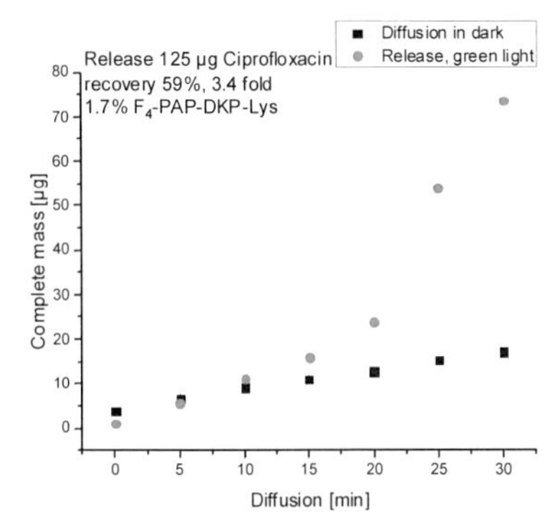

Figure 107. Light-induced release of encapsulated ciprofloxacin (a common antibiotic) from the hydrogel containing 17 g/L of 1 ("1.7% gel") in comparison to the passive diffusion ("leaking") of the cargo in darkness.

5.4.3.6 Chiral HPLC of F_4-PAP-DKP-Lys

The chiral HPLC runs were measured with the 1100 Series from AGILENT TECHNOLOGIES with an amylose-SA column (KSA99S05-2546WT) with 250×4.6 mm and 5 µm. The separation was performed with a 20 min gradient from 95% H_2O to 95% ACN with 0.1% TFA, flow rate 1 mL/min, slit = 4 nm, wavelength 256 nm or 330 nm for detection. In order to test the separation in this setting we chose Fmoc-(*R*)-Arg(Pbf)-OH and Fmoc-(*S*)-Arg(Pbf)-OH, which was separated by $\Delta t = 0.30$ min. The photochromic amino acid Boc-F_4-(*S*)-PAP-OH was *ee*=96% enantiopure. However, the deprotection of the Boc-groups with 50% TFA in CH_2Cl_2 the chiral HPLC showed double peaks indicating a racemization. The same is true for the gelator F_4-PAP-DKP-Lys **3**. This observation is not consistent with the literature[314], which is usually only possible with a strong base.[315] But a racemization can occur with an increased pre-activation time and similar to our amid coupling with HBTU and without the addition of HOBT.[316] Considering the CD-spectra of F_2-PAP-DKP-Lys described in 5.3.2.1, which was synthesized according to an analogue HBTU coupling procedure, a complete racemisation can be excluded.

It was not possible to grow crystals for an X-ray analysis and the determination of the absolute configuration by NMR with 6 mg Europium tris(3-(heptafluoropropylhydroxymethylene)-(+)-camphorate)[317] in deuterated acetone, THF or benzene lead to not analysable spectra. Therefore, it was not possible to verify the data of the chiral HPLC runs and distinguish between a potential racemization at the amid coupling step with HBTU and the cyclisation.

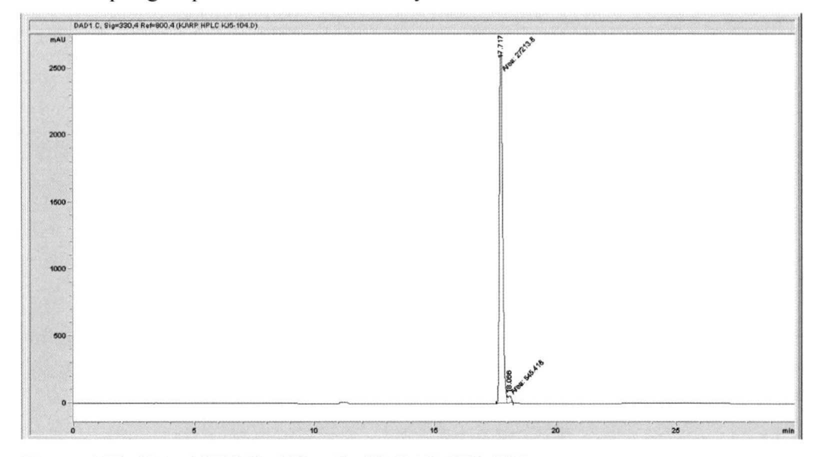

Figure 108. Chiral HPLC of Boc-F_4-(*S*)-PAP-OH, 330 nm.

5.5 Synthesis of tetra-*ortho*-fluoro-azobenzenes

2,6-Difluoro-4-bromoaniline 17

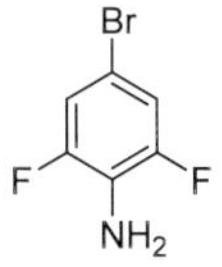

To a solution of 2,6-difluoroaniline (12.9 g, 100 mmol, 1.00 eq.) in acetonitrile (200 mL) was added NBS (17.8 g, 100 mmol, 1.00 eq.) at rt. The mixture was stirred overnight, and then diluted with water and *n*-hexane. The two phases were separated, and the organic phase was dried over $MgSO_4$, filtered, and concentrated under reduced pressure. The crude residue was purified by column chromatography (CH_2Cl_2:*n*-hexane 1:1) to give 10.0 g 2,6-difluoro-4-bromoaniline **17** (48.1 mmol, 48%) as a purple solid.[30a]

^{1}H NMR (300 MHz, $CDCl_3$): δ = 6.99 (dd, *J*=6.1, 1.3 Hz, 2 H), 3.73 (br s, 2 H) ppm. **^{13}C NMR (75 MHz, $CDCl_3$):** δ = 153.6, 150.3, 133.0, 114.7, 107.2 ppm. **HRMS (EI+):** *m/z* calcd for $C_6H_4BrF_2N$: 206.9490 Da [M], found: 206.9488 Da (Δ = 0.8 ppm). **IR (ATR):** $\tilde{\nu}$ = 3422 (m), 3326 (m), 3227 (w), 1642 (m), 1604 (m), 1582 (m), 1496 (s), 1427 (s), 1298 (m), 1241 (m), 1216 (w), 1151 (s), 963 (s), 868 (m), 839 (s), 760 (m), 717 (m), 632 (m), 599 (m), 565 (s), 507 (s), 462 (m) cm^{-1}.

4-Amino-3,5-difluorobenzonitrile 24

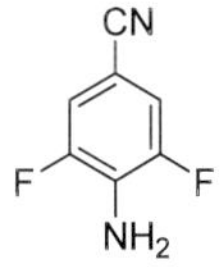

A mixture of 4-bromo-2,4-difluoroaniline **17** (48.7 g, 234 mmol, 1.00 eq.) and copper(I) cyanide (25.2 g, 281 mmol, 1.20 eq.)[318] in NMP (100 mL) was refluxed at 200 °C for 1.5 h (preheated silicone oil bath). After cooling down to rt, the reaction mixture was poured into an aqueous solution of 12% ammonium hydroxide (1000 mL), CH_2Cl_2 (300 mL) and active coal (5.00 g). The blue mixture was filtrated over Celite® (h=4 cm, Ø=4 cm) and the Celite® pad was washed with CH_2Cl_2 (2×50 mL). The solution was extracted with CH_2Cl_2 (3×200 mL). After the separation of the layers, the organic phase was washed with water (2×200 mL) and sat. aqueous $NaHCO_3$ (100 mL) and dried over Na_2SO_4. The solvent was removed under reduced pressure (800 mbar, 60 °C to remove CH_2Cl_2, then

< 50 mbar, 90 °C). The crude product was purified by recrystallization: First active coal (5.0 g) and water (100 mL) was added and the mixture was refluxed for 10 min (oil bath, 120 °C). Then more water was added (ca. 100 mL) until the product was dissolved. The hot solution was filtered off to remove undesirable side products and the active coal. The clear solution was cooled overnight in the fridge (4 °C).The crystalline product was filtered off, washed with water (2×20 mL), dried with the airstream of the pump (30 min), then dried under high vacuum to yield 27.2 g 4-amino-3,5-difluorobenzonitrile **24** (17.7 mmol, 76%) as a white solid.

TLC: R_f = 0.25 (CH_2Cl_2:cH 1:1). **^{1}H NMR (400 MHz, $CDCl_3$):** δ = 7.14 (dd, *J* = 6.0 Hz, 2.3 Hz, 2H), 4.30 (s, 1H) ppm. **^{13}C NMR (101 MHz, $CDCl_3$):** δ = 152.6 (d, *J* = 9 Hz), 150.2 (d, *J* = 8.9 Hz), 130.6 (t, *J* = 15.7 Hz), 118.9 (t, *J* = 3.5 Hz), 116.7 – 116.0 (m), 99.1 (t, *J* = 11.0 Hz) ppm. **^{1}H NMR (300 MHz, DMSO):** *δ* = 7.15 (dd, *J* = 6.0, 2.2 Hz, 2H), 4.27 (br s, 2H) ppm. **^{13}C NMR (101 MHz, DMSO):** *δ* = 150.6 (dd, *J* = 243.5, 8.9 Hz), 129.7 (t, *J* = 15.7 Hz), 118.1 (s), 115.6 (dd, *J* = 16.1, 8.5 Hz), 98.5 (t, *J* = 11.1 Hz) ppm. **^{19}F NMR (377 MHz, DMSO):** *δ* = −135.07 (s) ppm. **HRMS (EI+):** *m/z* calcd for $C_7H_4F_2N_2$: 154.0343 Da [M], found: 154.0344 Da (Δ = 0.6 ppm). **IR (ATR):** ṽ = 3375 (w), 3314 (w), 3204 (w), 2225 (w), 1648 (w), 1608 (w), 1579 (m), 1524 (m), 1511 (m), 1443 (m), 1337 (m), 1163 (m), 1120 (w), 977 (m), 876 (m), 862 (m), 728 (w), 664 (m), 614 (m), 537 (m), 470 (w), 458 (m), 413 (w) cm^{-1}. **elemental analysis calcd for $C_7H_4F_2N_2$ (%):** C: 54.55, H: 2.62, F: 24.65, N: 18.18, found: C: 54.56, H: 2.44, N: 18.29.

4-Amino-3,5-difluorobenzoic acid 25

HO O

F F

NH_2

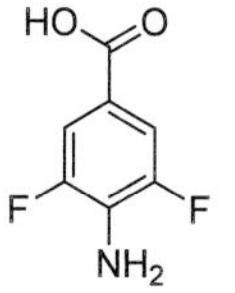

The substrate 4-amino-3,5-difluorobenzonitrile **24** (9.11 g, 59.1 mmol, 1.00 eq.) was suspended in NaOH (313 mL, 313 mmol, 5.30 eq.) and refluxed for 1 h to dissolve the substrate and stirred further at rt for 1 day. After cooling down to rt, conc. HCl was added until the solution reached pH 1. The precipitate was filtered off and dried by lyophilisation, yielding 9.45 g 4-amino-3,5-difluorobenzoic acid **25** (55.7 mmol, 94%).

TLC: R_f = 0.33 (1% formic acid, 30% cH in EtOAc). **^{1}H NMR (400 MHz, DMSO):** δ = 12.70 (s, 1H), 7.40 (d, J = 9.7 Hz, 2H), 6.04 (s, 2H) ppm. **^{13}C NMR (101 MHz, DMSO):** δ = 166.2 (t, J = 3.2 Hz), 151.1 (d, J = 9.3 Hz), 148.7 (d, J = 9.5 Hz), 130.7 (t, J = 16.6 Hz), 115.8 (t, J = 7.7 Hz), 112.4 (dt, J = 15.7, 7.8 Hz) ppm. **^{19}F NMR (377 MHz, DMSO):** δ = −135.99. **HRMS (EI+):** *m/z* calcd for $C_7H_5O_2N_1F_2$: 173.0288 Da [M], found: 173.0290 Da (Δ = 1.0 ppm). **IR (ATR):** $\tilde{\nu}$ = 3497 (vw), 3450 (vw), 3400 (w), 2836 (w), 2605 (w), 1862 (vw), 1693 (m), 1632 (m), 1587 (m), 1537 (w), 1449 (m), 1419 (m), 1336 (m), 1275 (m), 1239 (m), 1147 (w), 1082 (vw), 962 (w), 953 (w), 890 (w), 765 (m), 716 (w), 628 (w) 556 (w), 456 (w), 385 (vw) cm^{-1}.

(*E*)-1-(4-Bromo-2,6-difluorophenyl)-2-(2,6-difluorophenyl)diazene 78

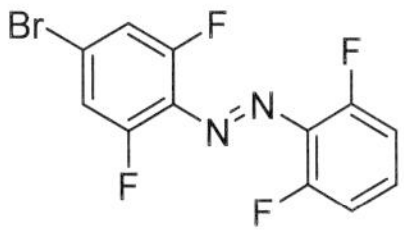

According to BLEGER *et al.*[2], Oxone® (1×$KHSO_5$, 0.5×$KHSO_4$, 0.5×K_2SO_4) (54.1 g, 176 mmol, 4.00 eq.) was dissolved in water (531 mL), added to 2,6-difluoroaniline **35** (5.69 g, 44.1 mmol, 1.00 eq.) in CH_2Cl_2 (177 mL). The solution was stirred vigorously at rt for 4 h. The layers were separated, the green organic layer washed with water (3×100 mL), dried over Na_2SO_4, filtered off and concentrated under reduced pressure (max. 40 °C). The crude precursor **11** (quant.) was instantly used without any further purification for the next reaction at the same day.

HRMS (EI+): *m/z* calcd for $C_6H_3ONF_2$: 143.0177 Da [M], found: 143.0176 Da (Δ = 1.0 ppm).

The nitrosobenzene (6.15 g, 43.0 mmol, 1.17 eq.) was suspended in glacial AcOH:PhMe:TFA 6:6:1 (147 mL) and 4-bromo-2,6-difluoroaniline **17** (7.65 g, 36.8 mmol 1.00 eq.) was added. The resulting mixture was stirred in darkness (covered with aluminium foil) at rt for 2 d. The solution was concentrated under reduced pressure after adding PhMe (150 mL) (to remove excess of the green nitrosobenzene). The crude product was purified by silica gel column chromatography with 20% CH_2Cl_2 in cH (R_f = 0.20). The solvent was evaporated under reduced pressure. Drying under high vacuum overnight resulted in 6.03 g (*E*)-1-(4-bromo-2,6-difluorophenyl)-2-(2,6-difluorophenyl)diazene **78** (18.1 mmol, 49%).

^{1}H NMR (400 MHz, DMSO): δ = 7.74 (d, J = 8.8 Hz, 2H), 7.63 (tt, J = 8.5, 6.1 Hz, 1H), 7.39 – 7.29 (m, 2H) ppm. **^{13}C NMR (101 MHz, DMSO):** δ = 156.0

(t, $J = 4.7$ Hz), 153.4 (dd, $J = 8.3$, 4.5 Hz), 133.5 (t, $J = 10.7$ Hz), 131.0 (t, $J = 9.7$ Hz), 129.9, 124.5 (t, $J = 12.5$ Hz), 117.1 (dd, $J = 23.7$, 3.6 Hz), 113.2 (dd, $J = 19.9$, 3.5 Hz) ppm. **^{19}F NMR (376 MHz, DMSO):** $\delta = -121.31$, -122.96 ppm. **HRMS (EI+):** *m/z* calcd for $C_{12}H_5N_2F_4$: 331.9572 Da [M], found: 331.9574 Da ($\Delta = 0.6$ ppm). **IR (ATR):** $\tilde{\nu}$ =3097 (vw), 1581 (m), 1482 (w), 1466 (m), 1415 (m), 1297 (w), 1242 (w), 1221 (w), 1198 (w), 1150 (w), 1081 (w), 1051 (m), 1021 (m), 892 (w), 861 (w), 844 (m), 788 (m), 742 (m), 622 (w), 585 (w), 513 (w), 417 (vw), 383 (w) cm^{-1}. **UV-Vis (MeCN):** $\lambda_{max} = 229$, 314, 457 nm.

2,2',6,6'-Tetrafluoro-azobenzene 79

According to BLEGER *et al.* [2, 30a], Oxone® (1×$KHSO_5$, 0.5×$KHSO_4$, 0.5×K_2SO_4) (9.53 g, 31.0 mmol, 4.00 eq.) was dissolved in water (94 mL), added to 2,6-difluoroaniline **35** (1.00 g, 7.75 mmol, 1.00 eq.) in CH_2Cl_2 (31 mL). The solution was stirred vigorously at rt for 4 h. The layers were separated, the green organic layer was washed with water (3×20 mL), dried over Na_2SO_4, filtered off and concentrated under reduced pressure (max. 40 °C). The crude 2,6-difluoro-nitrosobenzene **11** (1.11 g, 7.75 mmol, quant.) was used instantly without any further purification used for the next reaction. The 2,6-difluoro-nitrosobenzene **11** (1.11 g, 7.75 mmol, 1.20 eq.) was suspended in glacial AcOH:PhMe:TFA 6:6:1 (25 mL) and the 2,6-difluoroaniline **35** (833 mg, 6.46 mmol, 1.00 eq.). The resulting mixture was stirred in dark (covered with aluminium foil) at rt for 2 d. The solution was concentrated under reduced pressure after adding PhMe (100 mL) (to remove excess of the green nitrosobenzene). The crude product was purified by silica gel column chromatography with 20% CH_2Cl_2 in cH. Evaporation of combined fractions and drying under high vacuum overnight resulted in 508 mg 2,2',6,6'-tetrafluoro-azobenzene **79** (2.00 mmol, 31%).

^{1}H NMR (400 MHz, DMSO): $\delta = 7.62$ (tt, $J = 8.5$, 6.1 Hz, 2H), 7.35 (t, $J = 9.0$ Hz, 3H) ppm. **^{13}C NMR (101 MHz, DMSO):** $\delta = 155.9$ (d, $J = 4.1$ Hz), 153.3 (d, $J = 4.2$ Hz), 133.1 (t, $J = 10.7$ Hz), 130.7 (t, $J = 10.0$ Hz), 113.2 (dd, $J = 20.2$, 3.4 Hz) ppm. **^{19}F NMR (376 MHz, DMSO):** $\delta = -126.30$ ppm. **HRMS (EI+):** *m/z* calcd for $C_{12}H_6N_2F_4$: 254.0467 Da [M], found: 254.0468 Da ($\Delta = 0.2$ ppm). **IR (ATR):** $\tilde{\nu} = 3106$ (w), 1936 (vw), 1613 (m), 1586 (m), 1471 (s), 1284 (w), 1245 (m), 1219 (m), 1018 (s), 883 (w), 787 (s), 736 (s), 692 (w), 599 (m), 509 (m), 472 (m) cm^{-1}. **elemental analysis calcd for $C_{12}H_6F_4N_2$ (%):** C: 56.70, H: 2.38, N: 11.02, found: C: 56.70, H: 2.10, N: 11.15. **UV-Vis (MeCN):** λ_{max} = 226, 306, 454 nm.

5.5.1 Synthesis of *ortho*-F_4-4,4'-bis(hydroxymethyl)azobenzene

The following synthesis towards 2,2',6,6'-tetrafluoro-4,4'-bis(hydroxymethyl)azobenzene **29** *via* a MILLS coupling and symmetric oxidative coupling was initially performed by Mariam Lena Schulte in course of her bachelor thesis under the supervision of Johannes Karcher.[220] All described methods were optimised further to enable a gram scale preparation.

(4-Amino-3,5-difluorophenyl)methanol 26

OH

F F

NH_2

$LiAlH_4$ (2.61 g, 68.7 mmol, 1.20 eq.) was added to anhydrous THF (72 mL) in two-neck flask under argon. The substrate 4-amino-3,5-difluorobenzoic acid **25** (12.0 g, 57.3 mmol, 1.00 eq.) was dissolved in anhydrous THF (72 mL) and was added slowly at 0 °C. The mixture was stirred for 1 h at rt. Then additional $LiAlH_4$ (2.61 g, 68.7 mmol, 1.20 eq.) was added and the mixture was refluxed for 2 h. The reaction was quenched with EtOAc (50 mL) and with H_2O (10 mL) under argon stream. Aqueous solution of NaOH (1.0 M) was added until the solution reached pH 8-10 and the undesired precipitated salts were filtered off and washed with EtOAc (3×50 mL). The liquid was extracted with EtOAc (5×100 mL), dried over Na_2SO_4 and evaporated under reduced pressure. The crude product was purified by silica gel column chromatography with 1% Et_3N in EtOAc:cH 1:1 and 7.38 g of (4-amino-3,5-difluorophenyl)methanol **26** (46.4 mmol, 81%) were obtained.[319]

TLC: R_f = 0.35 (EtOAc:cH 1:1 + 1% Et_3N). **^{1}H NMR (400 MHz, DMSO):** δ = 6.83 (d, *J* = 9.7 Hz, 2H), 5.13 (t, *J* = 5.8 Hz, 1H), 5.03 (s, 2H), 4.34 (d, *J* = 5.8 Hz, 2H) ppm. **^{13}C NMR (101 MHz, DMSO):** δ = 151.0 (dd, *J* = 238.4, 9.3 Hz), 129.8 (t, *J* = 7.2 Hz), 123.8 (t, *J* = 16.8 Hz), 110.4 – 106.6 (m), 62.2 – 61.7 (m) ppm. **^{19}F NMR (376 MHz, DMSO):** δ = −135.90 ppm. **HRMS (FAB+):** *m/z* calcd for $C_7H_7O_1N_1F_2$: 159.0496 Da [M], found: 159.0.498 Da (Δ = 1.3 ppm). **IR (ATR):** ṽ = 3414 (w), 3176 (w), 2945 (w), 1645 (vw), 1594 (m), 1519 (m), 1472 (w), 1447 (m), 1361 (w), 1322 (m), 1158 (m), 1124 (w), 1033 (m), 1001 (m), 954 (m), 845 (m), 774 (m), 720 (m), 667 (m), 568 (m), 511 (w), 478 (w), 385 (w) cm^{-1}.

4-Amino-3,5-difluorobenzyl acetate 27

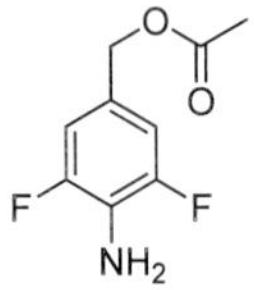

Ac_2O (2.57 mL, 27.2 mmol, 1.00 eq.) was added to a solution of (4-amino-3,5-difluorophenyl)methanol **26** (5.20 g, 32.7 mmol, 1.20 eq.) in pyridine (20 mL) and the reaction was stirred at 0 °C. After 2 h the reaction was stopped and the mixture was extracted with EtOAc (3×100 mL) and washed with brine (2×10 mL), then with $NaHCO_3$ (2×10 mL) and dried over anhydrous Na_2SO_4. The solvent was removed under reduced pressure. The product was purified by column chromatography with 20% EtOAc in cH ($R_f = 0.25$) and 1.42 g (4-amino-3,5-difluorophenyl)methanol (8.93 mmol, 33%) was recovered and 3.55 g 4-amino-3,5-difluorobenzyl acetate **27** (17.6 mmol, 64%) was obtained as a white solid.

^{1}H NMR (400 MHz, DMSO): δ = 6.94 (dd, J = 7.3, 2.3 Hz, 2H), 5.28 (s, 2H), 4.90 (s, 2H), 2.03 (s, 3H) ppm. **^{13}C NMR (101 MHz, DMSO):** δ = 170.3, 151.9 (d, J = 9.5 Hz), 149.5 (d, J = 9.6 Hz), 125.4 (t, J = 16.6 Hz), 122.4 (t, J = 8.3 Hz), 111.4 (dq, J = 14.6, 7.3 Hz), 64.7 (t, J = 2.0 Hz), 20.7 ppm. **^{19}F NMR (376 MHz, DMSO):** δ = −135.81 ppm. **HRMS (EI+):** m/z calcd for $C_9H_9O_2N_1F_2$: 201.0601 Da [M], found: 201.0602 Da (Δ = 0.5 ppm). **IR (ATR):** ṽ = 3369 (w), 1731 (m), 1646 (w), 1592 (m), 1527 (m), 1451 (m), 1381 (w), 1363 (m), 1228 (m), 1150 (m), 1024 (m), 953 (m), 847 (w), 722 (w), 665 (w), 633 (w), 601 (w), 573 (w), 452 (w) cm^{-1}.

2,2',6,6'-Tetrafluoro-4,4'-azobenzylalkohol-diacetate 28

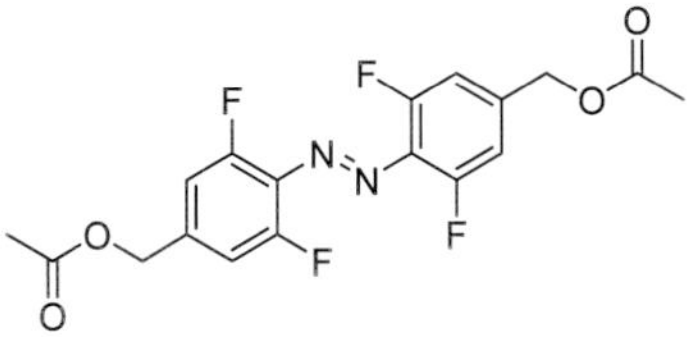

Route I, Mills-reaction: cf.[2]

Oxone® (1×$KHSO_5$, 0.5×$KHSO_4$, 0.5×K_2SO_4) (21.1 g, 68.6 mmol, 4.00 eq.) was dissolved in diH_2O (343 mL). A solution of 4-amino-3,5-difluorobenzyl acetate **27** (3.45 g, 17.2 mmol, 1.00 eq.) was dissolved in CH_2Cl_2 (172 mL) and added at rt. The biphasic solution was stirred vigorously (500 rpm) for 4 h at rt. The conversion was analysed by analytic HPLC every 30 min and the reaction was stopped after 4 h. The layers were separated, and the water phase was extracted with CH_2Cl_2 (3×62 mL). The green organic layer was washed with water (3×10 mL), dried over Na_2SO_4, filtered off and concentrated under reduced pressure (40 °C, 600 mbar). The white product was used instantly in the next reaction. This mixture consisting of 4-amino-3,5-difluorobenzyl acetate (est. 8.72 mmol, 1.28 eq.) and 3,5-difluoro-4-nitrosobenzyl acetate (est. 6.83 mmol, 1.00 eq.) (total 17.2 mmol, purity 90%) was dissolved in 34.2 mL AcOH:PhMe:TFA (6:6:1) and stirred for 4 d at rt. The reaction mixture was diluted with PhMe (100 mL) and the solvent was removed under reduced pressure. The crude product was purified by silica gel column chromatography with 30% EtOAc in cH (R_f = 0.25) and 2.16 g 2,2',6,6'-tetrafluoro-4,4'-azobenzylalkohol-diacetate **28** (5.42 mmol, yield 79%, purity 93%) was obtained as an orange solid. The side-product, which is presumably the (*E*)-1,2-bis(4-(acetoxymethyl)-2,6-difluorophenyl)diazene 1-oxide, was difficult to separate despite several attempts by recrystallization in 2-propanol or preparative HPLC.

Side-product: MS (FAB+): *m/z* calcd for $C_{18}H_{15}O_4N_2F_4$: 399.0968 Da [M+H], found: 399.1 Da; *m/z* calcd for $C_{18}H_{15}F_4N_2O_5$: 415.0912 Da [M+OH], found: 415.1 Da.

Route II, symmetric oxidative coupling: cf.[30a]

A flask charged with 4-amino-3,5-difluorobenzyl acetate **27** (0.200 g, 0.994 mmol, 1.00 eq.) and a freshly ground mixture of $KMnO_4$ (0.511 g, 3.23 mmol, 3.25 eq.) and $FeSO_4{\cdot}7H_2O$ (0.511 g, 1.84 mmol, 1.85 eq.) dissolved in CH_2Cl_2 (5 mL) was added. The mixture was refluxed overnight, filtered over Celite® and dried over Na_2SO_4. The crude product was purified by silica gel column chromatography in 30% EtOAc in cH (product: R_f = 0.25, substrate: R_f = 0.38) and 53 mg 4-amino-3,5-difluorobenzyl acetate (0.263 mmol, 27%) was recovered and 28 mg (0.070 mmol, 14%) 2,2',6,6'-tetrafluoro-4,4'-azobenzylalkohol-diacetate **28** was obtained as an orange solid.

Route III, symmetric DBU/NCS coupling: cf.[221]

DBU (4.94 mL, 33.1 mmol, 2.00 eq.) was added to a solution of 4-amino-3,5-difluorobenzyl acetate **27** (3.33 g, 16.5 mmol, 1.00 eq.) in dry CH_2Cl_2 (248 mL) was added under argon. The solution was stirred at rt for 5 min before being cooled down to −78 °C. Powdered NCS (4.42 g, 33.1 mmol, 2.00 eq.) was added to the reaction mixture under argon. The orange solution was stirred for 10 min at −78 °C before quenching by addition of a sat. u $NaHCO_3$. The organic layer was separated, washed sequentially with water (100 mL) and diluted aqueous solution of HCl (1.0 mol/L, 100 mL), dried over anhydrous Na_2SO_4, and concentrated to dryness *in vacuo*. The crude product was purified by silica gel column chromatography in 1% EtOAc in CH_2Cl_2 (R_f = 0.35) and 2.00 g 2,2',6,6'-tetrafluoro-4,4'-azobenzyl-alkohol-diacetate **28** (5.03 mmol, 61%) was obtained as an orange solid.

^{1}H NMR (400 MHz, DMSO): δ = 7.38 (d, J = 10.1 Hz, 4H), 5.17 (s, 4H), 2.14 (s, 6H) ppm. **^{13}C NMR (101 MHz, DMSO):** δ = 170.1, 155.9 (d, J = 4.6 Hz), 153.3 (d, J = 4.8 Hz), 142.7, 129.7, 111.7 (dd, J = 21.4, 3.1 Hz), 63.7, 20.6 ppm. **^{19}F NMR (376 MHz, DMSO):** δ = −125.37 ppm. **HRMS (FAB+):** *m/z* calcd for $C_{18}H_{15}O_4N_2F_4$: 399.0968 Da [M+H], found: 399.0970 Da (Δ = 0.5 ppm). **IR (ATR):** ṽ = 2936 (vw), 1738 (w), 1627 (w), 1577 (w), 1438 (w), 1383 (w), 1362 (w), 1234 (m), 1040 (m), 972 (w), 923 (w), 850 (m), 752 (vw), 665 (w), 586 (w), 529 (w), 455 (w), 386 (vw) cm^{-1}. **UV-Vis (MeCN):** λ_{max} = 229, 315, 454 nm.

2,2',6,6'-Tetrafluoro-4,4'-bis(hydroxymethyl)azobenzene 29

2,2',6,6'-Tetrafluoro-4,4'-azobenzylalkohol-diacetate **28** (1.98 g, 4.97 mmol, 1.00 eq.) was dissolved in a solution of conc. HCl (37%):methanol 1:100 (500 mL)[320] and the resulting solution stirred at 40 °C for 72 h. The reaction was diluted with PhMe (100 mL) and the solvent was removed under reduced pressure. The crude product was purified by silica gel column chromatography (R_f = 0.20 in 20% EtOAc in CH_2Cl_2) and the 1.56 g 2,2',6,6'-tetrafluoro-4,4'-bis(hydroxymethyl)azobenzene **29** (4.96 mmol, quant.) was obtained as an orange solid.

^{1}H NMR (400 MHz, DMSO): δ = 7.25 (d, J = 11.3 Hz, 4H), 5.58 (s, 2H), 4.60 (s, 4H) ppm. **^{13}C NMR (101 MHz, DMSO):** δ = 156.0 (d, J = 4.5 Hz), 153.4 (d, J = 4.6 Hz), 149.6 (t, J = 9.2 Hz), 129.0 (t, J = 9.9 Hz), 110.1 (dd, J = 20.8,

2.9 Hz), 61.7 ppm. **^{19}F NMR (376 MHz, DMSO):** δ = −126.49 ppm. **HRMS (FAB+):** *m/z* calcd for $C_{14}H_{11}O_2N_2F_4$: 315.0757 Da [M+H], found: 315.0756 Da (Δ = 0.3 ppm). **IR (ATR):** ṽ = 3286 (w), 1623 (m), 1574 (m), 1441 (m), 1359 (w), 1297 (w), 1200 (w), 1125 (w), 1067 (m), 1041 (m), 971 (m), 881 (w), 845 (m), 748 (w), 658 (w), 580 (w), 543 (w), 520(w) cm^{-1}. **UV-Vis (MeCN):** λ_{max} = 230, 318, 454 nm.

Isomerization of 2,2',6,6'-Tetrafluoro-4,4'-bis(hydroxymethyl)azobenzene **29**: After the irradiation for 30 min with 3 W LED 530 nm, 89% isomerized of *Z*-2,2',6,6'-tetrafluoro-4,4'-bis(hydroxymethyl)azobenzene (17 g/L in DMSO). The back-isomerization was possible and after 30 min with 3 W LED 410 nm 86% *E* in 17 g/L in DMSO was isomerized.

3,5-Difluoro-4-nitrosobenzyl alcohol 80[220]

The substrate 4-amino-3,5-difluorobenzyl alcohol **26** (0.100 g, 0.629 mmol, 1.00 eq.) was dissolved in 6 mL CH_2Cl_2:acetone 5:1 while Oxone® (0.775 g, 2.52 mmol, 4.00 eq.) was dissolved in water (7.5 mL). The solutions were combined and stirred for 4 h. The organic solvents were evaporated under reduced pressure. Due to the low solubility of the product in water, a beige solid precipitated. The solid was filtered off, washed with water and dried for 1 h under high vacuum to yield 33 mg of the crude 3,5-difluoro-4-nitrosobenzyl alcohol (0.191 mmol, 34%).[2, 220]

TLC: R_f = 0.65 (cH:EtOAc 1:1). **HRMS (EI+):** *m/z* calcd for $C_7H_5O_2N_1F_2$: 173.0288 Da [M], found: 173.0289 Da (Δ = 0.5 ppm). **IR (ATR):** ṽ = 3255 (w), 3068 (vw), 2922 (w), 2853 (w), 1712 (vw), 1628 (w), 1600 (w), 1445 (w), 1364 (w), 1278 (w), 1193 (w), 1041 (m), 988 (m), 835 (w), 799 (w), 717 (vw), 679 (vw), 586 (w), 538 (vw), 513 (vw), 423 (vw) cm^{-1}.

5.5.2 (*E*)-4-(4-Tetra-*ortho*-F_4-azobenzamido)naphthalene-2,6-dicarboxylate

The following synthesis towards (*E*)-4-(4-tetra-*ortho*-F_4-azobenzamido)naphthalene-2,6-dicarboxylate **46** was initially performed by Mariam Lena Schulte in course of her bachelor thesis under the supervision of Johannes Karcher.[220] The last amide coupling was optimised further and a full characterisation of the product was provided by Johannes Karcher.

3,5-Difluoro-4-nitrosobenzoic acid 31

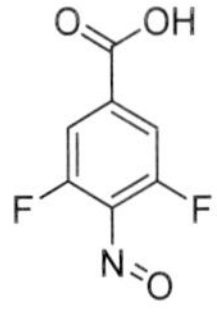

The substrate 4-amino-3,5-difluorobenzoic acid **25** (1.00 g, 5.78 mmol, 1.00 eq.) was dissolved in CH_2Cl_2 (115 mL) while Oxone® (7.11 g, 23.1 mmol, 4.00 eq.) was dissolved in water (46 mL). The solutions were combined and stirred for 2 h.[2] The product precipitates during the reaction. The organic solvent was evaporated under reduced pressure, the solid was filtered off and washed three times with water. Drying under high vacuum led to a yield of 883 mg 3,5-difluoro-4-nitrosobenzoic acid **31** (4.72 mmol, 82%).

TLC: $R_f = 0.23$ (30% MeOH in EtOAc). **^{1}H NMR (300 MHz, DMSO):** $\delta = 14.10$ (s, 1H), 7.85 (d, $J = 9.8$ Hz, 2H) ppm. **HRMS (EI+):** *m/z* calcd for $C_7H_3O_3N_1F_2$: 187.0081 Da [M], found: 187.0082 Da ($\Delta = 0.6$ ppm). **IR (ATR):** $\tilde{\nu} = 3388$ (vw), 3167 (w), 1714 (m), 1665 (m), 1619 (w), 1592 (m), 1440 (m), 1411 (m), 1340 (w), 1281 (m), 1199 (w), 1100 (vw), 1051 (m), 949 (vw), 877 (w), 824 (w), 772 (w), 727 (w), 681 (w), 628 (w), 578 (w), 560 (w), 513 (w), 431 (m) cm^{-1}.

(*E*)-Phenyl-*ortho*-F$_4$-azobenzoic acid 44

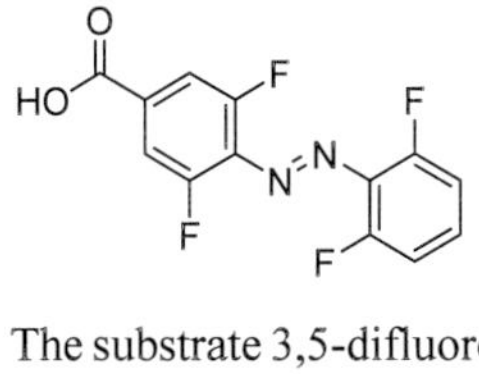

The substrate 3,5-difluoro-4-nitrosobenzoic acid **31** (0.883 g, 4.72 mmol, 1.00 eq.) was added to a solution of 2,6-difluoroaniline **35** (0.820 g, 6.35 mmol, 1.30 eq.) in 58 mL AcOH:PhMe:TFA (6:6:1) and stirred for 3 d at rt. PhMe (100 mL) was added and the mixture was evaporated under reduced pressure. The crude was purified by silica gel column chromatography. Initially, pure CH_2Cl_2 was used to remove side products. Then the product was eluted with 5% MeOH in CH_2Cl_2, followed by an increasing amount of MeOH. After the column chromatography the combined fractions (646 mg) were recrystallized from ACN (10 mL) and cH (5 mL) to yield 564 mg of (*E*)-phenyl-*ortho*-F_4-azobenoic acid **44** (1.89 mmol, 40%).[2]

TLC: $R_f = 0.3$ (EtOAc + 0.5% TFA). **^{1}H NMR (400 MHz, DMSO):** δ = 13.86 (s, 1H), 7.76 (d, J = 9.1 Hz, 2H), 7.67 (tt, J = 8.4, 6.1 Hz, 1H), 7.37 (t, J = 9.2 Hz, 2H) ppm. **^{13}C NMR (101 MHz, DMSO):** δ = 164.6, 155.8 (dd, J = 72.3, 3.9 Hz), 153.2 (dd, J = 71.3, 3.9 Hz), 134.4 (t, J = 9.0 Hz), 134.1 (t, J = 10.7 Hz), 133.2 (t, J = 10.6 Hz), 130.5 (t, J = 9.6 Hz), 113.8 (dd, J = 22.2, 3.0 Hz), 113.3 (dd, J = 19.8, 3.5 Hz) ppm. **^{19}F NMR (376 MHz, DMSO):** δ = −125.19, −125.36 ppm. **HRMS (FAB+):** *m/z* calcd for $C_{13}H_7O_2N_2F_4$: 299.0444 Da [M+H], found: 299.0442 Da (Δ = 0.6 ppm). **IR (ATR):** $\tilde{\nu}$ = 2818 (w), 2567 (w), 1699 (m), 1612 (w), 1577 (m), 1467 (m), 1432 (m), 1414 (m), 1330 (w), 1262 (m), 1242 (m), 1197 (m), 1063 (m), 1027 (m), 917 (w), 889 (m), 873 (w), 781 (m), 771 (m), 742 (m), 708 (w), 698 (w), 649 (vw), 589 (vw), 561 (w), 505 (w), 458 (w), 417 (w), 381 (w) cm^{-1}. **UV-Vis (MeCN):** λ_{max} = 310, 460 nm.

Dimethyl 4-nitronaphthalene-2,6-dicarboxylate 42

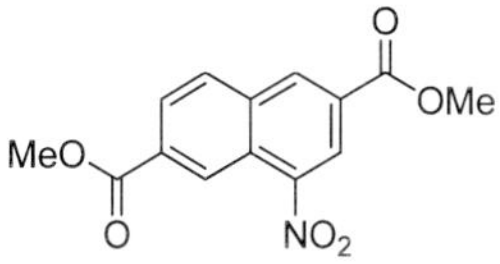

The substrate dimethyl-2,6-naphtalene-dicarboxylate **41** (2.00 g, 8.19 mmol, 1.00 eq.) was dissolved in conc. H_2SO_4 (200 mL). A solution of conc. HNO_3 (0.516 g, 8.19 mmol, 1.00 eq., 100%) in H_2SO_4 (20 mL) was added over 5 h by a syringe pump at 0 °C. Ice and sat. aqueous NH_4Cl (100 mL) solution were added and the product was filtered off and dried under high vacuum for 2 d to yield 2.03 g of a mixture with the desired product and the side product, which contains 2 nitro groups in relation to 4:1 (by NMR), yielding 1.62 g of pure dimethyl 4-nitronaphthalene-2,6-dicarboxylate **42** (5.62 mmol, 69%). The mixture could not be purified by silica gel column chromatography or recrystallization and was used in the next reaction without any further purification.[321]

TLC: $R_f = 0.2$ (developed cH:EtOAc (80:20) + 1% Net_3). **HPLC:** C_{18}, 30 min gradient 5-95% ACN +0.1% TFA in H_2O, elution 30.0 min. **^{1}H NMR (400 MHz, Pyridine-*d*$_5$):** δ = 9.71 – 8.84 (m, 1H), 8.73 (s, 1H), 8.26 (d, J = 34.4 Hz, 1H), 7.82 (d, J = 185.6 Hz, 1H), 7.22 (s, 1H), 5.10 (s, 3H), 3.91 (s, 1H) ppm. **^{13}C NMR (101 MHz, Pyridine-*d*$_5$):** δ = 166.6, 165.4, 148.2, 137.1, 133.3, 131.5, 130.5,

129.3, 128.3, 126.9, 126.4, 124.8, 53.4, 53.3 ppm. **HRMS (FAB+):** *m/z* calcd for $C_{14}H_{12}O_6N_1$: 290.0665 Da[M+H], found 290.0666 Da (Δ = 0.6 ppm). **IR (ATR):** $\tilde{\nu}$ = 3075 (vw), 2960 (vw), 2850 (vw), 1720 (m), 1605 (w), 1525 (w), 1436 (w), 1332 (w), 1271 (m), 1178 (w), 1101 (w), 986 (w), 909 (w), 845 (w), 762 (m), 703 (w), 479 (w) cm^{-1}.

dimethyl 4-aminonaphthalene-2,6-dicarboxylate 43

MeO OMe NH2 O O

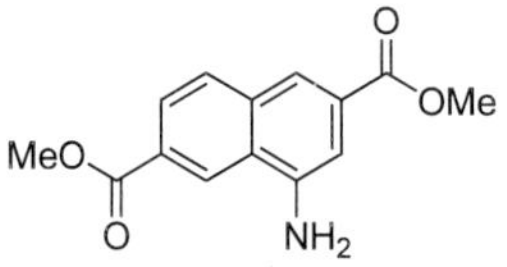

Dimethyl 4-nitronaphthalene-2,6-dicarboxylate **42** (1.90 g, 6.56 mmol, 1.00 eq.), 5% Pd on charcoal (0.076 mg, 4 wt%), $CHCl_3$ (150 mL) and MeOH (50 mL) were added under argon and the mixture was heated until the substrate was solved. The reaction mixture was stirred 3 d under hydrogen atmosphere at rt. The solution was filtered through Celite®, washed three times with MeOH and the solvent was evaporated under reduced pressure. Flash chromatography was used to purify the crude product. The solvent for elution was CH_2Cl_2:cH:EtOAc (40:40:20) and yielded 1.38 g dimethyl 4-aminonaphthalene-2,6-dicarboxylate **43** (5.33 mmol, 81%).[322]

TLC: R_f = 0.33 (CH_2Cl_2:cH:EtOAc 40:40:20 + 1% NEt_3). **^{1}H NMR (400 MHz, DMSO):** δ = 8.85 (s, 1H), 8.01 (d, *J* = 8.7 Hz, 1H), 7.93 (d, *J* = 8.6 Hz, 1H), 7.79 (s, 1H), 7.27 (s, 1H), 6.40 (s, 2H), 3.90 (d, *J* = 17.4 Hz, 6H) ppm. **^{13}C NMR (101 MHz, DMSO):** δ = 166.5 (d, *J* = 28.4 Hz), 147.0, 135.6, 129.9 (d, *J* = 12.7 Hz), 126.6, 125.3 (d, *J* = 22.9 Hz), 123.4, 116.6, 106.8, 52.2 (d, *J* = 7.1 Hz) ppm. **HRMS (EI+):** *m/z* calcd for $C_{14}H_{13}O_4N_1$: 259.0845 Da [M], found: 259.0843 Da (Δ = 0.7 ppm). **IR (ATR):** $\tilde{\nu}$ = 3460 (w), 3373 (w), 3249 (vw), 2950 (w), 2847 (vw), 1698 (m), 1639 (m), 1574 (w), 1506 (w), 1436 (m), 1396 (w), 1312 (w), 1274 (m), 1234 (m), 1214 (m), 1126 (m), 1103 (m), 1054 (w), 991 (w), 918 (w), 888 (w), 868 (w), 823 (w), 776 (m), 763 (m), 644 (vw), 485 (w), 434 (w) cm^{-1}.

Dimethyl *(E)*-4-(4-((2,6-difluorophenyl)diazenyl)-3,5-difluorobenzamido)naphthalene-2,6-dicarboxylate 46

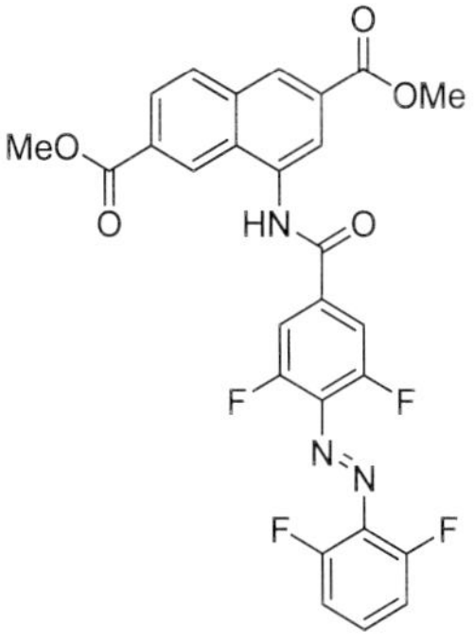

The substrate (*E*)-phenyl-*ortho*-F_4-azobenzoic acid **44** (0.396 g, 1.33 mmol, 1.00 eq.) was dissolved in thionyl chloride (16 mL) under argon and a few drops of dry DMF were added. The reaction mixture was refluxed for 6 h, cooled to rt and the solvent was evaporated under reduced pressure. The crude (*E*)-phenyl-*ortho*-F_4-azobenzoyl chloride **45** was dried under high vacuum for 1 h and was used in the next reaction without further purification.[323]

HRMS (EI+): *m/z* calcd for $C_{13}H_5O_1N_2Cl_1F_4$: 316.0027 Da [M], found: 316.0027 Da (Δ = 0.3 ppm). **IR (ATR):** $\tilde{\nu}$ = 3461 (vw), 3097 (w), 2648 (vw), 2574 (vw), 1945 (vw), 1740 (m), 1704 (w), 1613 (w), 1569 (s), 1470 (m), 1426 (m), 1332 (w), 1243 (m), 1206 (w), 1127 (w), 1060 (w), 1027 (m(w), 416 (w) cm^{-1}.

The substrate (*E*)-phenyl-*ortho*-F_4-azobenzoyl chloride **45** (0.344 g, 1.33 mmol, 1.00 eq.) was dissolved in dry PhMe (11 mL) under argon. A solution of dimethyl 4-aminonaphthalene-2,6-dicarboxylate **43** (0.344 g, 1.32 mmol, 1.00 eq.) in dry DMF (4 mL) and dry triethylamine (0.157 g, 1.56 mmol, 1.17 eq.) was added. The reaction mixture was stirred for 7 h at 60 °C under argon. The reaction was slowly cooled down to rt and stirring was continued overnight. The solution was evaporated under reduced pressure and the crude was purified *via* silica gel chromatography. Initially, pure CH_2Cl_2 was used to remove side products, then the product was eluted with 2% MeOH in CH_2Cl_2 and remaining substrate with 5% MeOH, followed by increasing amount until 30% MeOH in CH_2Cl_2. The solvent was evaporated yielding 508 mg dimethyl *(E)*-4-(4-tetra-*ortho*-F_4-azobenzamido)naphthalene-2,6-dicarboxylate **46** (9.42 mmol, 71%).[323]

TLC: $R_f = 0.8$ (5% MeOH in CH_2Cl_2). **HPLC:** C_{18}, 30 min gradient 20-95% ACN +0.1% TFA in H_2O +0.1% TFA, elution 31.0 min. **^{1}H NMR (400 MHz, THF-d_8):** δ = 10.18 (s, 1H), 8.84 (d, J = 1.2 Hz, 1H), 8.58 (t, J = 1.2 Hz, 1H), 8.44 (d, J = 1.5 Hz, 1H), 8.14 (dd, J = 3.7, 1.1 Hz, 2H), 7.94 (d, J = 9.3 Hz, 2H), 7.56 (ddd, J = 8.5, 5.9, 2.6 Hz, 1H), 7.22 (t, J = 8.8 Hz, 2H), 3.96 (s, 3H), 3.92 (s, 3H) ppm. **^{13}C NMR (101 MHz, THF-d_8):** δ = 167.1, 166.7, 164.2, 157.8 (dd, J = 65.1, 3.9 Hz), 155.2 (dd, J = 63.9, 3.9 Hz), 139.2 (t, J = 8.5 Hz), 136.6, 136.4, 134.5 (t, J = 10.7 Hz), 133.9 (t, J = 10.3 Hz), 132.7 (t, J = 9.9 Hz), 131.2, 131.1, 130.7, 130.7, 129.4, 127.0, 126.5, 123.9, 113.8 (dd, J = 20.1, 3.8 Hz), 113.4 (dd, J = 22.8, 2.9 Hz), 52.7 ppm. **^{19}F NMR (376 MHz, THF):** δ = −125.71, -125.99 ppm. **^{1}H NMR (300 MHz, DMSO):** δ = 11.04 (s, 1H), 8.71 (d, J = 11.9 Hz, 2H), 8.37 (d, J = 8.6 Hz, 1H), 8.24 (d, J = 1.6 Hz, 1H), 8.17 – 8.01 (m, 3H), 7.71 (ddd, J = 14.6, 8.5, 6.1 Hz, 1H), 7.42 (t, J = 9.3 Hz, 2H), 3.94 (d, J = 11.9 Hz, 6H) ppm. **HRMS (FAB+):** *m/z* calcd for $C_{27}H_{18}O_5N_3F_4$: 540.1183 Da [M+H], found: 540.1183 Da (Δ = 0.1 ppm). **IR (ATR):** $\tilde{\nu}$ = 3251 (vw), 3099 (vw), 2953 (vw), 1717 (m), 1646 (m), 1613 (w), 1576 (w), 1542 (m), 1440 (m), 1392 (w), 1338 (w), 1278 (m), 1244 (m), 1207 (m), 1127 (w), 1102 (w), 1049 (w), 1027 (m), 981 (w), 913 (w), 872 (w), 796 (m), 776 (m), 745 (w), 704 (w), 597 (w), 489 (w), 431 (w) cm^{-1}.

5.5.3 (*E*)-*ortho*-F_4-4-(*N*-Fmoc-aminomethylphenylazo)benzoic acid

The following synthesis towards (*E*)-*ortho*-F_4-4-(*N*-Fmoc-aminomethylphenylazo)benzoic acid **34** was performed by Anna-Lena Leistner in course of her bachelor thesis under the supervision of Johannes Karcher and are cited in verbatim.[227] The synthesis was optimised by Susanne Kirchner in her master studies.[209]

4-(Aminomethyl)-2,6-difluoroaniline 32

$LiAlH_4$ (0.74 g, 19.5 mmol, 3.00 eq.) was placed as dry solid in a three-necked-flask equipped with an additional funnel under argon atmosphere. 4-amino-3,5-difluoronitrile **24** (1.00 g, 6.49 mmol, 1.00 eq.) was also provided as solid in the additional funnel. 13 mL of dry THF were added to $LiAlH_4$ in the flask, while 19.5 mL were added to the substrate in the additional funnel through a septum. After the substrate was dissolved, the solution was added dropwise, and the resulting mixture was allowed to react for 2.5 h. 35 mL EtOAc followed by 50 mL water were used to quench the remaining $LiAlH_4$. After stirring for another 15 min, the solid components were filtered off. Subsequently the phases were separated, and the aqueous phase was extracted with EtOAc (3×30 mL). The combined organic layers were dried with anhydrous Na_2SO_4 and filtered off and the solvent was evaporated under reduced pressure. Silica gel column chromatography was used to purify the crude components. Initial elution with CH_2Cl_2 washed out remnants of the substrate followed by the side product of hydrolysis (4-amino-3,5-difluorobenzaldehyde, 90 mg, 0.57 mmol). The solvent for elution was changed to 2%, 5% 10% and 30% MeOH to elute 409 mg of 4-(aminomethyl)-2,6-difluoroaniline **32** (2.58 mmol, 40%) as beige solid.[324]

TLC: $R_f = 0.13$ (30% MeOH in CH_2Cl_2) **^{1}H NMR (300 MHz, DMSO):** $\delta = 6.78$ (dd, $J = 7.5, 2.3$ Hz, 2H), 5.07 (s, 2H), 4.09 (d, $J = 5.9$ Hz, 2H), 1.85 (s, 2H) ppm. **^{13}C NMR (101 MHz, DMSO):** $\delta = 152.2$ (d, $J = 9.3$ Hz), 149.8 (d, $J = 9.3$ Hz), 129.8 – 129.5 (m), 123.6 (t, $J = 16.9$ Hz), 109.9 (dd, $J = 14.5, 6.8$ Hz), 43.9 ppm **HRMS (EI+):** *m/z* calcd for $C_7H_8N_2F_2$: 158.0656 Da [M], found: 158.0657 Da ($\Delta = 0.3$ ppm). **IR (ATR):** $\tilde{\nu} = 3461$ (w), 3379 (w), 3288 (w), 2486 (w), 1650 (w), 1583 (m), 1519 (m), 1473 (m), 1446 (m), 1427 (m), 1378 (m), 1312 (m), 1277 (m), 1155 (m), 1039 (w), 964 (m), 820 (m), 719 (m), 665 (m), 592 (m), 565 (m), 513, (w), 473 (w) cm^{-1}.

O-(9H-Fluoren-9-yl)methyl-N-(4-amino-3,5-difluorobenzyl)carbamate 33

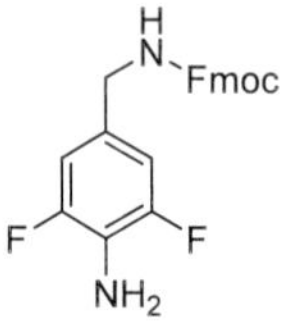

A suspension of 4-(aminomethyl)-2,6-difluoroaniline **32** (200 mg, 1.26 mmol, 1.0 eq.) and Na_2CO_3 (167 mg, 1.58 mmol, 1.25 eq.) in dioxane:H_2O 2:1 (9.9 mL) was cooled down to 0 °C. NaOH (0.1 mL, 2.5 M) was added to the mixture followed immediately by Fmoc-OSu (81.0 mg, 0.24 mmol, 0.19 eq.). The addition was repeated 5 times in total at 0 °C with a time interval of 10 min between each addition step, following the mixture was stirred for 1 h at rt. The product was precipitated by addition of water (9 mL) and dried under high vacuum to obtain 272 mg *O*-(9*H*-fluoren-9-yl)methyl-*N*-(4-amino-3,5-difluorobenzyl)carbamate **33** (0.72 mmol, 57%).[325]

^{1}H NMR (400 MHz, DMSO): δ = 7.86 (dd, *J* = 18.0, 7.4 Hz, 4H), 7.41 (td, *J* = 7.4, 1.2 Hz, 2H), 7.34 (td, *J* = 7.4, 1.2 Hz, 2H), 7.23 (t, *J* = 6.2 Hz, 1H), 6.79 (d, 2H), 6.27 (s, 2H), 5.05 (s, 2H), 4.00 (d, *J* = 6.2 Hz, 2H), 3.70 (dd, *J* = 39.7, 6.4 Hz, 1H) ppm. **^{13}C NMR (101 MHz, DMSO):** δ = 157.7, 152.2 (d, *J* = 9.4 Hz), 149.8 (d, *J* = 9.4 Hz), 139.4, 137.4, 128.9, 127.7 (t, *J* = 7.4 Hz), 127.3, 123.8 (t, *J* = 16.8 Hz), 121.4, 120.0, 109.6 (dd, *J* = 22.7 Hz), 63.8, 50.1, 42.8 ppm. **^{19}F NMR (376 MHz, DMSO):** δ = −135.82 ppm. **HRMS (FAB+):** *m/z* calcd for $C_{22}H_{19}O_2N_2F_2$: 381.1415 Da [M+H], found: 381.1413 Da (Δ = 0.3 ppm). **IR (ATR):** ṽ = 3473 (vw), 3377 (vw), 3331 (vw), 2937 (vw), 1684 (vw), 1643 (vw), 1591 (vw), 1524 (vw), 1448 (vw), 1360 (vw), 1328 (vw), 1258 (vw), 1144 (vw), 1115 (vw), 1041 (vw), 992 (vw), 956 (vw), 844 (vw), 781 (vw),757 (vw), 734 (vw), 672 (vw), 553 (vw), 500 (vw), 426 (vw) cm^{-1}.

(E)-ortho-F_4-4-(N-Fmoc-aminomethylphenylazo)benzoic acid 34

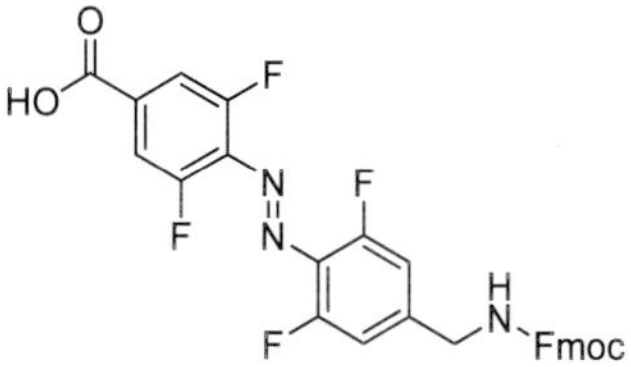

(9*H*-fluoren-9-yl)methyl-*N*-(4-amino-3,5-difluorobenzyl)carbamate **33** (175 mg, 0.46 mmol, 1.50 eq.) was dissolved in 4.6 mL AcOH:PhMe:TFA (6:6:1).[2] After adding 3,5-difluoro-4-nitrosobenzoic acid **31**, the solution was stirred for 3 d. The crude was purified by silica gel column chromatography with CH_2Cl_2 as initial solvent to elute impurities followed by increasing the polarity of the solvent stepwise (1%, 2%, 5%, 10% MeOH in CH_2Cl_2). Due to impurities the product was recrystallized afterwards from $CHCl_3$:cH 1:1 and 65 mg of *(E)-ortho*-F_4-4-(*N*-Fmoc-aminomethylphenylazo)benzoic acid **34** (118 µmol, 26%) was isolated as an orange solid.

^{1}H NMR (400 MHz, DMSO): δ = 13.87 (br s, 1H), 8.00 (t, J = 6.1 Hz, 2H), 7.89 (d, J = 7.5 Hz, 2H), 7.78 (d, J = 9.6 Hz, 2H), 7.71 (d, J = 7.4 Hz, 2H), 7.41 (t, J = 7.4 Hz, 2H), 7.33 (t, J = 7.4 Hz, 2H), 7.20 (d, J = 10.9 Hz, 2H), 4.43 (d, J = 6.7 Hz, 2H), 4.28 (t, J = 6.8 Hz, 2H), 4.24 (s, 1H) ppm. **^{13}C NMR (101 MHz, DMSO):** δ = 165.3, 156.5, 156.1 (d, J = 4.7 Hz), 155.4 (d), 153.5 (d, J = 4.4 Hz), 152.8 (d, J = 3.7 Hz), 143.8, 140.8, 139.0 – 138.7 (m), 132.5 – 132.2 (m), 129.2 (t, J = 9.9 Hz), 127.6, 127.0, 125.1, 124.6 – 124.4 (m), 120.1, 113.4 (d, J = 20.8 Hz), 111.2 (d, J = 20.8 Hz), 65.5, 46.8, 43.2 ppm. **HRMS (FAB+):** *m/z* calcd for $C_{29}H_{20}O_4N_3F_4$: 550.1390 Da [M+H], found: 550.1388 Da (Δ = 0.4 ppm). **IR (ATR):** $\tilde{\nu}$ = 3333 (vw), 2922 (vw), 1684 (w), 1625 (w), 1574 (w), 1478 (vw), 1440 (w), 1346 (w), 1263 (w), 1195 (vw), 1114 (vw), 1051 (w). 992 (vw), 967 (vw), 891 (vw), 845 (vw), 789 (vw), 757 (w), 739 (w), 621 (vw), 566 (vw), 503 (vw), 462 (vw), 426 (vw) cm^{-1}.

Fmoc-deprotection of *(E)-ortho*-F_4-4-(*N*-Fmoc-aminomethylphenylazo)benzoic acid

NHFmoc F F N N F F O OH 34

2 vol% Base, DMF rt, 20 or 60 min

NH$_2$ F F N N F F O OH 34d + R HN

degradation

+

HN F F N N F F HO O

F NH$_2$ F O OH 25 + R F R F not identified

degradation

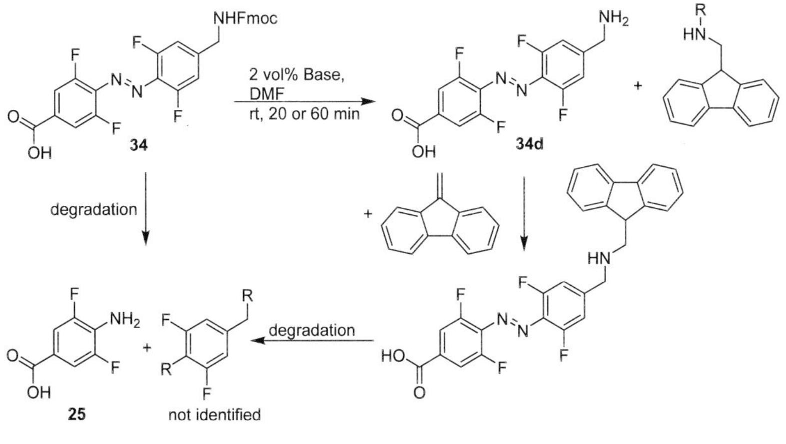

Figure 109. Overview over the proposed degradation pathway.

(E)-ortho-F_4-4-(*N*-Fmoc-aminomethylphenylazo)benzoic acid **34** (5.49 mg, 0.0100 mmol, 1.0 eq.) was dissolved in DMF (500 μL). A solution of a base (0.0500 mmol, 5.0 eq.) in DMF (500 μL) was added and quenched after 20 min or 60 min with AcOH (500 μL). The mixture was analyzed by HPLC under standard conditions, detection at 256 nm with a gradient 5-95% ACN with 0.1% TFA in H_2O with 0.1% TFA.

Table 18. Conversion I of Fmoc-deprotection, by HPLC, 256 nm.

Substance	eq	20 min reaction	60 min reaction
Fmoc-F4-Azo	1.00	23.7 min	-
DBU	5.00	13.9 min 90%, 20.3 min 4.6%	further degradation
1,2-Lutidine	5.00	stable	stable
Piperidin	5.00	13.9 min 60%, 14.4 min 14%, 22.3 6%, 23.7 min 16%	further degradation
Morpholin	5.00	14.0 min 5%, 23.7 min 90%	further degradation
DIPEA	5.00	stable	stable

A solution of *(E)-ortho*-F_4-4-(*N*-Fmoc-aminomethylphenylazo)benzoic acid **34** (1.37 mg, 0.0025 mmol, 1.0 eq.) was dissolved in DMF (500 μL, 5 mM). A solution of a 2 vol% base in DMF (500 μL, 4 vol%) was added and quenched after 20 min or 60 min with AcOH (100 μL). The mixture was analysed by HPLC with the absorption at 256 nm and a gradient 5-95% ACN with 0.1% TFA in H_2O with 0.1% TFA.

Table 19. Conversion II of Fmoc-deprotection, by HPLC, 256 nm.

substance	eq	m[mg]	c[mol/L]	20 min	60 min
Fmoc-F4-Azo	1.00	1.37	0.005	r=23.7 min	
2% DBU	53.6	20.4	0.268	degraded (6 p.)	degraded
2% Difluoroanilin	74.2	23.9	0.371	stable	stable
2% Morpholin	98.2	21.3	0.491	degraded (8 p.)	degraded
4.3% Octanthiol	100	36.5	0.500	stable	slow degradation

5.5.4 (*E*)-Phenyl-*ortho*-F4-*para*-azobenzoic acid

The synthesis until TFBA **40** was initially performed by Johannes Karcher in his master studies[243a] and reproduced by Tobias Bantle.[243b] The methods to 2-(4-amino-3,5-difluorophenyl)acetic acid **39** are cited from the master thesis of Susanne Kirchner, which included further optimisations.[209] All procedures were reproduced by undergraduate students under the supervision of Johannes Karcher to provide constant replenishment.

1,3-Difluoro-2-nitrobenzene 36

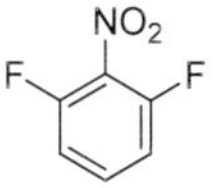

To a stirred suspension of sodium perborate tetrahydrate (66.5 g, 426 mmol) in glacial AcOH (100 mL) 2,6-difluoroaniline **35** (11.0 g, 85.2 mmol) in glacial acetic (50 mL) was added slowly at 80 °C. The reaction mixture was stirred at 80 °C for 1 h. The reaction mixture was then allowed to cool to rt and poured into water (100 mL). The aqueous phase was extracted with diethyl ether (2×100 mL) and the combined organic layers were washed with a sat. aqueous solution of $NaHCO_3$ (2×100 mL). The organic layer was dried with Na_2SO_4 and the solvent was removed *in vacuo*. The crude product was purified by silica gel column chromatography (cH:THF 9:1, $R_f = 0.18$) and subsequently recrystallized in cH to afford 3.84 g 1,3-difluoro-2-nitrobenzene **36** (24.1 mmol, 28%) as a yellow powder.[326]

^{1}H NMR (400 MHz, $CDCl_3$): $\delta = 7.53$ (tt, $J = 8.5$, 5.9 Hz, 1H), 7.14 – 7.07 (m, 2H) ppm. **^{13}C NMR (101 MHz, $CDCl_3$):** $\delta = 154.8$ (dd, $J = 261.6$, 1.9 Hz), 133.1 (t, $J = 9.6$ Hz), 130.0 (t, $J = 6.5$ Hz), 113.1 (dd, $J = 19.1$, 4.2 Hz) ppm. **^{19}F NMR (377 MHz, $CDCl_3$):** $\delta = -123.20$ ppm. **HRMS (EI+):** *m/z* calcd for $C_6H_3F_2NO_2$: 159.0126 Da [M], found: 159.0133 Da ($\Delta = 0.4$ ppm). **IR (ATR):** $\tilde{\nu} = 3104.8$ (w), 3076.6 (w), 2895.3 (w), 1612.6 (s), 1535.9 (s), 1477.1 (s), 1357.7 (s), 1297.2 (w), 1246.8 (m), 1026.2 (s), 852.7 (m), 789.5 (s), 750.8 (m), 691.8 (m), 589.0 (m), 513.2 (m), 408.1 (w) cm^{-1}. **elemental analysis calcd for $C_6H_3F_2NO_2$ (%):** C: 45.30, H: 1.90, F: 23.88, N: 8.80, O: 20.11, found: C: 45.16, H: 1.80, N: 8.61.

***tert*-Butyl 2-(3,5-difluoro-4-nitrophenyl)acetate 38**

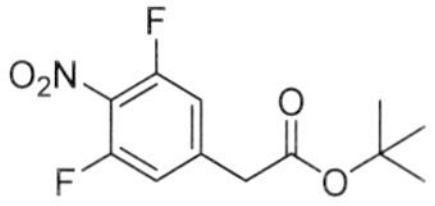

A solution of 2,6-difluoronitrobenzene **35** (6.00 g, 37.7 mmol) and *tert*-butyl chloroacetate **37** (9.66 g, 64.1 mmol) in anhydrous DMF (48 mL) was added dropwise over 1 h to a cold (−41 °C) suspension of KO^{t}Bu (14.8 g, 112 mmol) in anhydrous DMF (48 mL) under argon. The reaction mixture was stirred at (−41 °C for 1.5 h, quenched with 2 M HCl (50 mL) and extracted with cH (4×100 mL). The combined organic layers were washed with water (1×100 mL), sat. aqueous solution of NH_4Cl (2×100 mL) and brine (1×100 mL), dried with Na_2SO_4 and concentrated *in vacuo*. The crude product was purified by silica gel column chromatography (cH:CH_2Cl_2 2:1, $R_f = 0.25$) to yield 2.77 g *tert*-butyl 2-(3,5-difluoro-4-nitrophenyl)acetate **38** (10.1 mmol, 27%) as a yellow oil.[244b]

^{1}H NMR (400 MHz, $CDCl_3$): $\delta = 7.03$ (d, $J = 8.7$ Hz, 2H), 3.57 (s, 2H), 1.44 (s, 9H) ppm. **^{13}C NMR (101 MHz, $CDCl_3$):** $\delta = 168.4$, 154.6 (dd, $J = 261.3$, 2.4 Hz), 141.6, 114.0 (dd, $J = 20.8$, 3.0 Hz), 82.5, 42.15, 28.1 ppm. **^{19}F NMR (377 MHz, $CDCl_3$):** $\delta = -123.08$ ppm. **HRMS (EI+):** *m/z* calcd for $C_{12}H_{13}F_2NO_4$: 273.0807 Da [M], found. 273.0811 Da ($\Delta = 1.5$ ppm). **IR (ATR):** $\tilde{\nu} = 3067.0$ (w), 2980.9 (w), 2934.4 (w), 1727.6 (m), 1626.2 (m), 1537.4 (s), 1445.7 (m), 1364.0 (s), 1324.3 (m), 1219.0 (m), 1143.6 (s), 1056.5 (s), 946.4 (w), 866.0 (m), 840.8 (m), 753.5 (w), 591.3 (w), 551.3 (w) cm^{-1}.

2-(3,5-Difluoro-4-nitrophenyl)acetic acid 39

TFA (27 mL) was added to a 0 °C solution of *tert*-butyl 2-(3,5-difluoro-4- nitrophenyl)acetate **38** (2.70 g, 9.88 mmol) in CH_2Cl_2 (27 mL). Triisopropylsilane (1.56 g, 9.88 mmol) was added and the reaction mixture was allowed to warm up to rt and stirred for 3 h. PhMe (2×16 mL) was added to the reaction mixture and it was concentrated *in vacuo*. Recrystallization in 20 mL PhMe:cH 1:1 yielded 1.61 g of 2-(3,5-difluoro-4-nitrophenyl)acetic acid (7.39 mmol, 75%) as a white solid.[244b]

^{1}H NMR (300 MHz, $CDCl_3$): $\delta = 9.86$ (s, 1H), 7.07 (d, $J = 8.4$ Hz, 2H), 3.73 (s, 2H) ppm. **^{13}C NMR (101 MHz, $CDCl_3$):** $\delta = 175.2$, 154.7 (dd, $J = 262.0$, 2.4 Hz), 139.6, 114.3 (dd, $J = 20.8$, 3.3 Hz, C_{arom}-H), 40.5 ppm. **^{19}F NMR (377 MHz, $CDCl_3$):** $\delta = -122.26$ ppm. **HRMS (EI+):** *m/z* calcd for $C_8H_5F_2NO_4$: 217.0181 Da [M], found: 217.0184 Da ($\Delta = 1.4$ ppm). **IR (ATR):** $\tilde{\nu} = 2929.9$ (w), 2636.7 (w), 1709.0 (s), 1627.3 (m), 1594.6 (m), 1528.8 (s), 1445.5 (m), 1413.7 (m), 1366.0 (m), 1346.8 (s), 1230.8 (s), 1182.3 (w), 1056.0 (s), 914.6 (m), 860.3 (m), 839.8 (s), 762.3 (w), 742.9 (m), 697.6 (m), 651.5 (m), 612.0 (w), 589.0 (s), 545.8 (w), 508.1 (m), 456.0 (w) cm^{-1}. **elemental analysis calcd for $C_8H_5F_2NO_4$ (%):** C: 44.25, H: 2.32, F: 17.50, N: 6.45, O: 29.47, found: C: 44.02, H: 2.23, N: 6.30.

2-(4-Amino-3,5-difluorophenyl)acetic acid 39

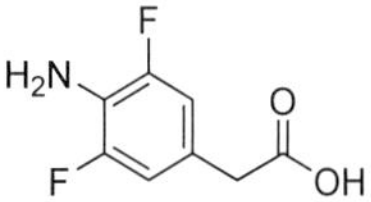

2-(3,5-difluoro-4-nitrophenyl)acetic acid (1.00 g, 4.61 mmol), 5% Pd/C (100 mg) and EtOH (50 mL) were added into a vial. The reaction mixture was flushed with hydrogen and stirred under hydrogen atmosphere (35 bar) at rt for 3 h. The crude was filtrated over Celite® and washed with MeOH (3×5 mL). The solvent was removed *in vacuo* and the product was dried under HV to yield 843 mg 2-(4-amino-3,5-difluorophenyl)acetic acid **39** (4.50 mmol, 98%) as a white solid.

^{1}H NMR (400 MHz, DMSO): δ = 12.31 (s, 1H), 6.81 (dd, J = 7.7, 2.0 Hz, 2H), 5.06 (s, 2H), 3.43 (s, 2H) ppm. **^{13}C NMR (101 MHz, DMSO):** δ = 172.7, 150.9 (dd, J = 238.0, 9.6 Hz), 123.9 (t, J = 16.8 Hz), 121.7 (t, J = 8.8 Hz), 112.1 (dd, J = 14.6, 7.4 Hz), 39.2 ppm. **^{19}F NMR (377 MHz, DMSO):** δ = −136.35 ppm. **HRMS (EI+):** *m/z* calcd for $C_8H_7F_2NO_2$: 187.0439 Da [M], found: 187.0446 Da (Δ = 3.7 ppm). **IR (ATR):** $\tilde{\nu}$ = 3352.3 (w), 3283.4 (w), 3186.7 (w), 2794.9 (w), 2502.7 (w), 1683.2 (m), 1597.7 (m), 1515.1 (m), 1445.5 (s), 1333.7 (w), 1302.2 (s), 1274.9 (m), 1219.6 (m), 1166.2 (m), 1141.7 (s), 989.0 (s), 941.7 (m), 892.1 (m), 850.4 (s), 797.9 (s), 715.0 (w), 640.3 (m), 601.4 (m), 551.1 (w), 504.8 (w), 438.2 (m) cm^{-1}. **elemental analysis calcd for $C_8H_7F_2NO_2$ (%):** C: 51.43, H: 3.77, F: 20.30, N: 7.48, O: 17.10, found: C: 51.33, H: 3.84, N: 7.45. **UV-Vis (MeCN):** λ_{max} = 237 nm.

(*E*)-Phenyl-*ortho*-F_4-*p*-azobenzoic acid, TFBA 40

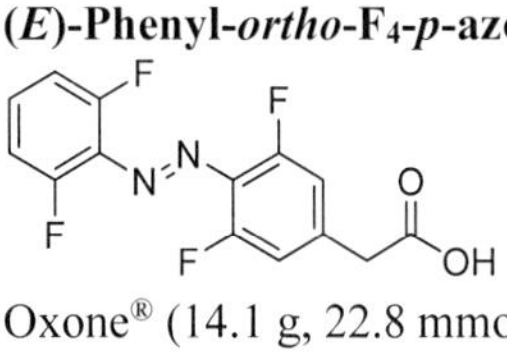

Oxone® (14.1 g, 22.8 mmol, 2.00 eq.) dissolved in diH_2O (137 mL) was added to 2,6-difluoroaniline **35** (1.47 g, 11.4 mmol, 1.00 eq.) dissolved in CH_2Cl_2 (46 mL).[2] The mixture was stirred vigorously at rt for 4 h. The layers were separated, the organic layer washed with water (3×100 mL) and concentrated under reduced pressure. Without any further purification the crude 1.51 g 1,3-difluoro-2-nitrosobenzene **11** (10.5 mmol, 92%) was instantly used in the next reaction.

The 1,3-difluoro-2-nitrosobenzene **11** (1.51 g, 10.5 mmol, 1.14 eq.) was suspended in AcOH:PhMe:TFA 6:6:1 (79 mL) and 2-(4-amino-3,5-difluorophenyl)acetic acid (1.72 g, 9.19 mmol, 1.00 eq.) was added.[2] The resulting mixture was stirred at rt for 3 d. The solution was diluted with water (100 mL), and extracted with EtOAc (3×100 mL, the organic phase was dried over $MgSO_4$, filtered off, and concentrated under reduced pressure. The resulting mixture was purified by column chromatography in 5% MeOH in CH_2Cl_2 (R_f = 0.13) to yield 1.40 g (*E*)-phenyl-*ortho*-F_4-*p*-azobenzoic acid **40** (4.48 mmol, 49%) as an orange solid.

^{1}H NMR (500 MHz, DMSO): δ = 12.67 (s, 1H), 7.62 (tt, *J* = 8.5, 6.1 Hz, 1H), 7.40 – 7.33 (m, 2H), 7.31 (d, *J* = 10.6 Hz, 2H), 3.76 (s, 2H) ppm. **^{13}C NMR (126 MHz, DMSO):** δ = 171.4, 155.5 (dd, *J* = 35.1, 4.5 Hz), 153.5 (dd, *J* = 35.1, 4.5 Hz), 141.7 (t, *J* = 10.7 Hz), 133.0 (t, *J* = 10.5 Hz), 130.8, 129.1 (t, *J* = 9.9 Hz), 114.4 (dd, *J* = 20.5, 3.3 Hz), 113.2 (dd, *J* = 19.8, 3.8 Hz), 39.78 (t, *J* = 21.0 Hz, HSQC) ppm. **^{19}F NMR (471 MHz, DMSO):** δ = −122.06 (dd, *J* = 10.4, 6.2 Hz), −122.22 (d, *J* = 11.3 Hz) ppm. **HRMS (ESI−):** *m/z* calcd for $C_{14}H_7F_4N_2O_2^-$: 311.0449 Da $[M]^-$, found: 311.0450 Da (Δ = 0.3 ppm). **IR (ATR):** $\tilde{\nu}$ = 3107 (w), 1936 (w), 1735 (w), 1612 (m), 1586 (s), 1470 (s), 1284 (w), 1245 (s), 1218 (m), 1057 (w), 1016 (w), 883 (w), 786 (s), 735 (s), 692 (w), 599 (m), 509 (m), 472 (m), 379 (w) cm^{-1}. **UV-Vis (MeCN):** λ_{max} = 226, 306, 456 nm.

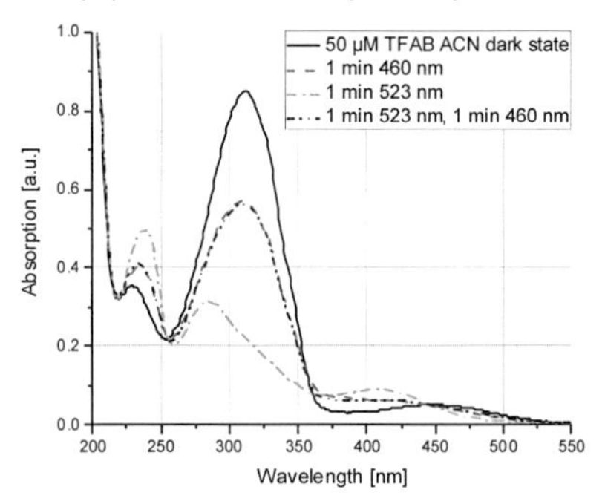

Figure 110. UV-Vis of 50 µM TFAB in ACN in dark and after isomerization with 460 nm and 523 nm.

5.6 Cl_4-PAP-DKP-Lys_2

5.6.1 (*S*)-Boc-Phe(3,5-Cl_2-4-NH_2)-OMe

The following syntheses towards (*S*)-Boc-Phe(3,5-Cl_2-4-NH_2)-OMe **51** were performed by Anna-Lena Leistner in her bachelor thesis under the supervision of Johannes Karcher and are cited in verbatim.[227] All procedures were reproduced by undergraduate students under the supervision of Johannes Karcher to provide constant replenishment.

***L*-Phe-(4-NO_2)-OH, (*S*)-2-amino-3-(4-nitrophenyl)propanoic acid 81**

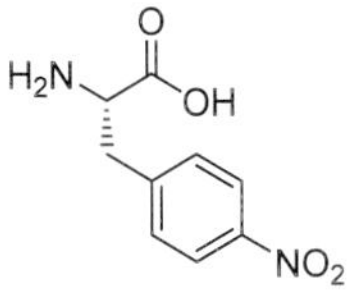

S-Phenylalanine (40.0 g, 0.242 mol, 1.00 eq.) was dissolved in 96% sulfuric acid (H_2SO_4, 121 mL). The solution was stirred vigorously while cooling down to 0 °C with an ice water bath. When the internal temperature decreased below +5 °C, a mixture of 65% nitric acid (HNO_3, 33.7 mL, 2.00 eq.) and concentrated 26.5 mL H_2SO_4 (1.4:1.1 v/v) were added drop by drop over a period of 4 h while stirring was continued. The internal temperature did not exceed +10 °C. After the addition was finished, the reaction mixture was stirred for an additional 2.5 h in the ice bath (below +10 °C). The reaction mixture was poured on 400 mL of an ice-water-mixture and neutralized with 25% aqueous NH_3. The crude precipitates and the reaction mixture were stirred at rt for further 45 min and then filtered to separate out the precipitate. The precipitate was washed with water (3×20 mL). To purify the crude product the precipitate was recrystallized from 600 mL water, filtered off and dried under high vacuum, yielding 36.9 g of *L*-Phe(pNO_2)-OH **81** (153.6 mmol, 73%).[327]

^{1}H NMR (300 MHz, D_2O/KOH): δ = 8.12 (d, *J* = 8.9 Hz, 2H), 7.39 (d, *J* = 8.8 Hz, 2H), 3.50 (t, *J* = 1.3 Hz, 1H), 3.11 – 2.83 (m, 2H) ppm. **^{13}C NMR (75 MHz, D_2O/KOH):** δ = 178.9, 144.0, 143.5, 127.5, 120.8, 54.5, 37.9 ppm. **HRMS (EI+):** *m/z* calcd for $C_9H_{10}O_4N_2$: 210.0635 Da [M], found 210.0634 Da (Δ = 0.2 ppm). **IR (ATR):** $\tilde{\nu}$ = 3181 (vw), 2872 (w), 1697 (vw), 1610 (w), 1567 (w), 1513 (w), 1441 (w), 1417 (w), 1344 (m), 1311 (w), 1175 (vw), 1140 (vw), 1105 (w), 1069 (vw), 1012 (vw), 946 (vw), 877 (w), 862 (w), 814 (vw), 767 (vw), 744 (w), 696 (w), 653 (w), 523 (w), 491 (w), 416 (vw) cm^{-1}.

L-Phe-(4-NO_2)-OMe·HCl, methyl (*S*)-2-amino-3-(4-nitrophenyl)propanoate hydrochloride 82

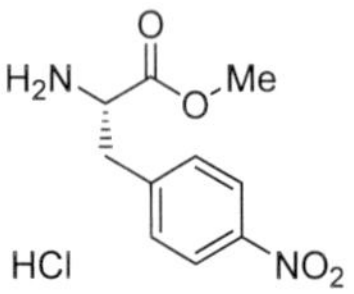

The solution of L-Phe-($p$$NO_2$)-OH **81** (46.3 g, 220 mmol, 1.00 eq.) in MeOH (173 mL) was cooled down to 0 °C. Subsequently, $SOCl_2$ (119.5 mL, 1.65 mol, 7.50 eq.) was slowly added and the mixture was stirred overnight while allowing to warm up to rt. The solvent was removed under reduced pressure. The remaining solid was dissolved again in a minimal amount of MeOH and the solution was added to stirred vigorously diethyl ether (the final ratio of MeOH:ether, 3:7). White solid precipitated, was filtered off and washed with small amounts of diethyl ether to yield 28.6 g of L-Phe-($p$$NO_2$)-OMe·HCl **82** (110 mmol, 50%).[328]

TLC: R_f = 0.55 (30% MeOH in EtOAc). **^{1}H NMR (400 MHz, DMSO):** δ = 8.94 (s, 2H), 8.18 (d, J = 8.7 Hz, 2H), 7.59 (d, J = 8.7 Hz, 2H), 4.35 (dd, J = 7.6, 5.7 Hz, 1H), 3.66 (s, 3H), 3.39 (dd, J = 13.9, 5.7 Hz, 1H), 3.29 (dd, J = 13.9, 7.6 Hz, 1H) ppm. **^{13}C NMR (101 MHz, DMSO):** δ = 169.0, 146.8, 143.2, 131.0, 123.5, 52.8, 52.7, 35.3 ppm. **HRMS (EI+):** *m/z* calcd for $C_{10}H_{13}O_4N_2$: 225.0870 Da [M], found 225.0869 Da (Δ = 0.2 ppm). **IR (ATR):** $\tilde{\nu}$ = 2914 (w), 1741 (m), 1601 (w), 1549 (w), 1491 (m), 1450 (w), 1345 (m), 1238 (m), 1185 (w), 1146 (m), 1107 (w), 1059 (w), 981 (w), 950 (w), 932 (w), 868 (m), 859 (m), 844 (w), 812 (w), 751 (m), 700 (m), 655 (w), 528 (vw), 508 (w), 491 (w), 405 (w) cm^{-1}.

Boc-L-Phe(4-NH_2)-OMe, methyl (*S*)-2-((*tert*-butoxycarbonyl)amino)-3-(4-nitrophenyl)propanoate 83

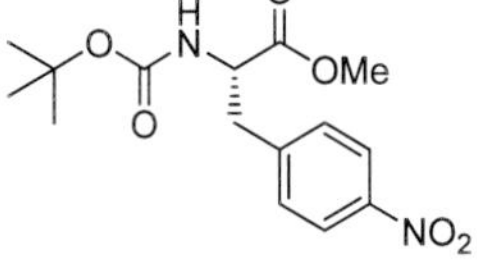

To the mixture of L-Phe-($p$$NO_2$)-OMe·HCl **82** (28.6 g, 110 mmol, 1.00 eq.) in 225 mL of aqueous $NaHCO_3$ (20.3 g, 242 mmol, 2.20 eq.) was added dropwise a solution of $(Boc)_2O$ (26.4 g, 121 mmol, 1.10 eq.) in dioxane (225 mL). The mixture was stirred at rt for 24 h, then concentrated to half of its original volume (by complete evaporation of dioxane) and extracted with EtOAc (3×100 mL). The organic layer was washed with 5% aqueous $KHSO_4$ (10-50vol% compared organic phase), 5% $NaHCO_3$ (10vol% compared to organic phase), and brine (10vol% compared to organic phase). After drying over anhydrous Na_2SO_4 and filtration through cotton, the solvent was removed under reduced pressure. The Boc-*L*-Phe(*p*NH_2)-OMe **83** was dried under high vacuum over night and was obtained as a yellow solid (33.8 g, 95%).[329]

TLC: R_f = 0.70 (EtOAc). **^{1}H NMR (300 MHz, $CDCl_3$):** δ = 8.14 (d, J = 8.5 Hz, 2H), 7.30 (d, J = 8.3 Hz, 2H), 5.08 (d, J = 8.1 Hz, 1H), 4.62 (d, J = 7.2 Hz, 1H), 3.71 (d, J = 3.7 Hz, 3H), 3.23 (dd, 1H), 3.11 (dd, 1H), 1.38 (s, J = 3.7 Hz, 9H) ppm. **^{13}C NMR (75 MHz, $CDCl_3$)**: δ = 171.7, 155.0, 147.2, 144.2, 130.3, 123.7, 80.4, 54.2, 52.6, 38.5, 28.3 ppm. **HRMS (FAB+):** *m/z* calcd for $C_{15}H_{21}O_6N_2$: 325.1400 Da [M+H], found 325.1400 Da (Δ = 0.2 ppm). **IR (ATR):** $\tilde{\nu}$ = 3357 (w), 1728 (m), 1687 (m), 1606 (w), 1518 (m), 1369 (w), 1343 (m), 1270 (m), 1230 (w), 1155 (m), 1102 (w), 1057 (w), 1033 (w), 1013 (w), 994 (w), 971 (w), 858 (w), 840 (w), 775 (w), 752 (w), 699 (w), 607 (w), 551 (w), 514 (w), 437 (vw) cm^{-1}.

Boc-*L*-Phe(4-NH_2)-OMe, methyl (*S*)-3-('*para*-aminophenyl)-2-[(*tert*-butoxycarbonyl)amino]propanoate 50

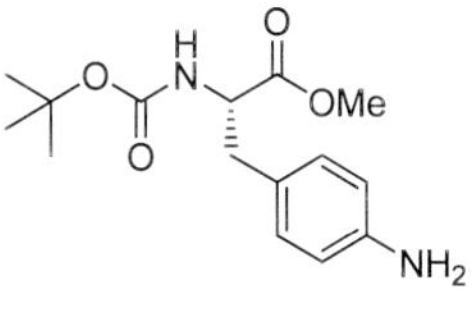

Boc-*L*-Phe(*p*NO_2)-OMe **83** (35.2 g, 109 mmol, 1.00 eq.) was dissolved in 200 mL MeOH in a three-necked flask and 5% Pd on charcoal (701 mg) was added. The atmosphere in the flask was exchanged three times with argon/vacuum. Next, vacuum was applied to the flask and hydrogen was added by usage of a balloon. The solution was stirred vigorously. TLC was used to survey the progress of the reaction. Subsequently 701 mg of Pd/C were added to ensure complete conversion to the amine. After 2 h of stirring the reaction mixture was filtered through Celite® to remove the remaining catalyst and charcoal. The solvent was removed under reduced pressure and the product was dried under high vacuum overnight to yield 29.3 g of Boc-*L*-Phe(*p*NH_2)-OMe **50** (99.5 mmol, 91%).[330]

TLC: R_f = 0.3 (cH:EtOAc 1:1). **^{1}H NMR (300 MHz, $CDCl_3$):** δ = 6.89 (d, J = 7.7 Hz, 2H), 6.62 (d, J = 7.9 Hz, 2H), 4.97 (d, J = 8.3 Hz, 1H), 4.49 (q,

J = 7.4 Hz, 1H), 3.69 (s, 3H), 3.13 – 2.79 (d, 2H), 1.41 (s, 9H) ppm. **^{13}C NMR (75 MHz, $CDCl_3$):** δ = 172.7, 157.1, 145.1, 130.2, 126.0, 115.6, 79.9, 54.7, 52.2, 37.5, 28.4 ppm. **HRMS (EI+):** *m/z* calcd for $C_{15}H_{22}O_4N_2$: 294.1574 Da [M], found 294.1572 Da (Δ = 0.3 ppm). **IR (ATR):** ṽ = 3373 (w), 2983 (vw), 1739 (m), 1688 (m), 1609 (w), 1512 (m), 1430 (w), 1366 (w), 1296 (w), 1274 (w), 1219 (m), 1166 (m). 1098 (w), 1056 (w), 1027 (w), 976 (w), 819 (w), 785 (w), 733 (w), 700 (w), 534 (m), 486 (w), 404 (w) cm^{-1}.

Boc-Phe(3,5-Cl_2-4-NH_2)-OMe, methyl-(*S*)-3-(4'-amino-3',5'-dichlorophenyl)-2-[(*tert*-butoxycarbonyl)amino]propanoate 51

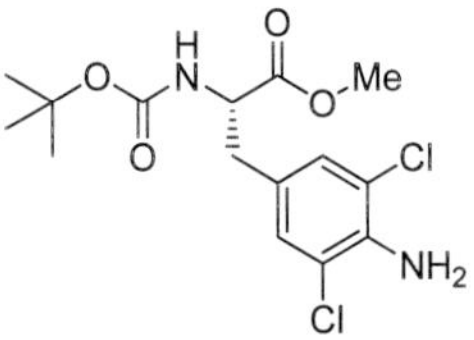

N-Chlorosuccinimide (16.7 g, 125 mmol, 2.40 eq.) was added to a solution[331] of (*S*)-Boc-Phe(4-NH_2)-OMe **50** (15.3 g, 34.0 mmol, 1.00 eq.) in DMF (520 mL). The solution was stirred under argon atmosphere at 55 °C for 2 h. The reaction was quenched by addition of sat. aqueous $NaHCO_3$ (115 mL) and the excess of DMF and water was evaporated under reduced pressure. To the mixture sat. aqueous $NaHCO_3$ (200 mL) was added and extracted with EtOAc (3×200 mL), washed two times with brine (100 mL) and dried over anhydrous Na_2SO_4. After filtration, the solvent was removed under reduced pressure. The crude product was purified *via* silica gel column chromatography (dry load). Starting the column with pure cH (0.50 L) to remove traces of DMF and eluting the product with 1% Et_3N, 20% EtOAc in cH (R_f = 0.20). The combined fractions were evaporated, dried under high vacuum overnight and 11.9 g (32.7 mmol, 63%) (*S*)-Boc-Phe(4-NH_2-3,5-Cl_2)-OMe **51** was obtained as a white solid.

TLC: R_f = 0.2 (1% Et_3N, 20% EtOAc in cH); R_f= 0.4 (1% Et_3N, 30% EtOAc in cH). **^{1}H NMR (300 MHz, $CDCl_3$):** δ = 6.94 (s, 2H), 5.01 (d, J = 8.1 Hz, 1H), 4.48 (q, J = 6.5 Hz, 1H), 4.38 (s, 2H), 3.73 (s, 3H), 2.94 (qd, J = 14.0, 5.8 Hz, 2H), 1.43 (s, 9H) ppm. **^{13}C NMR (75 MHz, $CDCl_3$):** δ = 172.1, 163.2, 139.2, 128.7, 126.3, 119.6, 80.2, 54.5, 52.5, 37.14, 28.4 ppm. **HRMS (FAB+):** *m/z* calcd for $C_{15}H_{20}O_4N_2Cl_2$: 362.0800 Da [M+H], found 362.0802 Da (Δ = 0.3 ppm). **IR (ATR):** ṽ = 3349 (w), 2979 (vw), 1734 (w), 1683 (m), 1619 (w), 1587 (vw), 1522 (w), 1489 (w), 1437 (w), 1413 (vw), 1366 (w), 1350 (w), 1299 (w), 1252 (w), 1210 (w), 1162 (m), 1055 (w), 1015 (w), 998 (w), 975 (vw), 886 (vw), 866 (vw), 849 (w), 777 (w), 609 (w) cm^{-1}.

(*S*)-4-(2'-[(*tert*-Butoxycarbonyl)amino]-3'-methoxy-3'-oxopropyl)-2,6-dichloro-benzenediazonium tetrafluoroborate 84

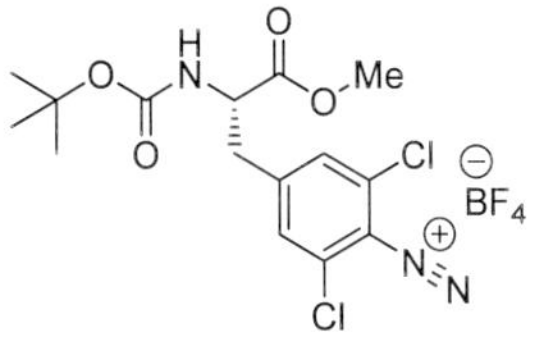

Boc-*L*-(4-amino-3,5-Cl)Phe-OMe **51** (695 mg, 1.92 mmol, 1.00 eq.) was solved in 4.4 mL EtOAc and cooled down to 0 °C. $NOBF_4$ (234 mg, 2.00 mmol, 0.96 eq.) was added in small portions. The solution was stirred for 1 h at 0 °C and was subsequently added to 44 mL of cooled diethyl ether, which was stirred vigorously until the complete reaction mixture was added. The product precipitated and the supernatant solvent was drained. After washing the solid with small amounts of diethyl ether the solid was dried overnight *in vacuo* and was shielded from light to prevent degradation. The diazonium salt was obtained as yellowish solid (708 mg, 1.53 mmol, 80%). cf.[37]

MS (FAB+): *m/z* calcd for $C_{15}H_{18}Cl_2N_3O_4$: 374.07 Da [M+H], found 374.10 Da.

5.6.2 Synthesis of Cl_4-PAP-DKP-Lys_2

(*S*)-sym-Cl_4-$(Boc)_2$-PAP-OMe, dimethyl 3,3'-(((*E*)-diazene-1,2-diyl)bis(3,5-dichloro-4,1-phenylene))(2*S*,2'*S*)-bis(2-((*tert*-butoxycarbonyl)amino)propanoate) 52

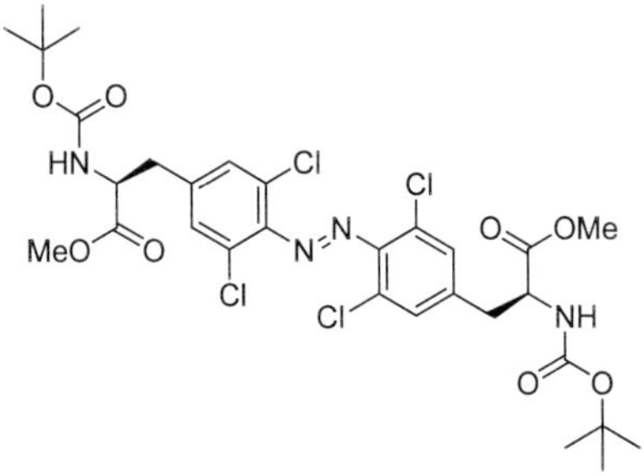

To a solution of (*S*)-Boc-Phe(3,5-Cl_2-4-NH_2)-OMe **51** (11.8 g, 32.5 mmol, 1.00 eq.) dissolved in dry CH_2Cl_2 (500 mL) was added DBU (9.89 g, 65.0 mmol, 2.00 eq.).[221] The solution was stirred at rt for 5 min before being cooled down to −78 °C. Then NCS (8.68 g, 65.0 mmol, 2.00 eq.) was added as a solid to the reaction mixture. The red solution was stirred for 10 min at −78 °C before quenching by addition of a sat. aqueous $NaHCO_3$. The organic layer was separated, washed sequentially with water (100 mL) and 1 M HCl (100 mL), dried over anhydrous Na_2SO_4, and concentrated to dryness *in vacuo*. The crude product was purified by silica gel column chromatography with 5% EtOAc in CH_2Cl_2 (R_f = 0.30) to yield 8.16 g (*S*)-sym-Cl_4-$(Boc)_2$-PAP-OMe **52** (11.3 mmol, 70%) as an orange solid.

^{1}H NMR (400 MHz, DMSO): δ = 7.58 (s, 4H), 7.38 (d, *J* = 8.6 Hz, 2H), 4.33 (ddd, *J* = 10.7, 8.5, 4.6 Hz, 2H), 3.67 (s, 6H), 3.20 – 2.84 (m, 4H), 1.31 (s, 18H) ppm. **^{13}C NMR (101 MHz, DMSO):** δ = 171.9, 155.4, 144.9, 141.8, 130.7, 125.8, 78.5, 54.1, 52.1, 39.5, 35.4, 28.1 ppm. **HRMS (FAB+):** *m/z* calcd for $C_{30}H_{36}O_8N_4Cl_4$: 720.1287 Da [M], found: 720.1288 Da (Δ = 0.2 ppm). **IR (ATR):** ṽ = 3344 (vw), 2974 (vw), 1732 (w), 1678 (w), 1590 (vw), 1520 (w), 1438 (vw), 1399 (vw), 1367 (vw), 1348 (w), 1296 (w), 1272 (w), 1251 (w), 1230 (w), 1160 (w), 1057 (w), 1016 (w), 976 (w), 953 (vw), 926 (vw), 852 (vw), 815 (w), 759 (vw), 740 (vw), 621 (vw), 597 (w), 542 (vw), 436 (vw) cm^{-1}. **UV-Vis (MeCN):** λ_{max} = 297, 461 nm.

Stability of (*S*)-sym-Cl_4-$(Boc)_2$-PAP-OMe 52

The (*S*)-sym-Cl_4-$(Boc)_2$-PAP-OMe **52** (10.0 mg, 0.014 mmol, 1.00 eq.) was dissolved in a solution of 50 vol% morpholine or 20 vol% piperidine in ACN (1.4 mL) and quenched after 20 min with 2 mol/L HCl in aqueous brine (4.5 mL). The precipitate was filtrated, washed with water (3×1 mL) and dried under airstream. The crude was analysed by ^1H NMR and HPLC with the absorption at 220 nm and a gradient 5-95% H_2O to ACN with 0.1% TFA.

The sample in 50 vol% morpholine showed no change by ^1H NMR and HPLC, the sample in 20 vol% piperidine showed a degradation with estimated 88% substrate and 12% side-product by HPLC.

(*S*)-sym-Cl_4-$(Boc)_2$-PAP-OH, (2*S*,2'*S*)-3,3'-(((*E*)-diazene-1,2-diyl)bis(3,5-dichloro-4,1-phenylene))bis(2-((*tert*-butoxycarbonyl)amino)propanoic acid) 53

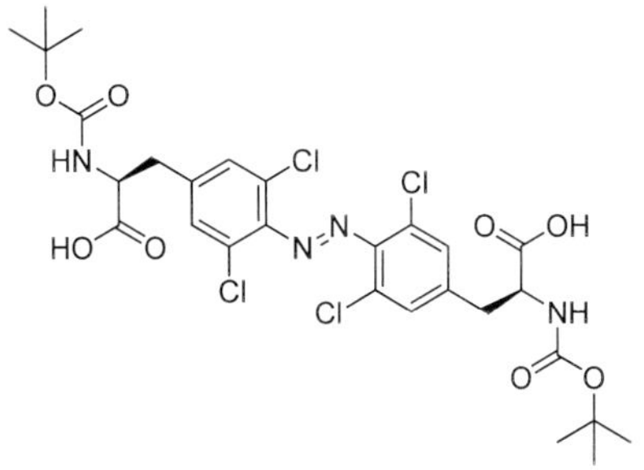

To a solution of (*S*)-sym-Cl_4-$(Boc)_2$-PAP-OMe **52** (2.05 g, 2.84 mmol, 1.00 eq.) in dioxane (41 mL) LiOH (4.08 g, 170 mmol, 60.0 eq.) in H_2O (41 mL) was added and the reaction mixture was stirred for 15 min at rt. The reaction was acidified by adding aqueous solution of HCl (2 mol/L, 93 mL). The product was extracted with EtOAc (2×100 mL), washed with brine (2×100 mL) and the combined organic layers were concentrated *in vacuo*. The crude product was purified by silica gel column (H=7 cm) chromatography (start in pure CH_2Cl_2 $R_f = 0.0$, stepwise gradient 1% MeOH; 2% MeOH; 20% MeOH; 30% MeOH with 1% FA in CH_2Cl_2; $R_f = 0.20$ in 2% MeOH 1% FA in CH_2Cl_2) to yield 1.67 g (2.41 mmol, 85%) (*S*)-sym-Cl_4-$(Boc)_2$-PAP-OH **53** as an red solid.

^1H NMR (400 MHz, DMSO): $\delta = 7.56$ (s, 4H), 7.12 (d, $J = 8.3$ Hz, 2H), 4.18 (td, $J = 9.5$, 8.7, 4.4 Hz, 2H), 3.16 (dd, $J = 13.7$, 4.5 Hz, 2H), 2.90 (dd, $J = 13.6$, 10.4 Hz, 2H), 1.32 (s, 18H) ppm. **^{13}C NMR (101 MHz, DMSO):** $\delta = 172.8$, 155.3, 144.8, 142.4, 130.6, 125.7, 78.2, 54.4, 35.7, 28.1 ppm. **HRMS (FAB+):** *m/z* calcd for $C_{28}H_{32}O_8N_4Cl_4$: 692.0974 Da [M], found: 692.0974 Da ($\Delta = 0.1$ ppm). **IR**

(ATR): $\tilde{\nu}$ = 3352 (vw), 2979 (vw), 1682 (w), 1591 (vw), 1516 (w), 1445 (vw), 1394 (w), 1367 (w), 1249 (w), 1159 (w), 1054 (w), 1025 (w), 848 (w), 803 (w), 567 (w), 463 (w), 435 (w) cm^{-1}.

sym-Cl_4-PAP-OMe, dimethyl 3,3'-(((*E*)-diazene-1,2-diyl)bis(3,5-dichloro-4,1-phenylene))(2*S*,2'*S*)-bis(2-aminopropanoate) trifluoroacetate 84

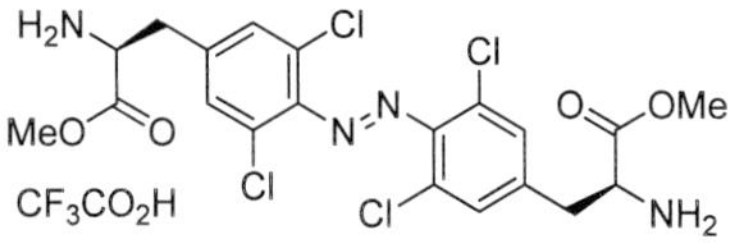

The sym-Cl_4-$(Boc)_2$-PAP-OMe **52** (3.30 g, 4.57 mmol, 1.00 eq.) was dissolved in CH_2Cl_2 (49.5 mL) yielding orange solution. Trifluoroacetic acid (49.5 mL) and 1 vol% Triisopropylsilane (1.00 mL, 4.57 mmol, 1.00 eq.) was added at rt. The mixture was stirred for 1 h at rt. The reaction mixture was diluted with PhMe (100 mL) and evaporated under reduced pressure. The crude product was purified by silica gel column chromatography: H=8 cm, dry load, start in pure CH_2Cl_2 R_f = 0.0, stepwise gradient 1% MeOH (100 mL); 2% MeOH (100 mL); elution in 10-30% MeOH with 1% TFA in CH_2Cl_2 (R_f = 0.13 in 10% MeOH with 1% TFA in CH_2Cl_2). Evaporation of combined fractions and drying under high vacuum overnight resulted in 3.41 g sym-Cl_4-PAP-OMe·TFA **84** (4.55 mmol, quant.) as a red solid.

^{1}H NMR (400 MHz, DMSO): δ = 8.70 (s, 6H), 7.64 (s, 4H), 4.51 (t, J = 6.7 Hz, 2H), 3.77 (s, 6H), 3.24 (ddd, J = 44.4, 14.2, 6.6 Hz, 4H) ppm. **^{13}C NMR (101 MHz, DMSO):** δ = 169.0, 145.5, 138.8, 131.0, 126.1, 52.9, 52.5, 39.5, 34.5 ppm. **HRMS (FAB+):** *m/z* calcd for $C_{20}H_{21}O_4N_4Cl_4$: 521.0317 Da [M], found: 521.0316 Da (Δ = 0.1 ppm). **IR (ATR):** $\tilde{\nu}$ = 2958 (vw), 1748 (w), 1656 (w), 1534 (w), 1435 (w), 1405 (vw), 1296 (vw), 1179 (m), 1133 (m), 1062 (w), 992 (vw), 946 (vw), 881 (w), 837 (w), 818 (w), 801 (w), 721 (w), 648 (vw), 610 (vw), 519 (vw), 418 (vw) cm^{-1}. **UV-Vis (MeCN):** λ_{max} = 294, 457 nm.

sym-Cl$_4$-(Boc)Lys$_2$-(Boc)PAP-OMe, dimethyl 2,2'-(((2*S*,2'*S*)-3,3'-(((*E*)-diazene-1,2-diyl)bis(3,5-dichloro-4,1-phenylene))bis(2-((*tert*-butoxycarbonyl)amino)propanoyl))bis(azanediyl))(2*S*,2'*S*)-bis(6-((*tert*-butoxycarbonyl)amino)hexanoate) 54

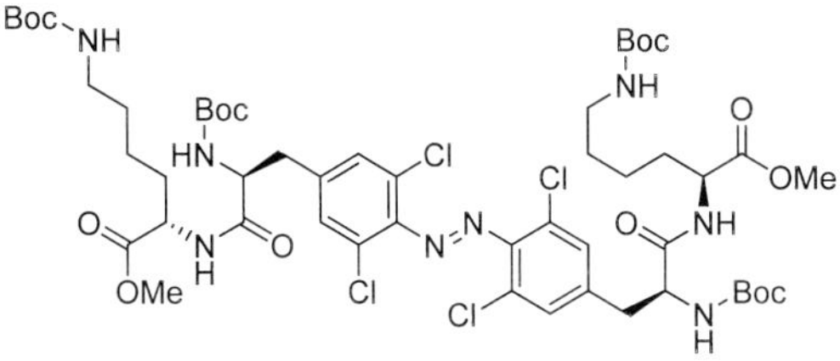

Sym-Cl$_4$-(Boc)$_2$-PAP-OH **53** (1.60 g, 2.30 mmol, 1.00 eq.), HBTU (1.75 g, 4.61 mmol, 2.00 eq.) and DIPEA (1.0 mL, 5.76 mmol, 2.50 eq.) were dissolved in anhydrous DMF (8.0 mL) and stirred for 10 min at rt under argon. The same amount of DIPEA (1.0 mL, 5.76 mmol, 2.50 eq.) was added together with solid *ω*-*N*-Boc-lysine methyl ester hydrochloride (H-Lys(Boc)-OMe·HCl, 1.38 g, 4.63 mmol, 2.01 eq.). The reaction mixture was stirred further at rt under argon and the reaction progress was followed by TLC. After 2 h full conversion of the starting material was observed. The reaction mixture was quenched with sat. aqueous NH_4Cl solution (200 mL) and extracted once with EtOAc (200 mL). The organic layer was washed with sat. aqueous NH_4Cl solution (3×200 mL), brine (1×200 mL), dried with anhydrous Na_2SO_4 and the solvent was removed under reduced pressure. The crude product was purified by silica gel column chromatography with an initial elution of 40% EtOAc in cH, then the product was eluted with EtOAc:cH 1:1 (R_f = 0.2). Evaporation of combined orange fractions and drying under high vacuum resulted in 1.59 g sym-Cl$_4$-(Boc)PAP-(Boc)Lys-OMe **54** (1.35 mmol, 59%) as red solid.

TLC: R_f = 0.2 (cH:EtOAc 1:1). **^{1}H NMR (400 MHz, DMSO):** δ = 8.36 (d, *J* = 7.5 Hz, 2H), 7.62 (s, 3H), 7.55 (d, *J* = 95.8 Hz, 1H), 7.03 (d, *J* = 8.8 Hz, 2H), 6.79 (t, *J* = 5.6 Hz, 2H), 4.26 (td, *J* = 9.5, 7.5, 4.4 Hz, 4H), 3.63 (s, 6H), 3.06 (dd, *J* = 13.7, 3.9 Hz, 2H), 2.91 (d, *J* = 6.4 Hz, 4H), 2.78 (dd, *J* = 13.6, 10.7 Hz, 2H), 1.81 – 1.55 (m, 4H), 1.34 (d, *J* = 26.3 Hz, 44H) ppm. **^{13}C NMR (101 MHz, DMSO):** δ = 172.4, 171.2, 155.5, 155.2, 144.7, 142.3, 130.6, 125.7, 78.2, 77.3, 54.9, 51.9, 51.9, 36.6, 30.6, 29.0, 28.2, 28.0, 22.6 ppm. **HRMS (ESI+):** *m/z* calcd for $[C_{52}H_{77}O_{14}N_8Cl_4]^+$: 1177.4308 Da $[M+H]^+$, found: 1177.4276 Da (Δ = 2.7 ppm). **IR (ATR):** ṽ = 3331 (w), 2934 (vw), 1749 (w), 1686 (m), 1655 (m), 1520 (m), 1463 (w), 1445 (w), 1366 (w), 1342 (w), 1290 (m), 1249 (w), 1222 (w), 1167 (m), 1043 (w), 1023 (w), 1001 (w), 918 (w), 872 (w), 845 (vw), 801 (w), 759 (vw), 640 (w), 434 (vw), 400 (vw) cm^{-1}. **UV-Vis (MeCN):** λ_{max} = 308 nm.

sym-Cl$_4$-PAP-Lys-OMe, dimethyl 2,2'-(((2*S*,2'*S*)-3,3'-(((*E*)-diazene-1,2-diyl)bis(3,5-dichloro-4,1-phenylene))bis(2-aminopropanoyl))bis(azanediyl))(2*S*,2'*S*)-bis(6-aminohexanoate) bis(trifluoroacetate) 54d

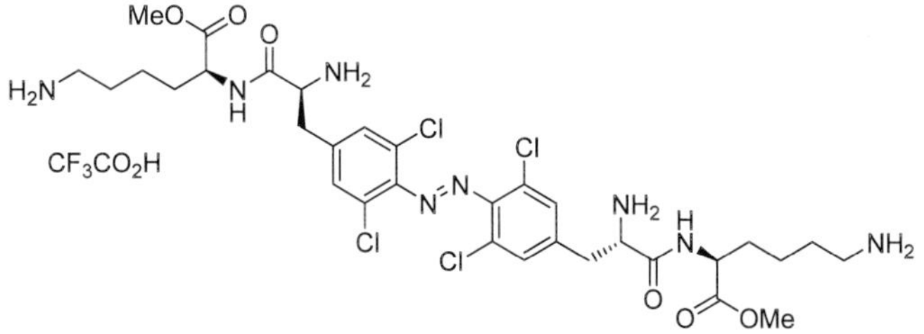

Sym-Cl$_4$-(Boc)PAP-(Boc)Lys-OMe **54** (1.50 g, 1.27 mmol, 1.00 eq.) was dissolved in CH_2Cl_2 (60 mL) yielding an orange solution. Trifluoroacetic acid (TFA, 60 mL) and 1 vol% Triisopropylsilane (1.2 mL) was added at rt. The mixture was stirred for 1 h at rt. The reaction mixture was diluted with PhMe (100 mL) and evaporated under reduced pressure. After evaporation to dryness 2.19 g crude sym-Cl$_4$-PAP-Lys-OMe·TFA$_2$ **54d** (1.27 mmol, quant.) was obtained as TFA salt and was used without further purification in the next reaction.

^{1}H NMR (400 MHz, DMSO): δ = 9.04 (d, *J* = 7.4 Hz, 2H), 8.33 (s, 6H), 7.82 (s, 6H), 7.61 (s, 4H), 4.34 (td, *J* = 8.0, 5.4 Hz, 2H), 4.22 (s, 2H), 3.67 (s, 6H), 3.24 (dd, *J* = 14.2, 4.8 Hz, 2H), 3.06 (dd, *J* = 14.2, 8.1 Hz, 2H), 2.78 (q, *J* = 6.7 Hz, 4H), 1.78 (dt, *J* = 13.9, 7.3 Hz, 2H), 1.67 (dd, *J* = 9.1, 5.4 Hz, 2H), 1.61 – 1.49 (m, 4H), 1.46 – 1.30 (m, 4H) ppm. **^{13}C NMR (101 MHz, DMSO):** δ = 171.9, 167.9, 158.4 (q, *J* = 34.2 Hz), 145.5, 138.8, 131.1, 126.1, 116.3 (d, *J* = 294.7 Hz), 52.8, 52.2, 52.0, 38.5, 35.8, 30.4, 26.6, 22.2 ppm. **HRMS (FAB+):** *m/z* calcd for $C_{32}H_{45}O_6N_8Cl_4$: 777.2216 Da [M+H], found: 777.2215 Da (Δ = 0.1 ppm). **IR (ATR):** ṽ = 3078 (w), 2873 (w), 2958 (w), 1740 (w), 1665 (m), 1555 (m), 1439 (w), 1405 (w), 1356 (vw), 1137 (s), 884 (w), 840 (w), 797 (m), 723 (m), 703 (m), 598 (w), 517 (w), 465 (w), 435 (w), 412 (vw) cm^{-1}. **UV-Vis (MeCN):** λ_{max} = 253, 295 nm

Cl_4-PAP-DKP-Lys_2, (3*S*,3'*S*,6*S*,6'*S*)-6,6'-((((*E*)-diazene-1,2-diyl)bis(3,5-dichloro-4,1-phenylene))bis(methylene))bis(3-(4-aminobutyl)piperazine-2,5-dione) bis(trifluoroacetate) 4

The crude sym-Cl_4-PAP-Lys-OMe·TFA_2 **54d** (2.09 g, 1.27 mmol, 1.00 eq.) was dissolved in 2-butanol (113 mL). It was mixed with glacial AcOH (568 μL, 9.93 mmol, 7.82 eq.), *N*-methylmorpholine (378 μL, 3.44 mmol, 2.63 eq.) and *N,N*-diisopropyl-*N*-ethylamine (681 μL, 3.91 mmol, 3.08 eq.). The resulting mixture was refluxed for 2 h (oil bath, external heating 120 °C), then cooled down. Next, approximately 60% of the solvent was removed under reduced pressure. Cooling down to rt resulted in red gel like precipitation. The precipitate was filtered off, washed with small amount of cold 2-butanol (2×10 mL) on the filter and dried on the filter. The residue was re-dissolved in small amount of AcOH (20 mL) and dropped into stirred CH_2Cl_2 (200 mL) to precipitate the product. Then the precipitate was filtered of, washed with CH_2Cl_2 (3×20 mL) and dried under high vacuum to obtain 0.977 g Cl_4-PAP-DKP-Lys_2 **4** (1.04 mmol, 82%). For the analytics the product was purified by HPLC with following settings: start in 5% ACN with 0.1%TFA in diH_2O with 0.1%TFA for 3 min, then switch within 1 min to isocratic 20% ACN in diH_2O with 0.1%TFA, 15 mL/min, detection at 274 nm, C_{18}-column, 30 °C, retention 11.3 min, c.a. 200 mg of the sample in TFA:diH_2O 1:5 (2 mL) was injected as a gel.

^{1}H NMR (400 MHz, DMSO): δ = 8.29 (s, 2H), 8.25 (s, 2H), 7.68 (s, 6H), 7.51 (s, 4H), 4.34 (s, 2H), 3.83 (s, 2H), 3.09 (ddd, *J* = 62.1, 13.8, 5.0 Hz, 4H), 2.67 (s, 4H), 1.41 (p, *J* = 7.6 Hz, 8H), 1.19 – 0.83 (m, 4H) ppm. **^{13}C NMR (101 MHz, DMSO):** δ = 167.5, 166.6, 145.0, 140.9, 131.2, 125.7, 54.7, 53.4, 39.5, 38.6, 36.9, 31.9, 26.6, 20.5 ppm. **^{1}H NMR (500 MHz, acetic acid-*d*$_4$):** δ = 7.52 (s, 4H), 4.66 (td, *J* = 5.4, 1.6 Hz, 2H), 4.20 (td, *J* = 4.8, 2.2 Hz, 2H), 3.34 (d, *J* = 5.5 Hz, 4H), 3.12 (t, *J* = 7.6 Hz, 4H), 1.84 – 1.72 (m, 4H), 1.50 – 1.36 (m, 8H) ppm. **^{13}C NMR (126 MHz, acetic acid-*d*$_4$):** δ = 179.0, 171.8, 170.5, 148.1, 140.5, 132.9, 128.7, 57.0, 55.7, 41.2, 40.1, 34.7, 27.8, 22.7 ppm. **HRMS (FAB+):** *m/z* calcd for $C_{30}H_{37}O_4N_8Cl_4$: 713.1692 Da [M+H], found: 713.1690 Da (Δ = 0.2 ppm). **IR (ATR):** $\tilde{\nu}$ = 2950 (vw), 2030 (vw), 1671 (w), 1548 (vw), 1436 (vw), 1404 (vw), 1331 (vw), 1199 (w), 1129 (w), 836 (vw), 798 (vw), 721 (vw), 517 (vw), 458 (vw), 433 (vw), 406 (vw) cm^{-1}. **UV-Vis (H_2O):** λ_{max} = 300, 450 nm.

Cl$_4$-Boc-PAP-OMe, (*S*,*E*)-2-((*tert*-butoxycarbonyl)amino)-3-(3,5-dichloro-4-((2,6-dichlorophenyl)diazenyl)phenyl)propanoate 85

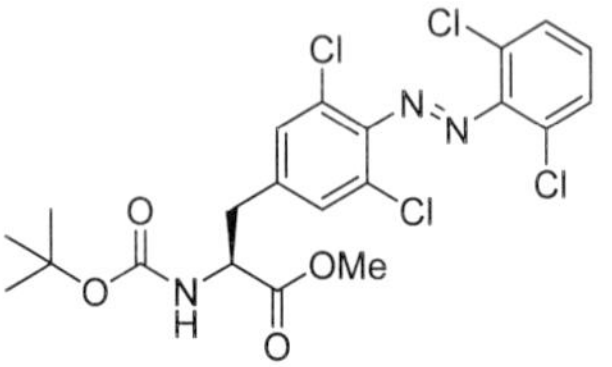

To a solution of methyl (*S*)-3-(4-amino-3,5-dichlorophenyl)-2-((*tert*-butoxycarbonyl)amino)propanoate (0.200 g, 0.551 mmol, 1.00 eq.) and 2,6-dichloroaniline (0.357 g, 2.20 mmol, 4.00 eq.) in dry CH_2Cl_2 (15 mL) DBU (0.822 mL, 5.50 mmol, 10.0 eq.) was added under argon.[221] The solution was stirred at rt for 5 min before being cooled down to −78 °C. The NCS (0.735 g, 5.50 mmol, 10.0 eq.) was added as a solid to the reaction mixture. The orange solution was stirred for 10 min at −78 °C before quenching by addition of a sat. aqueous $NaHCO_3$. The organic layer was separated, washed sequentially with water (10 mL) and 1 mol/L HCl (10 mL), dried over anhydrous Na_2SO_4, and concentrated to dryness *in vacuo*. The crude product was purified by silica gel column chromatography (20% CH_2Cl_2 in cH R_f of (*E*)-1,2-bis(2,6-dichlorophenyl)diazene = 0.30 then switched to 20% EtOAc in cH R_f = 0.30) to yield 187 mg of a mixture. Part of this mixture decomposed in deuterated DMSO while measuring a NMR and the column was repeated with the same conditions to yield 47 mg methyl (*S*,*E*)-2-((*tert*-butoxycarbonyl)amino)-3-(3,5-dichloro-4-((2,6-dichlorophenyl)diazenyl)phenyl)propanoate **85** (0.0902 mmol, 16%) as a red solid. The attempt to reproduce this procedure failed 3 times and was not optimised further.

^{1}H NMR (300 MHz, DMSO): δ = 8.02 (s, 1H), 7.47 (s, 2H), 7.37 (d, *J* = 8.6 Hz, 1H), 6.47 (s, 2H), 4.39 – 4.20 (m, 1H), 3.65 (s, 3H), 3.09 (dd, *J* = 13.7, 4.5 Hz, 1H), 2.90 – 2.75 (m, 1H), 1.31 (s, 9H) ppm. **^{13}C NMR (75 MHz, DMSO):** δ = 172.1, 155.3, 141.5, 138.1, 133.4, 132.6, 132.2, 130.0, 129.6, 113.2, 78.4, 54.2, 52.0, 35.5, 28.1 ppm. **HRMS (FAB+):** *m/z* calcd for $C_{21}H_{22}O_4N_3Cl_4$: 520.0364 Da [M+H], found: 520.0363 Da (Δ = 0.3 ppm). **IR (ATR):** ṽ =3326 (vw), 2978 (vw), 2269 (vw), 1704 (w), 1597 (vw), 1547 (vw), 1453 (w), 1366 (w), 1273 (w), 1209 (w), 1154 (w), 1096 (vw), 1047 (w), 1021 (w), 1000 (w), 961 (vw), 813 (w), 795 (w), 634 (vw), 481 (vw), 426 (vw), 394 (vw) cm^{-1}.

L-Ac-Phe(pNO₂)-OH, (S)-2-acetamido-3-(4-nitrophenyl)propanoic acid 86

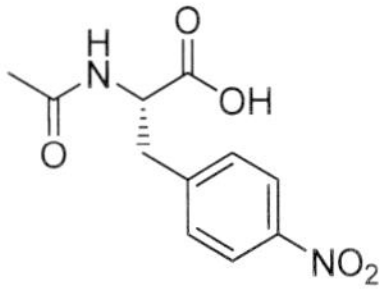

Finely powdered *L*-Phe(pNO$_2$)-OH **81** (13.8 g, 84.0 mmol, 1.00 eq.) was dissolved in Ac_2O (80 mL) in a 250 mL round-bottom flask equipped with a magnetic stirring bar at 25 °C in water bath. To this 2,6-lutidine (0.089 g, 0.83 mmol, 0.01 eq.), 4-DMAP (0.101 g, 0.83 mmol, 1.00 eq.) and $NaHCO_3$ (13.9 g, 165 mmol, 2.00 eq.) were added.[332] The reaction was stirred at 100 °C for 2 h and checked with thin layer chromatography (substrate $R_f = 0.00$, product $R_f = 0.30$ in 10% AcOH in CH_2Cl_2) for completion. The resulting mixture was purified by column chromatography (10% AcOH in CH_2Cl_2, $R_f = 0.30$) to yield 8.17 g (*S*)-2-acetamido-3-(4-nitrophenyl)propanoic acid **86** (32.4 mmol, 39%) white solid.

^{1}H NMR (400 MHz, DMSO): δ = 12.79 (s, 1H), 8.27 (d, *J* = 8.2 Hz, 1H), 8.15 (d, *J* = 8.7 Hz, 2H), 7.51 (d, *J* = 8.7 Hz, 2H), 4.49 (ddd, *J* = 9.8, 8.2, 4.8 Hz, 1H), 3.19 (dd, *J* = 13.8, 4.9 Hz, 1H), 2.97 (dd, *J* = 13.8, 9.8 Hz, 1H), 1.77 (s, 3H) ppm. **^{13}C NMR (101 MHz, DMSO):** δ = 172.8, 169.3, 146.3, 146.2, 130.5, 123.3, 52.9, 22.3 ppm. **HRMS (EI+):** *m/z* calcd for $C_{11}H_{12}O_5N_2$: 252.0746 Da [M], found: 252.0748 Da (Δ = 0.5 ppm). **IR (ATR):** ṽ = 3314 (vw), 2931 (vw), 1728 (w), 1599 (w), 1523 (w), 1442 (w), 1405 (w), 1344 (m), 1256 (vw), 1221 (w), 1183 (w), 1126 (w), 1105 (w), 1014 (vw), 991 (vw), 969 (vw), 876 (w), 845 (w), 791 (vw), 749 (vw), 736 (w), 724 (vw), 696 (w), 637 (w), 592 (w), 519 (w), 498 (w), 396 (vw) cm^{-1}.

L-Ac-Phe(pNH₂)-OH, (S)-2-acetamido-3-(4-aminophenyl)propanoic acid 87

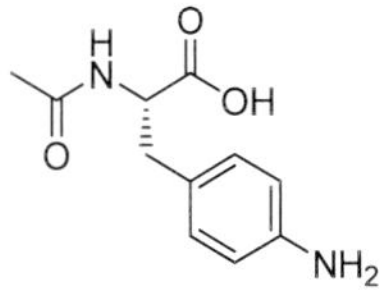

(*S*)-2-acetamido-3-(4-nitrophenyl)propanoic acid **86** (8.05 g, 31.9 mmol, 1.00 eq.) was mixed with 5% Pd on charcoal (81 mg) in a 1000 mL three-neck-flask. The flask was purged three times with argon, then MeOH (130 mL) was added and the mixture was hydrogenated at rt (2 L hydrogen under atmospheric pressure – 1 atm) until total substrate consumption. The temperature of the flask initially increased slightly (to c.a. 45 °C) and after ca. 30 min stirring decreased to rt and stirred for additional 60 min. The substrate consumption was confirmed by TLC (substrate $R_f = 0.30$ in 10% AcOH in CH_2Cl_2), hydrogen was removed by vacuum and the flask was flushed three times with argon. The mixture was filtered over Celite®, washed with methanol (2×150 mL), and the combined colourless organic fractions were evaporated to yield 7.61 g (*S*)-2-acetamido-3-(4-aminophenyl)propanoic acid **87** (31.9 mmol, quant.) as white solid.[332]

^{1}H NMR (400 MHz, DMSO): δ = 8.07 (d, *J* = 8.0 Hz, 1H), 6.86 (d, *J* = 8.2 Hz, 2H), 6.45 (d, *J* = 8.3 Hz, 2H), 4.26 (ddd, *J* = 9.3, 7.9, 5.0 Hz, 1H), 2.83 (dd, *J* = 13.8, 5.0 Hz, 1H), 2.63 (dd, *J* = 13.9, 9.3 Hz, 1H), 1.78 (s, 3H) ppm. **^{13}C NMR (101 MHz, DMSO):** δ = 173.5, 169.2, 147.1, 129.5, 124.5, 113.8, 54.1, 36.2, 22.4 ppm. **HRMS (EI+):** *m/z* calcd for $C_{11}H_{14}O_3N_2$: 222.1004 Da [M found:], 222.1006 Da (Δ = 0.8 ppm). **IR (ATR):** $\tilde{\nu}$ = 3293 (vw), 2568 (vw), 1625 (w), 1593 (w), 1548 (w), 1513 (w), 1398 (w), 1297 (w), 1209 (vw), 1126 (vw), 1021 (vw), 862 (vw), 819 (vw), 710 (vw), 678 (vw), 641 (vw), 614 (vw), 555 (w), 528 (w), 495 (w), 435 (vw), 383 (vw) cm^{-1}.

(*S*,*E*)-2-acetamido-3-(4-(phenyldiazenyl)phenyl)propanoic acid 88

The nitrosobenzene **7** (5.30 g, 49.5 mmol, 1.45 eq.) was suspended in AcOH (150 mL) and the (*S*)-2-acetamido-3-(4-aminophenyl)propanoic acid **87** (7.61 g, 34.2 mmol, 1.00 eq.) was added.[2] The resulting mixture was stirred at rt for 3 d. The reaction mixture was diluted with PhMe (200 mL) and was concentrated under reduced pressure. The resulting mixture was purified by column chromatography (start in pure CH_2Cl_2 and eluted with 3% MeOH in CH_2Cl_2 $R_f = 0.15$) to yield the corresponding non-symmetrical 1.41 g (*S*,*E*)-2-acetamido-3-(4-(phenyldiazenyl)phenyl)propanoic acid **88** (4.54 mmol, 13%) as an orange solid.

^{1}H NMR (500 MHz, DMSO): δ = 12.76 (s, 1H), 8.26 (d, *J* = 8.2 Hz, 1H), 7.92 – 7.84 (m, 2H), 7.82 (d, *J* = 8.4 Hz, 2H), 7.64 – 7.50 (m, 3H), 7.45 (d, *J* = 8.4 Hz, 2H), 4.49 (ddd, *J* = 9.6, 8.1, 5.0 Hz, 1H), 4.36 (ddd, *J* = 9.9, 8.2, 4.9 Hz, 1H), 3.19 – 3.11 (m, 1H), 2.95 (dd, *J* = 13.8, 9.7 Hz, 3H) ppm. **^{13}C NMR (126 MHz, DMSO):** δ = 173.0, 169.2, 152.0, 150.7, 141.8, 131.4, 130.2, 129.5,

122.5, 119.9, 53.2, 36.7, 22.4 ppm. **HRMS (FAB+):** *m/z* calcd for $C_{17}H_{18}O_3N_3$: 312.1348 Da [M+H], found: 312.1349 Da (Δ = 0.2 ppm). **IR (ATR):** $\tilde{\nu}$ = 3325 (vw), 2456 (vw), 1701 (w), 1618 (w), 1547 (w), 1439 (w), 1376 (vw), 1275 (w), 1238 (w), 1117 (vw), 1062 (vw), 930 (vw), 824 (w), 764 (w), 720 (vw), 688 (w), 667 (w), 597 (w), 557 (w), 544 (w), 467 (w), 440 (w), 393 (vw) cm^{-1}.

5.6.3 Photophysical properties Cl_4-PAP-DKP-Lys_2

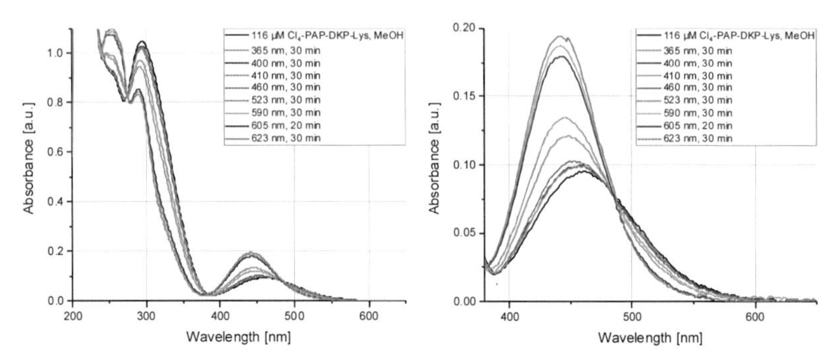

Figure 111. UV-Vis of 116 µM Cl_4-PAP-DKP-Lys_2 in MeOH, irradiation 30 min.

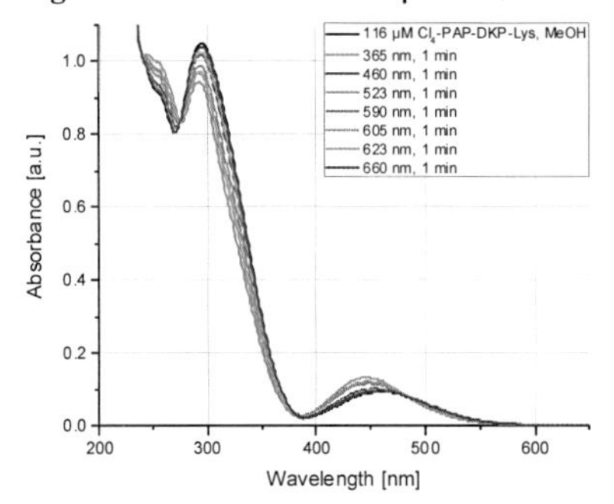

Figure 112. UV-Vis of 116 µM Cl_4-PAP-DKP-Lys_2 in MeOH, irradiation 1 min.

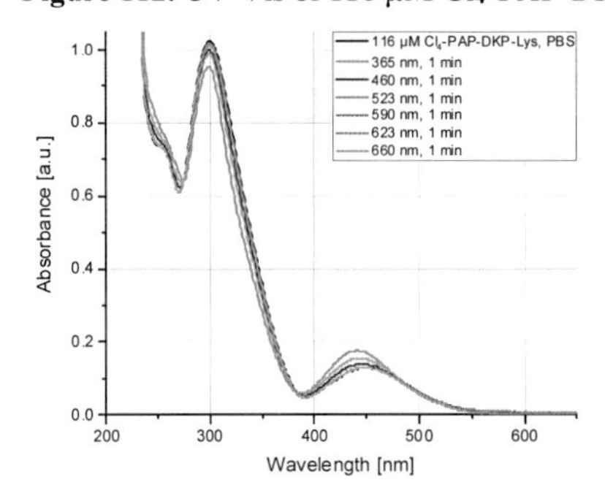

Figure 113. UV-Vis of 116 µM Cl_4-PAP-DKP-Lys_2 in MeOH, irradiation 1 min.

The 'dark state' refers to the *E*-isomer obtained by thermal equilibration during the purification procedure (crystallization from 2-butanol) during synthesis of the Cl_4-PAP-DKP-Lys_2 and its further storage in absence of light. Daylight irradiation results in decrease of the *trans*-isomer ratio in the mixture. By irradiation of the samples with various LED diodes (3 W) from the range 365 – 660 nm, until the photostationary state (PSS) was achieved after 30 min in MeOH.

For the HPLC-based quantification of the *Z*- and *E*-isomer ratio, we have chosen $\lambda = 274$ nm as the isosbestic point with a medium absorption 0.6342 ± 0.0005 au for the 116 μM solution of Cl_4-PAP-DKP-Lys_2 in PBS. The isosbestic point of 116 μM Cl_4-PAP-DKP-Lys_2 in MeOH was at 275 nm with an absorption of 0.831 ± 0.004 au.

Table 20. PSS of 1 mM Cl_4-PAP-DKP-Lys_2 in H_2O, detection 274 nm.

Wavelength	*E* in H_2O	Time	Power
nm	%	min	W
365	19	5	3
380	18	5	3
400	9	1	3
410	**9**	**1**	**3**
420	9	1	3
430	10	1	3
450	14	1	3
470	20	1	3
490	20	10	3
530	39	10	3
590	64	60	3
623	**74**	**60**	**10**
660	-	10	10

A sample of 1 mM Cl_4-PAP-DKP-Lys_2 in H_2O was irradiated for the given time in Table 20 at the given wavelength with a distance of 3 cm at 25 °C inside of a steel block. 25 μL of each solution was injected in the HPLC under following conditions: run of 20 min, gradient 10% H_2O to 40% ACN with 0.1% TFA (4 nm slit, wavelength 274 nm). The sample irradiated with 623 nm contained 74% *Z* and 26% *E*-isomer, and the sample irradiated with 410 nm – 9% *Z* and 91% *E*-isomer, whereas the non-irradiated sample contained over 99% *E*. The identical experiment in PBS buffer showed no significant difference. LED with 660 nm with 10 W showed a slow isomerization, but the PSS was not achieved and in comparison, with the 623 nm LED it was not a useful choice.

Figure 114. Kinetic of isomerization of 1 mM Cl_4-PAP-DKP-Lys_2 in H_2O.

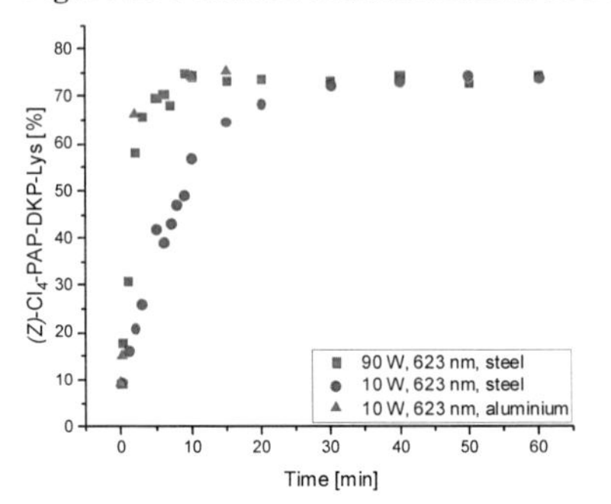

The slow isomerization rate and increased irradiation time of 60 min of the 1 mM Cl_4-PAP-DKP-Lys_2 in H_2O was compared in different setups (Figure 114). An increase of the light power of 1×LED with 700 lm (datasheet of LED Engin) to 9×LEDs, reduced the time to 9 min at 25 °C with a distance of 2 cm.

In general, a decrease of the concentration of the gelator would decrease the required irradiation time to reach the PSS. The same is true for a decreased distance between the LED and the sample, but this would compromise in present setup the cooling efficiency. Another possibility is an increased reflection of the surface inside the irradiation chamber, because the brushed steel in our setup has low reflection in comparison with a polished aluminium surface. Therefore, an identical sample was wrapped in aluminium foil and irradiated with 1×LED with a distance of 3 cm and the PSS was reached after 10 min. In this case an efficient cooling of the sample was not possible.

5.6.3.1 Measurement of the lifetime of Cl_4-PAP-DKP-Lys

The *Z*-isomers of azobenzenes are thermally less stable than the respective *E*-forms and they undergo spontaneous thermal equilibration. However, the rate of the equilibration depends on the substitution pattern on both aromatic rings of the azobenzene system and can span from microseconds to years.[298b] The half-life of an 623 nm irradiated sample of Cl_4-PAP-DKP-Lys_2 was measured at rt (25 °C). In water a half-life of $t_{1/2}$=60.2 ± 0.4 d and in DMSO a half-life of $t_{1/2}$=23.7 ± 0.1 d was observed (Figure 115). This decrease of thermal stability in less polar solvents was previously described by Bleger *et al.* for *ortho*-fluoro-azobenzenes[30a] and predicted by quantum chemical calculations.[333]

In the literature, the reported half-life of a cross-linked peptide with diazene-1,2-diylbis(3,5-dichloro-4,1-phenylene)bis(2-chloroacetamide) with the sequence

WGEACAREAAAREAACRQ at 37 °C was reported with 3.5 h.[30c] The symmetric 2,6,2',6'-tetrachloroazobenzene at 37 °C in DMSO was reported with $t_{1/2} > 20$ h.[37]

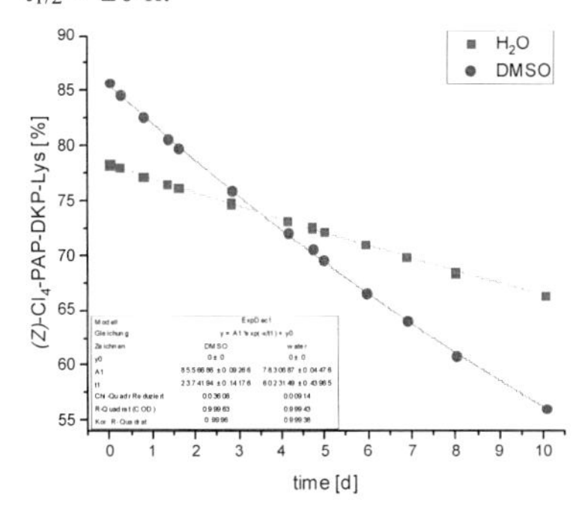

Figure 115. Lifetime of Cl_4-PAP-DKP-Lys_2 in H_2O and DMSO.

5.6.4 Gelation properties of Cl_4-PAP-DKP-Lys_2

The procedure to prepare the hydrogels and measure the gel-to-sol transition temperature was performed according to the procedure described previously (5.4.3 Gelation properties of F_4-PAP-DKP-Lys). The gelation properties of Cl_4-PAP-DKP-Lys_2 in PBS are shown in Table 21. The gel-to-sol transition of a 0.5% hydrogel is 92 ± 2 °C. The hydrogels with 1, 1.5 and 2% are stable at 100 °C by inversion of the vial. The 1.5 and 2% hydrogels can be heated up to 100 °C for 5 min and grabbed with a tweezer and shaken vigorously without a change of the gel structure. The hydrogels with a gel-to-sol transition of 100 °C can be turned into fluid upon boiling the hydrogel inside the closed vial with a heat gun. All given temperature refers to a closed vial to eliminate any loss of water and ensure a reproducible setup with constant concentration.

Table 21. Gelation properties of Cl_4-PAP-DKP-Lys_2 in PBS.

Composition x mg in 500 µL PBS	Approx. C. %	Gel to sol T. °C	deviation T. °C	reproduced n
1.0	0.2	66	9	4
1.5	0.3	77	3	3
2.0	0.4	82	5	3
2.5	0.5	91	2	6
5.0	1.0	100	0	6
7.5	1.5	100	0	6
10.0	2.0	100	0	6

The gelation at rt after boiling the gel is depending on the concentration of the hydrogel. For all hydrogels with 1% or higher the gelation progress was fast enough to invert the vial typically after 15 min. The hydrogel with 0.5% typically required an hour and lower concentrations up to 0.3% required a prolonged incubation at rt overnight. The concentration of 0.2% and lower required up to 7 d to form a hydrogel, which was tested by inversion of the vial.

The decreased concentration of the gelator below 0.5% leads to an increased fatigue of the hydrogel upon several boiling and re-gelation cycles. We observed 33% side-product by HPLC in a diluted 1.0 mol/L solution in PBS, after one boiling cycle. This degradation is dominant in PBS buffer and can be prevented in other buffers. A 1.0 mmol/L solution Cl_4-PAP-DKP-Lys_2 in diH_2O was 3% and in DMSO 2%. For the gelation isotonic buffers, like aqueous 0.9% NaCl solution or isotonic Ringer's solution (RS), is comparable to PBS buffer. The formation of the side-product in aqueous 0.9% NaCl was 3% and in the Ringer's solution 5% upon one boiling cycle. An isotonic buffer containing 10 mM Tris·HCl with 147 mM NaCl (iso. Tris) is not suitable for gelation, but showed also a reduced degradation of 6%. In Table 22 all results are listed about the stability of Cl_4-PAP-DKP-Lys_2 in different buffers.

Table 22. Stability of Cl_4-PAP-DKP-Lys_2 in different buffers.

Cl_4-PAP-DKP-Lys		degradation	boiling cycle	buffer	condition
mmol/L	%	%			after 24 h
1	97	3	1	RS	sol
1	95	5	1	0.9% NaCl	sol
1	94	6	1	iso. Tris	sol
1	67	33	1	PBS	sol
2	95	5	1	RS	slightly viscous
2	96	4	1	0.9% NaCl	viscous sol
2	97	3	1	iso. Tris	sol
2	97	3	1	diH_2O	sol
1	98	2	1	DMSO	solution
1	97	3	2	RS	sol
1	95	5	2	0.9% NaCl	sol
1	94	6	2	iso. Tris	sol
1	66	34	2	PBS	sol
2	95	5	2	RS	slightly viscous
2	92	8	2	0.9% NaCl	viscous sol
2	96	4	2	iso. Tris	sol
2	96	4	2	diH_2O	sol

5.6.4.1 Light-induced gel-to-sol transition

Samples of the hydrogels (0.5-2 wt% of Cl_4-PAP-DKP-Lys_2 in PBS buffer pH 7.4, prepared according to the Table 21) in 1.5-mL-vials (crimp top, 12×32 mm) have been irradiated with red light (623 nm, 10 or 120 W LED diodes, device see Figure 116) at rt (25 °C). The hydrogel sample with a concentration of 0.5% an above remained as a gel and the isomerization rate was only between 21% to 55% of the *Z*-isomer by HPLC. Without the cooling it was possible to switch a 0.5% hydrogel within 10 min to a sol, but as a prove of principle and broad scope of future applications, a light-induced gel-to-sol transition at rt is preferred.

Table 23. Isomerization of Cl_4-PAP-DKP-Lys_2 4 hydrogel in PBS.

Cl_4-PAP-DKP-Lys %	*Z*-isomer %	Time min	Intensity W	condition
0.5	31	30	120	remained gel, upon shaking sol
1	25	30	120	remained gel
1	31	60	120	remained gel
1	55	180	10	remained gel
2	53	60	120	remained gel
2	47	120	120	remained gel
2	21	60	120	remained gel
2	26	120	120	remained gel

Figure 116. Setup for the irradiation (90 or 120 W) of the hydrogel.

For further tests, a concentration of 0.2% Cl_4-PAP-DKP-Lys_2 in 4 different aqueous solutions was prepared and irradiated with 10 W LED, see

Table 24. A light-induced gel-to-sol transition is possible with this setup and with isotonic 0.9% NaCl or isotonic Ringer's solution.

Table 24. Isomerization of Cl_4-PAP-DKP-Lys_2 hydrogel in various buffer.

Z-isomer %	Side-product %	Time min	Intensity W	Buffer	condition before	condition after
72	1	30	10	diH_2O	sol	sol
68	2	30	10	iso. Tris	sol	sol
63	2	30	10	0.9% NaCl	gel (weak)	sol
60	3	30	10	RS	gel (weak)	sol

5.7 Phenylalanine-containing cyclic dipeptides

Methyl *N*^{6}-(*tert*-butoxycarbonyl)-*N*^{2}-((*tert*-butoxycarbonyl)-*L*-phenylalanyl)-*L*-lysinate 89

The (*tert*-butoxycarbonyl)-*L*-phenylalanine (5.00 g, 18.9 mmol, 1.00 eq.), HBTU (7.15 g, 18.9 mmol, 1.00 eq.) and DIPEA (8.2 mL, 47.1 mmol, 2.50 eq.) were dissolved in anhydrous DMF (40 mL) and stirred for 10 min at rt under argon. The same amount of DIPEA (8.2 mL, 47.1 mmol, 2.50 eq.) was added together with *ω*-*N*-Boc-lysine methyl ester hydrochloride (H-Lys(Boc)-OMe·HCl (5.65 g, 19.0 mmol, 1.01 eq.) in DMF (40 mL). The reaction mixture was stirred further at rt under argon and the reaction progress was followed by TLC. After 2 h, full conversion of the starting material was observed. The reaction mixture was quenched with sat. aqueous NH_4Cl solution (200 mL) and extracted once with EtOAc (200 mL). The organic layer was washed with sat. aqueous NH_4Cl solution (3×200 mL), brine (200 mL), dried over anhydrous Na_2SO_4 and the solvent was evaporated under reduced pressure. The crude product was purified by silica gel column chromatography (cH:EtOAc 3:1 to cH:EtOAc 2:1) to yield 8.87 g methyl *N*^{6}-(*tert*-butoxycarbonyl)-*N*^{2}-((*tert*-butoxycarbonyl)-*L*-phenylalanyl)-*L*-lysinate **89** (17.5 mmol, 93%) as a white solid.

^{1}H NMR (300 MHz, DMSO): δ = 8.26 (d, J = 7.5 Hz, 1H), 7.27 (d, J = 4.4 Hz, 4H), 7.19 (t, J = 4.3 Hz, 1H), 6.89 (d, J = 8.6 Hz, 1H), 6.77 (t, J = 5.7 Hz, 1H), 4.22 (t, J = 6.3 Hz, 2H), 3.61 (s, 3H), 3.03 – 2.82 (m, 3H), 2.78 – 2.62 (m, 1H), 1.65 (dd, J = 17.5, 7.0 Hz, 2H), 1.32 (d, J = 22.1 Hz, 22H) ppm. **^{13}C NMR (75 MHz, DMSO):** δ = 172.5, 172.0, 155.6, 155.2, 138.1, 129.2, 128.0, 126.2, 78.0, 77.4, 51.9, 37.3, 30.7, 29.1, 28.3, 28.1, 22.6 ppm. **HRMS (FAB+):** *m/z* calcd for $C_{26}H_{42}O_7N_3$: 508.3023 Da [M+H], found: 508.3025 Da (Δ = 0.4 ppm). **IR (ATR):** $\tilde{\nu}$ =3364 (w), 2983 (vw), 2953 (vw), 2871 (vw), 1737 (w), 1679 (w), 1664 (w), 1516 (w), 1443 (w), 1389 (vw), 1365 (w), 1323 (w), 1250 (w), 1214 (w), 1164 (w), 1130 (w), 1052 (w), 1011 (w), 894 (vw), 860 (vw), 779 (vw), 757 (w), 700 (w), 660 (w), 607 (w), 564 (w), 523 (w), 407 (vw) cm^{-1}.

Methyl *L*-phenylalanyl-*L*-lysinate trifluoroacetate 90

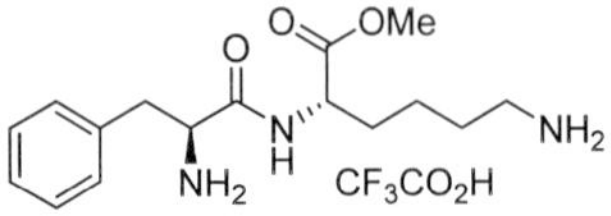

The methyl N^6-(*tert*-butoxycarbonyl)-N^2-((*tert*-butoxycarbonyl)-*L*-phenylalanyl)-*L*-lysinate **89** (8.70 g, 17.1 mmol, 1.00 eq.) was dissolved in CH_2Cl_2 (171 mL) yielding a solution. Trifluoroacetic acid (TFA, 171 mL) and 1 vol% Triiso-propylsilane (3.5 mL, 17.1 mmol, 1.00 eq.) was added at rt and the mixture was stirred for 1 h at rt. The reaction mixture was diluted with PhMe (200 mL) and evaporated under reduced pressure. After evaporation to dryness 10.1 g methyl *L*-phenylalanyl-*L*-lysinate trifluoroacetate **90** (17.1 mmol, quant.) was obtained and was used without further purification in the next reaction.

^{1}H NMR (300 MHz, DMSO): δ = 8.90 (d, *J* = 7.5 Hz, 1H), 8.51 – 8.12 (m, 3H), 7.81 (s, 3H), 7.41 – 7.20 (m, 5H), 4.29 (td, *J* = 8.2, 5.5 Hz, 1H), 4.07 (d, *J* = 3.9 Hz, 1H), 3.11 (dd, *J* = 14.0, 5.8 Hz, 1H), 2.95 (dd, *J* = 14.0, 7.6 Hz, 1H), 2.76 (dd, *J* = 14.2, 5.9 Hz, 2H), 1.86 – 1.43 (m, 4H), 1.34 (q, *J* = 7.7 Hz, 2H) ppm. **^{1}H NMR (400 MHz, acetic acid-*d*$_4$):** δ = 7.48 – 7.09 (m, 5H), 4.65 (t, *J* = 7.0 Hz, 1H), 4.58 (dd, *J* = 9.0, 4.8 Hz, 1H), 3.74 (s, 3H), 3.32 (dd, *J* = 14.4, 6.4 Hz, 1H), 3.23 (dd, *J* = 14.2, 7.3 Hz, 1H), 3.08 (t, *J* = 6.6 Hz, 2H), 1.96 – 1.85 (m, 1H), 1.81 – 1.68 (m, 3H), 1.49 (dd, *J* = 17.4, 8.6 Hz, 2H) ppm. **^{13}C NMR (101 MHz, acetic acid-*d*$_4$):** δ = 179.0, 173.5, 170.8, 135.3, 131.2, 130.4, 129.2, 56.3, 54.1, 53.8, 41.2, 38.5, 32.3, 27.9, 23.6 ppm. **HRMS (FAB+):** *m/z* calcd for $C_{16}H_{26}O_3N_3$: 308.1974 Da [M+H], found: 308.1973 Da (Δ = 0.2 ppm). **IR (ATR):** ṽ = 2957 (w), 1665 (s), 1438 (w), 1135 (s), 838 (w), 798 (m), 746 (w), 722 (m), 702 (m), 598 (w), 518 (w), 437 (w) cm^{-1}.

Phe-DKP-Lys, (3*S*,6*S*)-3-(4-aminobutyl)-6-benzylpiperazine-2,5-dione 5[171]

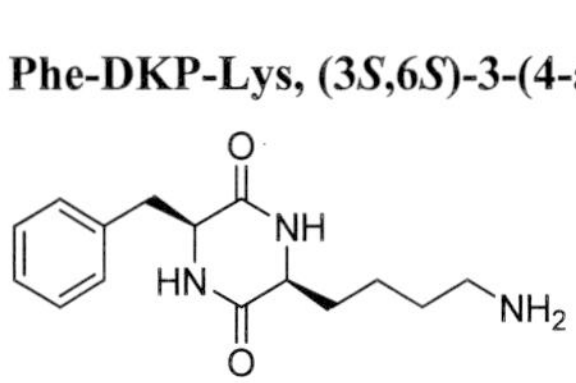

The crude methyl *L*-phenylalanyl-*L*-lysinate trifluoroacetate **90** (8.78 g, 16.4 mmol, 1.00 eq.) was dissolved in 2-butanol (693 mL). It was mixed with glacial AcOH (3.5 mL, 60.5 mmol, 3.69 eq.), *N*-methylmorpholine (2.3 mL, 20.9 mmol, 1.28 eq.) and *N,N*-diisopropyl-*N*-ethylamine (DIPEA, 4.2 mL, 23.8 mmol, 1.45 eq.). The resulting mixture was refluxed for 2 h (oil bath, external heating 120 °C), then cooled down. Next, approximately 60% of the solvent was removed under reduced pressure. Cooling down to rt resulted in opaque gel precipitation, which was filtered off, washed with small amounts of cold ACN (2×10 mL) and dried on the filter. The crude product was purified by repeating this recrystallization procedure with acetonitrile to yield 3.91 g (3*S*,6*S*)-3-(4-aminobutyl)-6-benzylpiperazine-2,5-dione **5** (14.2 mmol, 86%) as a white solid. To obtain analytic purity it was necessary to further purify the product by HPLC with a gradient from 10-20% ACN with 0.1% TFA in diH_2O with 0.1% TFA, retention 13.5 min.

^{1}H NMR (400 MHz, DMSO): δ = 8.19 (d, *J* = 2.2 Hz, 1H), 8.03 (d, *J* = 2.3 Hz, 1H), 7.75 (s, 3H), 7.33 – 7.18 (m, 3H), 7.18 – 7.08 (m, 2H), 4.29 – 4.10 (m, 1H), 3.62 – 3.55 (m, 1H), 3.15 (dd, *J* = 13.4, 3.6 Hz, 1H), 2.83 (dd, *J* = 13.4, 5.0 Hz, 1H), 2.64 – 2.51 (m, 2H), 1.23 (dt, *J* = 15.5, 7.6 Hz, 2H), 1.00 (ddd, *J* = 12.5, 9.9, 4.8 Hz, 1H), 0.69 (dddd, *J* = 24.4, 17.1, 9.2, 5.0 Hz, 3H) ppm. **^{13}C NMR (101 MHz, DMSO):** δ = 166.8, 166.1, 158.3 (q, *J* = 31.2 Hz), 136.2, 130.5, 128.0, 126.7, 117.2 (d, *J* = 299.6 Hz), 55.3, 53.5, 38.7, 38.0, 32.7, 26.6, 20.5 ppm. **HRMS (EI+):** *m/z* calcd for $C_{15}H_{21}O_2N_3$: 275.1634 Da [M], found: 275.1636 Da (Δ = 0.7 ppm). **IR (ATR):** ṽ = 2894 (vw), 1666 (m), 1546 (vw), 1454 (w), 1337 (vw), 1206 (w), 1181 (w), 1132 (w), 835 (w), 798 (w), 758 (w), 721 (w), 705 (w), 584 (vw), 515 (vw), 494 (vw), 444 (w), 413 (vw) cm^{-1}.

5-(*tert*-butyl) 1-Methyl (*tert*-butoxycarbonyl)-*L*-phenylalanyl-*L*-glutamate 91

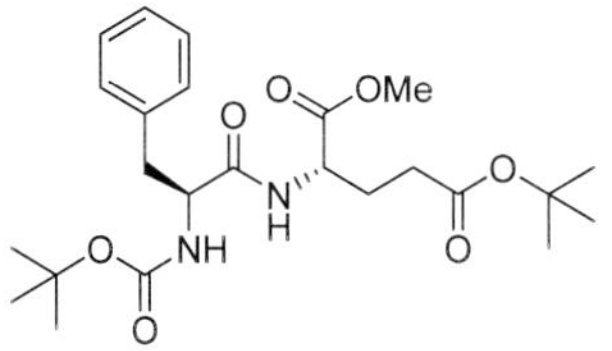

L-Boc-Phe-OH (5.00 g, 18.5 mmol, 1.00 eq.), *L*-H-Glu(OtBu)-OMe.HCl, (4.83 g, 19.0 mmol, 1.01 eq.) and HOBT (3.06 g, 22.6 mmol, 1.20 eq.) were added to a flame-dried round-bottom flask with a stir bar. Then DMF (94 mL, over 4Å molecular sieve) and DIPEA (13.1 mL, 75.4 mmol, 4.00 eq.) were added followed by EDC·HCl (3.97 g, 20.7 mmol, 1.10 eq.). The reaction mixture was stirred further at rt and the reaction progress was followed by TLC. After 48 h, full conversion of the starting material was observed, and the solvent was removed under reduced pressure. The reaction mixture was quenched with sat. aqueous NH_4Cl solution (10 mL) and extracted with EtOAc (3×100 mL). Combined organic layers were washed with sat. aqueous NH_4Cl solution (3×10 mL), brine (1×10 mL), dried over anhydrous Na_2SO_4 and the solvent was removed under reduced pressure. The crude product was purified by silica gel column chromatography (30% EtOAc in cH $R_f = 0.30$) to yield 12.5 g 5-(*tert*-butyl) 1-methyl (*tert*-butoxycarbonyl)-*L*-phenylalanyl-*L*-glutamate **91** (17.4 mmol, 92%) as a transparent fluid with high viscosity.

^{1}H NMR (400 MHz, DMSO): δ = 8.30 (d, *J* = 7.8 Hz, 1H), 7.27 (d, *J* = 4.4 Hz, 4H), 7.19 (q, *J* = 4.2 Hz, 1H), 6.94 (d, *J* = 8.5 Hz, 1H), 4.33 (ddd, *J* = 9.4, 7.7, 4.9 Hz, 1H), 4.17 (ddd, *J* = 10.3, 8.4, 4.3 Hz, 1H), 3.62 (s, 3H), 2.93 (dd, *J* = 13.8, 4.4 Hz, 1H), 2.72 (dd, *J* = 13.8, 10.3 Hz, 1H), 2.28 (t, *J* = 7.6 Hz, 2H), 1.96 (td, *J* = 13.4, 7.8 Hz, 1H), 1.78 (ddd, *J* = 16.5, 14.2, 7.6 Hz, 1H), 1.39 (s, 9H), 1.30 (s, 9H) ppm. **^{13}C NMR (101 MHz, DMSO):** δ = 172.1, 171.5, 155.3, 138.1, 129.2, 128.0, 126.2, 79.8, 78.0, 55.6, 52.0, 50.9, 37.2, 30.8, 28.2, 27.8, 26.3 ppm. **^{1}H NMR (300 MHz, $CDCl_3$):** δ = 7.28 – 7.08 (m, 5H), 6.52 (d, *J* = 7.8 Hz, 1H), 4.91 (s, 1H), 4.48 (td, *J* = 7.8, 4.7 Hz, 1H), 4.29 (d, *J* = 8.0 Hz, 1H), 3.64 (s, 3H), 3.01 (t, *J* = 6.4 Hz, 2H), 2.25 – 1.95 (m, 3H), 1.88 – 1.71 (m, 1H), 1.35 (d, *J* = 5.7 Hz, 18H) ppm. **^{13}C NMR (75 MHz, $CDCl_3$):** δ = 172.1, 171.9, 171.2, 155.4, 136.6, 129.5, 128.8, 127.1, 80.9, 80.4, 55.8, 52.6, 51.8, 38.2, 31.3, 28.4, 28.2, 27.5 ppm. **HRMS (FAB+):** *m/z* calcd for $C_{24}H_{37}O_7N_2$: 465.260 Da [M], found: 465.2599 Da (Δ = 0.4 ppm). **IR (ATR):** ṽ = 3306 (w), 2978 (w), 1727 (m), 1657 (m), 1498 (m), 1454 (w), 1366 (m), 1249 (m), 1151 (s), 1020 (w), 848 (w), 752 (m), 699 (m), 501 (w) cm^{-1}.

(*S*)-4-((*S*)-2-Amino-3-phenylpropanamido)-5-methoxy-5-oxopentanoic acid 92

CF_3CO_2H

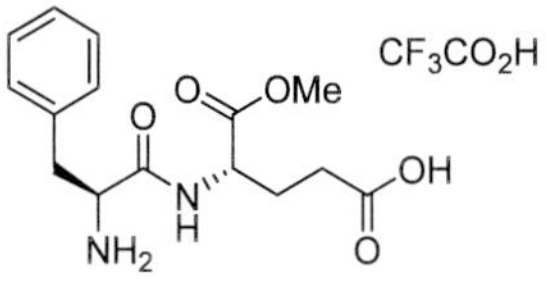

The 5-(*tert*-butyl) 1-methyl (*tert*-butoxycarbonyl)-*L*-phenylalanyl-*L*-glutamate **91** (8.00 g, 17.2 mmol, 1.00 eq.) was dissolved in CH_2Cl_2 (172 mL). Trifluoroacetic acid (TFA, 172 mL) and 1 vol% Triisopropylsilane (3.4 mL, 17.2 mmol, 1.00 eq.) was added at rt. The mixture was stirred for 1 h at rt. The reaction mixture was diluted with PhMe (100 mL) and evaporated under reduced pressure. After evaporation to dryness, 7.87 g (*S*)-4-((*S*)-2-amino-3-phenylpropanamido)-5-methoxy-5-oxopentanoic acid **92** (17.2 mmol, quant.) was obtained as TFA salt.

^{1}H NMR (400 MHz, DMSO): δ = 8.89 (d, *J* = 7.6 Hz, 1H), 8.21 (s, 3H), 7.54 – 7.30 (m, 2H), 7.30 – 7.07 (m, 3H), 4.37 (ddd, *J* = 9.1, 7.6, 5.2 Hz, 1H), 4.06 (s, 1H), 3.63 (s, 3H), 3.10 (dd, *J* = 14.0, 5.9 Hz, 1H), 2.95 (dd, *J* = 14.0, 7.6 Hz, 1H), 2.31 (t, *J* = 7.5 Hz, 2H), 1.99 (dtd, *J* = 13.2, 7.7, 5.3 Hz, 1H), 1.80 (ddt, *J* = 14.0, 9.0, 7.2 Hz, 1H) ppm. **^{13}C NMR (101 MHz, DMSO):** δ = 173.6, 171.5, 168.3, 158.1 (d, *J* = 32.6 Hz), 134.8, 129.5, 128.6, 127.2, 116.8 (d, *J* = 297.3 Hz), 53.3, 52.2, 51.3, 36.9, 29.7, 26.3 ppm. **HRMS (EI+):** *m/z* calcd for $C_{15}H_{21}O_5N_2$: 309.1450 Da [M+H], found: 309.1450 Da (Δ = 0.1 ppm). **IR (ATR):** ṽ = 2957 (w), 1661 (s), 1438 (m), 1176 (s), 1133 (s), 837 (w), 798 (m), 746 (w), 722 (m), 700 (m), 597 (w), 518 (vw), 441 (vw) cm^{-1}.

Phe-DKP-Glu, 3-((2*S*,5*S*)-5-benzyl-3,6-dioxopiperazin-2-yl)propanoic acid 55[171]

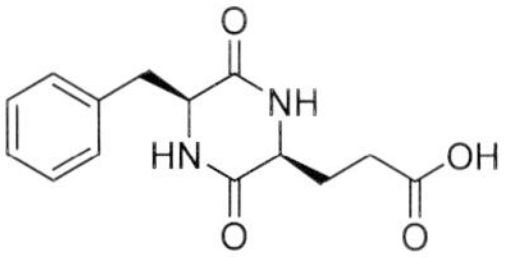

The crude (*S*)-4-((*S*)-2-amino-3-phenylpropanamido)-5-methoxy-5-oxopentanoic acid **92** as a TFA salt (7.20 g, 17.1 mmol, 1.00 eq.) was dissolved in 2-butanol (720 mL). It was mixed with glacial AcOH (3.60 mL, 62.9 mmol, 3.69 eq.), *N*-methylmorpholine (2.40 mL, 21.7 mmol, 1.28 eq.) and *N,N*-diisopropyl-*N*-ethylamine (4.32 mL, 24.8 mmol, 1.45 eq.). The resulting mixture was refluxed for 2 h (oil bath, external heating 120 °C), then cooled down. The solvent was removed under reduced pressure and the crude product was purified by recrystallization in isopropanol (R_f = 0.20 in 30% MeOH, 5% FA in EtOAc) to yield 3.90 g 3-((2*S*,5*S*)-5-benzyl-3,6-dioxopiperazin-2-yl)propanoic acid **55** (14.1 mmol, 83%) as a white solid.

TLC: R_f = 0.2 (developed in30% MeOH, 5% FA in EtOAc). **^{1}H NMR (400 MHz, DMSO):** δ = 11.91 (s, 1H), 8.19 (d, *J* = 2.2 Hz, 1H), 8.08 (d, *J* = 2.2 Hz, 1H), 7.25 (t, *J* = 7.3 Hz, 2H), 7.17 (dd, *J* = 12.5, 7.1 Hz, 3H), 4.19 (s, 1H), 3.74 – 3.62 (m, 1H), 3.13 (dd, *J* = 13.5, 3.9 Hz, 1H), 2.84 (dd, *J* = 13.4, 4.9 Hz, 1H), 1.66 (ddd, *J* = 9.7, 6.4, 4.1 Hz, 2H), 1.33 – 1.18 (m, 1H), 1.05 (ddd, *J* = 15.8, 13.1, 6.4 Hz, 1H) ppm. **^{13}C NMR (101 MHz, DMSO):** δ = 173.8, 166.6, 166.3, 136.0, 130.3, 128.0, 126.8, 55.3, 53.0, 38.0, 28.6, 28.5 ppm. **HRMS (EI+):** *m/z* calcd for $C_{14}H_{16}O_4N_2$: 276.1110 Da [M], found: 276.1108 Da (Δ = 0.7 ppm). **IR (ATR):** ṽ = 3031 (vw), 1738 (w), 1670 (m), 1471 (w), 1454 (vw), 1409 (vw), 1336 (w),

1312 (w), 1273 (w), 1176 (w), 1098 (w), 1035 (vw), 993 (vw), 856 (w), 821 (w), 777 (w), 750 (vw), 701 (w), 643 (vw), 587 (vw), 491 (vw), 454 (m), 423 (w) cm^{-1}.

5.8 Photochromic tyrosine based on spiropyrans

L-Boc-Tyr-OH, (*tert*-butoxycarbonyl)-*L*-tyrosine 93

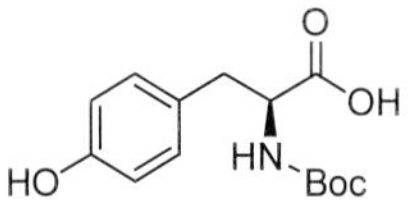

To a solution of *L*-Tyrosine (40.0 g, 220.8 mmol, 1.00 eq.) in 1:1 dioxane:water (500 mL) was added triethylamine (46.2 mL, 331 mmol, 1.50 eq.). The reaction flask was cooled to 0 °C with an ice bath and di-*tert*-butyl-dicarbonate (53.0 g, 242 mmol, 1.10 eq.) was added. After 1 h, the ice bath was removed, and the reaction mixture was stirred at ambient temperature for 20 h. The reaction mixture was then concentrated under reduced pressure and the residue diluted with water and EtOAc. The aqueous layer was acidified to pH 1 with 1.0 mol/L HCl and back extracted with EtOAc. The organic layer was washed with brine, dried over Na_2SO_4 and evaporated to give the 61.6 g (*tert*-butoxycarbonyl)-*L*-tyrosine **93** (219 mmol, 99%) as a white solid, which was used without further purifications.[334]

^{1}H NMR (300 MHz, DMSO): δ = 12.51 (s, 1H), 9.21 (s, 1H), 7.02 (d, *J* = 8.3 Hz, 2H), 6.65 (d, *J* = 8.5 Hz, 2H), 4.00 (ddd, *J* = 13.0, 8.4, 3.2 Hz, 1H), 2.88 (dd, *J* = 13.8, 4.6 Hz, 1H), 2.69 (dd, *J* = 13.8, 10.1 Hz, 1H), 1.32 (s, 9H) ppm. **^{13}C NMR (75 MHz, DMSO):** δ = 173.8, 155.9, 155.5, 130.0, 128.0, 115.0, 78.0, 55.6, 35.7, 28.2 ppm. **HRMS (FAB+):** *m/z* calcd for $C_{14}H_{19}O_5N_1$: 281.1258 Da [M], found: 281.1257 Da (Δ = 0.4 ppm). **IR (ATR):** ṽ = 2976 (vw), 1681 (w), 1614 (vw), 1513 (w), 1445 (vw), 1393 (vw), 1367 (w), 1221 (w), 1153 (w), 1104 (vw), 1052 (vw), 1023 (vw), 826 (vw), 779 (vw), 538 (vw), 491 (vw), 427 (vw), 388 (vw) cm^{-1}.

(*S*)-2-((*tert*-Butoxycarbonyl)amino)-3-(3-formyl-4-hydroxyphenyl)propanoic acid 94

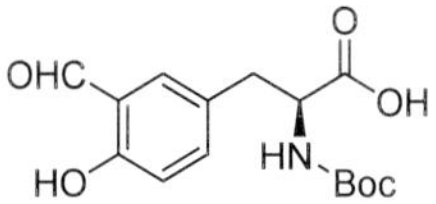

L-Boc-Tyr-OH **93** (57.0 g, 203 mmol, 1.00 eq.) in 600 mL of $CHCl_3$ (stabilized with 1% EtOH) was heated to 65 °C. A solution of NaOH (32.5 g, 812 mmol, 4.00 eq.,) in 114 mL of water was added to the hot reaction mixture (65 °C) and the mixture was refluxed for 1.5 h. After cooling, the solvent was removed under reduced pressure and the reaction mixture was diluted with water and EtOAc (100 mL), and the aqueous layer was acidified to pH 1 with 1 mol/L HCl and back extracted with EtOAc (3×100 mL). The organic layer was washed with brine (3×25 mL), dried over Na_2SO_4 and concentrated under reduced pressure. The product was purified by silica gel column chromatography and (5% MeOH, 1% AcOH in $CHCl_3$) afforded 17.9 g (*S*)-2-((*tert*-butoxycarbonyl)amino)-3-(3-formyl-4-hydroxyphenyl)propanoic acid **94** (57.8 mmol, 28%) as a pale-yellow oil. Unreacted starting material *L*-Boc-Tyr-OH was recovered.[334]

^{1}H NMR (400 MHz, DMSO): δ = 10.57 (s, 1H), 10.23 (s, 1H), 7.52 (d, *J* = 2.3 Hz, 1H), 7.40 (dd, *J* = 8.5, 2.4 Hz, 1H), 7.12 (d, *J* = 8.5 Hz, 1H), 6.91 (d, *J* = 8.4 Hz, 1H), 4.03 (ddd, *J* = 10.2, 8.4, 4.5 Hz, 1H), 2.95 (dd, *J* = 13.8, 4.6 Hz, 1H), 2.75 (dd, *J* = 13.9, 10.3 Hz, 1H), 1.30 (s, 9H) ppm. **^{13}C NMR (101 MHz, DMSO):** δ = 191.3, 173.5, 159.5, 155.5, 137.3, 129.2, 129.0, 121.9, 117.1, 78.1, 55.2, 35.4, 28.1 ppm. **HRMS (EI+):** *m/z* calcd for $C_{15}H_{19}O_6N_1$: 309.1212 Da [M], found: 309.1215 Da (Δ = 0.8 ppm). **IR (ATR):** ṽ = 2978 (vw), 1653 (m), 1591 (w), 1486 (w), 1394 (w), 1368 (w), 1280 (w), 1245 (w), 1149 (m), 1055 (w), 1025 (w), 835 (w), 770 (w), 578 (w), 457 (w) cm^{-1}.

1,2,3,3-Tetramethyl-3*H*-indol-1-ium iodide 95

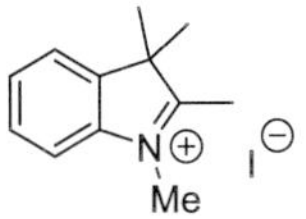

A solution of 2,3,3-trimethylindolenine (10.0 g, 62.8 mmol, 1.00 eq.) and iodomethane (10.7 g, 75.4 mmol, 1.20 eq.) in PhMe (200 mL) was heated at 80 °C for 48 h. The reaction mixture was cooled to rt and the solvent was concentrated under reduced pressure. The residue was suspended in *n*-hexane (100 mL), sonicated for 10 min and filtered off to afford 15.6 g 1,2,3,3-tetramethyl-3H-indol-1-ium iodide **95** (51.7 mmol, 82%) as a pale red solid.[335]

^{1}H NMR (400 MHz, DMSO): δ = 8.01 – 7.87 (m, 1H), 7.88 – 7.77 (m, 1H), 7.68 – 7.56 (m, 2H), 3.97 (d, *J* = 1.1 Hz, 3H), 2.77 (s, 3H), 1.53 (s, 6H) ppm. **^{13}C NMR (101 MHz, DMSO):** δ = 196.0, 142.1, 141.6, 129.3, 128.8, 123.3, 115.1, 53.9, 34.7, 21.7, 14.2 ppm. **HRMS (EI+):** *m/z* calcd for $C_{12}H_{16}N_1$: 174.1277 Da [M], found: 174.1278 Da (Δ = 0.6 ppm).**IR (ATR):** ṽ = 2963 (vw), 1629 (vw), 1608 (vw), 1480 (w), 1455 (w), 1392 (w), 1357 (vw), 1295 (vw), 1235 (vw), 1170 (vw), 1123 (vw), 1092 (vw), 1033 (vw), 1020 (vw), 989 (w), 938 (w), 884 (vw), 817 (vw), 776 (m), 679 (vw), 629 (vw), 598 (vw), 571 (vw), 541 (w),

456 (w) cm^{-1}. **elemental analysis calcd for $C_{12}H_{16}I_1N_1$ (%):** C: 47.86, H: 5.36, I: 42.14, N: 4.65, found: C: 48.08, H: 5.32, N: 4.75.

(Me)Spiro-Tyr(Boc)-OH, (2*S*)-2-((*tert*-butoxycarbonyl)amino)-3-(1',3',3'-trimethylspiro[chromene-2,2'-indolin]-6-yl)propanoic acid 96

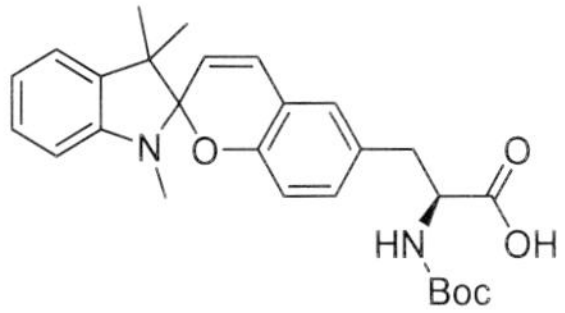

1,2,3,3-tetramethyl-3*H*-indol-1-ium iodide **95** (6.92 g, 23.0 mmol, 1.00 eq.),[336] (*S*)-2-((*tert*-butoxycarbonyl)amino)-3-(3-formyl-4-hydroxyphenyl)propanoic acid **94** (7.10 g, 23.0 mmol, 1.00 eq.), and *N*,*N*-Diisopropylethylamine (8.00 mL, 46.0 mmol, 2.00 eq.) were dissolved in EtOH (230 mL) and heated to 80 °C. After 2 h (TLC $R_f = 0.25$ in 1% AcOH in cH:EtOAc 1:1), the solvent was evaporated under reduced pressure. The crude product was purified by silica gel column chromatography (1% AcOH in cH:EtOAc 1:1) and additionally recrystallised in decalin (ca.50 mg/mL) to yield 7.09 g (2*S*)-2-((*tert*-butoxycarbonyl)amino)-3-(1',3',3'-trimethylspiro[chromene-2,2'-indolin]-6-yl)propanoic acid **96** (15.3 mmol, 66%).

^{1}H NMR (300 MHz, DMSO): δ = 7.12 – 6.92 (m, 6H), 6.76 (t, *J* = 7.4 Hz, 1H), 6.55 (t, *J* = 8.1 Hz, 2H), 5.72 (d, *J* = 10.2 Hz, 1H), 4.04 (tt, *J* = 8.0, 3.9 Hz, 1H), 2.93 (dd, *J* = 13.8, 4.6 Hz, 1H), 2.78 – 2.68 (m, 1H), 2.63 (s, 3H), 1.31 (s, 9H), 1.19 (s, 3H), 1.07 (s, 3H) ppm. **^{13}C NMR (75 MHz, DMSO):** δ = 173.8, 155.4, 152.5, 147.9, 145.0, 136.3, 130.6, 129.7, 129.4, 127.4, 121.4, 119.2, 118.9, 118.1, 113.9, 106.7, 103.8, 77.9, 55.4, 51.3, 33.7, 28.6, 28.2, 25.7 ppm. **HRMS (FAB+):** *m/z* calcd for $C_{27}H_{32}O_5N_2$: 464.2306 Da [M], found: 464.2306 Da (Δ = 0.1 ppm). **IR (ATR):** ṽ = 2923 (vw), 1697 (w), 1607 (w), 1485 (w), 1382 (w), 1365 (w), 1302 (w), 1252 (w), 1161 (m), 1126 (w), 1053 (w), 1021 (w), 966 (w), 927 (w), 821 (w), 741 (w), 648 (vw), 569 (vw), 550 (vw), 458 (vw) cm^{-1}. **UV-Vis (MeCN):** λ_{max} = 223, 295 nm.

(Me)Spiro-Tyr(Boc)-Lys(Boc)-OMe, methyl *N*^{6}-(*tert*-butoxycarbonyl)-*N*^{2}-((2*S*)-2-((*tert*-butoxycarbonyl)amino)-3-(1',3',3'-trimethylspiro[chromene-2,2'-indolin]-6-yl)propanoyl)-*L*-lysinate 97

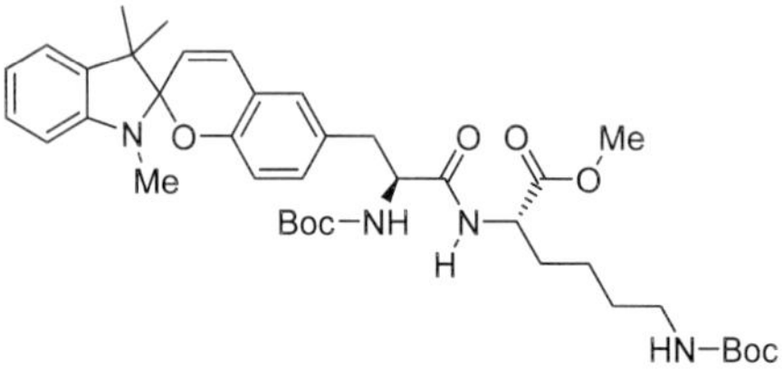

((2*S*)-2-((*tert*-butoxycarbonyl)amino)-3-(1',3',3'-trimethylspiro[chromene-2,2'-indolin]-6-yl)propanoic acid **96** ((Me)Spiro-Tyr(Boc)-CO_2H, 3.43 g, 7.39 mmol, 1.00 eq.), HBTU (2.80 g, 7.39 mmol, 1.00 eq.) and DIPEA (3.22 mL, 18.5 mmol, 2.50 eq.) were dissolved in anhydrous DMF (21.0 mL) and stirred for 15 min at rt under argon. Then DIPEA (1.3 mL, 7.39 mmol, 1.00 eq.) was added together with solid ω-*N*-Boc-lysine methyl ester hydrochloride (H-Lys(Boc)-OMe (1.92 g, 7.39 mmol, 1.00 eq.). The reaction mixture was stirred at rt under argon and the reaction progress was followed by TLC. After 3 h, full conversion of the starting material was observed. The reaction mixture was quenched with sat. aqueous NH_4Cl solution (100 mL) and extracted once with EtOAc (425 mL). The organic layer was washed with sat. aqueous NH_4Cl solution (3×100 mL), brine (1×100 mL), dried with anhydrous Na_2SO_4 and the solvent was removed under reduced pressure. The crude product was purified by silica gel column chromatography (EtOAc:cH) to yield 2.75 g methyl *N*^{6}-(*tert*-butoxycarbonyl)-*N*^{2}-((2*S*)-2-((*tert*-butoxycarbonyl)amino)-3-(1',3',3'-trimethylspiro[chromene-2,2'-indolin]-6-yl)propanoyl)-*L*-lysinate **97** (3.89 mmol, 53%).

^{1}H NMR (300 MHz, DMSO): δ = 8.23 (d, *J* = 7.5 Hz, 1H), 7.15 – 6.97 (m, 4H), 6.97 (d, *J* = 4.3 Hz, 1H), 6.93 (s, 1H), 6.90 – 6.80 (m, 1H), 6.76 (t, *J* = 7.3 Hz, 2H), 6.56 (dd, *J* = 10.2, 7.9 Hz, 2H), 5.74 (d, *J* = 10.2 Hz, 1H), 4.30 – 4.09 (m, 2H), 3.61 (s, 3H), 2.87 (dq, *J* = 8.3, 5.4, 4.1 Hz, 3H), 2.63 (s, 3H), 1.76 – 1.54 (m, 2H), 1.36 (s, 12H), 1.32 – 1.25 (m, 10H), 1.19 (s, 3H), 1.08 (s, 3H) ppm. **^{13}C NMR (75 MHz, DMSO):** δ = 172.50, 171.92, 155.53, 155.15, 152.46, 147.86, 136.33, 130.67, 129.64, 129.39, 127.52, 127.40, 121.41, 119.19, 118.85, 118.03, 113.86, 106.73, 103.78, 77.96, 77.32, 55.55, 51.87, 51.81, 51.29, 51.25, 36.73, 30.66, 29.06, 28.55, 28.26, 28.12, 22.59, 19.90 ppm. **HRMS (FAB+):** *m/z* calcd for $C_{39}H_{55}O_8N_4$: 707.4020 Da [M+H], found: 707.4021 Da (Δ = 0.2 ppm). **IR (ATR):** ṽ = 2980 (w), 1741 (w), 1665 (w), 1587 (w), 1534 (w), 1479 (w), 1458 (w), 1367 (w), 1304 (w), 1255 (w), 1153 (m), 1019 (w), 975 (w), 796 (w), 760 (w), 702 (w), 596 (vw), 567 (vw), 518 (vw), 456 (w), 388 (vw) cm^{-1}.

5.9 3-Arylazopyridinium derivatives

(*E*)-*N*,*N*-Diethyl-4-(pyridin-3-yldiazenyl)aniline 98

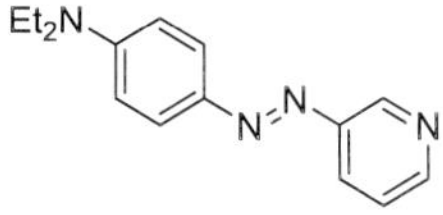

A two-phase solution of aqueous $NaNO_2$ (2.71 g, 39.3 mmol, 1.20 eq.) and diethylaniline (4.69 g, 32.7 mmol, 1.00 eq.) in H_2O (39.3 mL) was added dropwise over 1 h to a stirred solution of 3-aminopyridine **69** (3.70 g, 39.3 mmol, 1.20 eq.) in aqueous HCl (49 mL, 1.0 mol/L, 49.1 mmol, 1.50 eq.) at −5 °C.[301, 315] The mixture was neutralised with 1 mol/L aqueous NaOH. The resulting precipitate was isolated by filtration and washed with water (3×10 mL). The compound was purified by silica gel column chromatography in EtOAc:cH 1:1 (R_f = 0.20) and 5.34 g (*E*)-*N*,*N*-diethyl-4-(pyridin-3-yldiazenyl)aniline **98** (21.0 mmol, 64%) was obtained as an orange solid.

^{1}H NMR (500 MHz, DMSO): δ = 8.96 (dd, *J* = 2.4, 0.8 Hz, 1H), 8.59 (dd, *J* = 4.6, 1.6 Hz, 1H), 8.05 (ddd, *J* = 8.2, 2.4, 1.6 Hz, 1H), 7.80 (d, *J* = 9.2 Hz, 2H), 7.53 (ddd, *J* = 8.2, 4.7, 0.8 Hz, 1H), 6.82 (d, *J* = 9.2 Hz, 2H), 3.47 (q, *J* = 7.0 Hz, 4H), 1.15 (t, *J* = 7.0 Hz, 6H) ppm. **^{13}C NMR (126 MHz, DMSO):** δ = 150.6, 149.9, 148.0, 145.5, 142.2, 126.1, 125.6, 124.3, 111.0, 44.1, 12.5 ppm. **HRMS (FAB+):** *m/z* calcd for $C_{15}H_{18}N_4$: 254.1526 Da [M], found: 254.1527 Da (Δ = 0.2 ppm). **IR (ATR):** ṽ = 3055 (vw), 2963 (w), 1595 (m), 1557 (w), 1513 (m), 1467 (w), 1448 (w), 1423 (w), 1397 (m), 1376 (m), 1358 (m), 1311 (m), 1265 (m), 1192 (m), 1138 (m), 1092 (m), 1071 (m), 1015 (m), 923 (w), 823 (m), 788 (w), 725 (w), 702 (m), 629 (w), 611 (w), 542 (w), 510 (w), 495 (w), 415 (w) cm^{-1}. **elemental analysis calcd for $C_{15}H_{18}N_4$ (%):** C: 70.84, H: 7.13, N: 22.03, found: C: 70.71, H: 7.08, N: 21.83. **UV-Vis (ACN):** λ_{max} = 273, 432 nm.

(*E*)-1-(2-(*tert*-Butoxy)-2-oxoethyl)-3-((4-(diethylamino)phenyl)diazenyl)pyridin-1-ium bromide 56

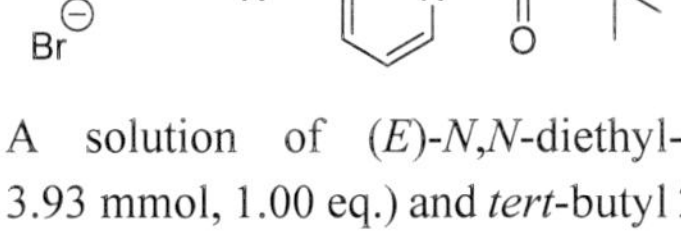

A solution of (*E*)-*N*,*N*-diethyl-4-(pyridin-3-yldiazenyl)aniline **98** (1.00 g, 3.93 mmol, 1.00 eq.) and *tert*-butyl 2-bromoacetate (920 mg, 4.72 mmol, 1.20 eq.) in PhMe[337] (9.8 mL) was stirred at 60 °C for 72 h. After cooling, the solvent was evaporated under reduced pressure. The mixture was dissolved in MeOH (5 mL) and precipitated with diethyl ether (50 mL). The precipitate was filtered of and washed with diethyl ether (3×10 mL) dried under high vacuum to afford 1.71 g (*E*)-1-(2-(*tert*-butoxy)-2-oxoethyl)-3-((4-(diethylamino)phenyl)diazenyl)pyridin-1-ium bromide **56** (3.80 mmol, 97%) as a red solid.

^{1}H NMR (400 MHz, DMSO): δ = 9.48 (s, 1H), 8.98 (d, *J* = 6.1 Hz, 1H), 8.86 (ddd, *J* = 8.5, 2.1, 1.1 Hz, 1H), 8.28 (dd, *J* = 8.5, 5.9 Hz, 1H), 7.86 (d, *J* = 9.2 Hz, 2H), 6.91 (d, *J* = 9.3 Hz, 2H), 5.65 (s, 2H), 3.53 (q, *J* = 7.0 Hz, 4H), 1.48 (s, 9H), 1.18 (t, *J* = 7.0 Hz, 6H) ppm. **^{13}C NMR (101 MHz, DMSO):** δ = 165.2, 152.3, 150.4, 144.5, 142.2, 141.2, 136.2, 128.3, 127.0, 111.7, 83.8, 60.9, 44.5, 27.7, 12.6 ppm. **^{1}H NMR (300 MHz, $CDCl_3$):** δ = 9.59 (d, *J* = 5.9 Hz, 1H), 9.08 (s, 1H), 8.67 (d, *J* = 8.4 Hz, 1H), 8.06 (dd, *J* = 8.6, 6.1 Hz, 1H), 7.87 (d, *J* = 9.5 Hz, 2H), 6.71 (d, *J* = 8.9 Hz, 2H), 6.14 (s, 2H), 3.50 (q, *J* = 7.2 Hz, 4H), 1.52 (s, 9H), 1.31 – 1.20 (m, 6H) ppm. **HRMS (FAB+):** *m/z* calcd for $C_{21}H_{29}O_2N_4$: 369.2285 Da [M], found: 369.2287 Da (Δ = 0.5 ppm). **IR (ATR):** ṽ = 2974 (w), 1734 (m), 1600 (m), 1552 (vw), 1522 (w), 1486 (w), 1387 (m), 1346 (m), 1317 (w), 1274 (w), 1241 (m), 1191 (w), 1140 (m), 1075 (m), 1006 (w), 844 (w), 825 (m), 791 (w), 746 (w), 674 (w), 579 (vw), 540 (vw), 496 (w), 449 (w) cm^{-1}.

(*E*)-1-(Carboxymethyl)-3-((4-(diethylamino)phenyl)diazenyl)pyridin-1-ium trifluoroacetate 56OH

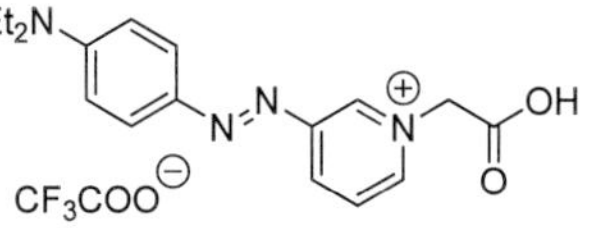

Et_2N N N ⊕N OH O $CF_3COO^{\ominus}$

(*E*)-1-(2-(*tert*-butoxy)-2-oxoethyl)-3-((4-(diethylamino)phenyl)diazenyl)pyridin-1-ium bromide **56** (1.26 g, 3.54 mmol, 1.00 eq.) was dissolved in CH_2Cl_2 (18 mL) yielding orange solution. Trifluoroacetic acid (TFA, 18 mL) and 1 vol% Triisopropylsilane (0.35 mL) was added at rt. The mixture was stirred for 1 h at rt. Then, PhMe (20 mL) was added and the mixture was co-evaporated to dryness. The crude was dissolved in MeOH (10 mL) and precipitated by dropwise addition of the solution into a stirred solution of diethyl ether (100 mL). The precipitate was filtered of and washed with diethyl ether (3×10 mL), dried under high vacuum to afford 1.33 g (*E*)-1-(carboxymethyl)-3-((4-(diethylamino)phenyl)diazenyl)pyridin-1-ium trifluoroacetate **56OH** (3.11 mmol, 88%) as a red solid.

^{1}H NMR (300 MHz, DMSO): δ = 13.70 (s, 1H), 9.49 (s, 1H), 9.01 (d, *J* = 5.9 Hz, 1H), 8.83 (d, *J* = 8.4 Hz, 1H), 8.28 (dd, *J* = 8.5, 5.7 Hz, 1H), 7.86 (d, *J* = 8.7 Hz,

2H), 6.89 (d, $J = 8.7$ Hz, 2H), 5.67 (s, 2H), 3.52 (d, $J = 7.0$ Hz, 4H), 1.17 (s, 6H) ppm. **^{13}C NMR (75 MHz, DMSO):** $\delta = 167.7$, 158.5 (q, $J = 37.3$ Hz), 152.3, 150.5, 144.5, 142.3, 141.1, 136.2, 128.3, 127.1, 115.4 (d, $J = 289.8$ Hz), 111.7, 60.9, 44.5, 26.7 ppm. **HRMS (EI+):** *m/z* calcd for $C_{17}H_{21}O_2N_4$: 313.1659 Da [M], found: 313.1658 Da ($\Delta = 0.2$ ppm). **IR (ATR):** $\tilde{\nu} = 3406$ (vw), 2975 (w), 1734 (w), 1600 (w), 1552 (vw), 1522 (w), 1487 (vw), 1387 (w), 1347 (w), 1274 (w), 1242 (w), 1191 (w), 1141 (m), 1075 (w), 1006 (w), 844 (w), 825 (w), 791 (w), 746 (w), 674 (w), 540 (vw), 496 (vw), 450 (vw) cm^{-1}. **UV-Vis (ACN):** $\lambda_{max} = 285$, 505 nm.

(E)-4-(Pyridin-3-yldiazenyl)phenol 99

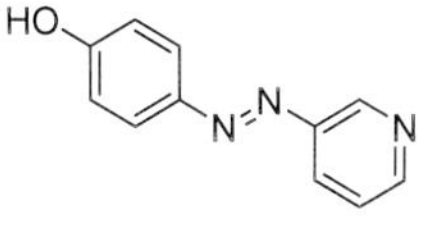

A mixture of aqueous $NaNO_2$ (2.63 g, 38.1 mmol, 1.20 eq.) and phenol (3.00 g, 31.8 mmol, 1.00 eq.) in H_2O (42 mL) with 6.4 mL 1 mol/L aqueous NaOH (1.06 mmol, 0.20 eq.) was added dropwise over 1 h[301, 338] to a stirred solution of 3-aminopyridine **69** (3.59 g, 38.1 mmol, 1.20 eq.) in aqueous 1.75 mol/L HCl (35 mL, 61.0 mmol, 1.92 eq.) at 0 °C and the reaction was stirred for 1 h. The mixture was neutralised with aqueous $NaHCO_3$ and three times extracted with EtOAc (100 mL). The combined organic phase was washed aqueous $NaHCO_3$ (50 mL), with water (50 mL) and brine (50 mL). The organic phase was dried over anhydrous Na_2SO_4 and the solvent was removed under reduced pressure. The compound was purified by column chromatography in EtOAc:cH 1:1. The solvent was removed under reduced pressure, then dried under high vacuum and 4.19 g of the orange *(E)*-4-(pyridin-3-yldiazenyl)phenol **99** (21.0 mmol, 66%) was obtained.

TLC: $R_f = 0.51$ (EtOAc). **^{1}H NMR (400 MHz, DMSO):** $\delta = 10.46$ (s, 1H), 9.03 (d, $J = 2.2$ Hz, 1H), 8.67 (dd, $J = 4.7$, 1.5 Hz, 1H), 8.08 (ddd, $J = 8.2$, 2.4, 1.6 Hz, 1H), 7.83 (d, $J = 8.8$ Hz, 2H), 7.55 (dd, $J = 8.2$, 4.7 Hz, 1H), 6.96 (d, $J = 8.9$ Hz, 2H) ppm. **^{13}C NMR (101 MHz, DMSO):** $\delta = 161.6$, 151.1, 147.5, 145.9, 145.3, 126.6, 125.2, 124.4, 116.0 ppm. **HRMS (EI+):** *m/z* calcd for $C_{11}H_9N_3O$: 199.0747 Da [M], found: 199.0746 Da ($\Delta = 0.8$ ppm). **IR (ATR):** $\tilde{\nu} = 2599$ (w), 1583 (m), 1472 (w), 1433 (w), 1414 (m), 1318 (w), 1279 (m), 1252 (m), 1180 (w), 1139 (m), 1100 (m), 1040 (w), 1021 (w), 951 (w), 842 (m), 817 (w), 800 (m), 767 (m), 725 (w), 696 (m), 626 (w), 557 (m), 529 (w), 493 (m), 437 (w), 411 (vw), 387 (vw) cm^{-1}. **UV-Vis (MeCN):** $\lambda_{max} = 245$, 349 nm.

(*E*)-1-(2-(*tert*-Butoxy)-2-oxoethyl)-3-((4-hydroxyphenyl)diazenyl)pyridin-1-ium bromide 57

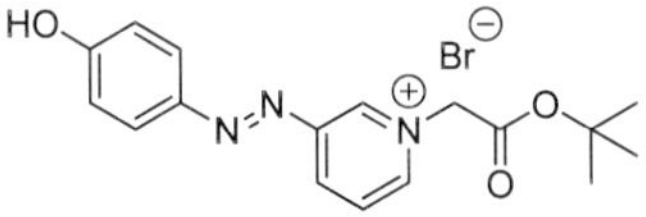

A solution of *(E)*-4-(pyridin-3-yldiazenyl)phenol **99** (1.00 g, 5.02 mmol, 1.0 eq.) and *tert*-butyl 2-bromoacetate (0.979 g, 5.02 mmol, 1.00 eq.) in 25 mL DMF[337] was stirred at 60 °C 24 h. After cooling, the solvent was evaporated under reduced pressure. The residue was dissolved in MeOH (20 ml) and precipitated in diethyl ether (200 mL). The precipitate was filtered of and washed with diethyl ether (3×10 mL), dried under high vacuum to afford 1.63 g (*E*)-1-(2-(*tert*-butoxy)-2-oxoethyl)-3-((4-hydroxyphenyl)diazenyl)pyridin-1-ium bromide **57** (4.12 mmol, 82%) as a red solid.

^{1}H NMR (400 MHz, DMSO): δ = 10.87 (s, 1H), 9.67 (s, 1H), 9.17 (d, J = 6.1 Hz, 1H), 8.96 (dd, J = 8.4, 1.0 Hz, 1H), 8.39 (dd, J = 8.4, 6.0 Hz, 1H), 7.92 (d, J = 8.9 Hz, 2H), 7.05 (d, J = 8.9 Hz, 2H), 5.74 (s, 2H), 1.48 (s, 9H) ppm. **^{13}C NMR (101 MHz, DMSO):** δ = 165.1, 163.5, 149.5, 146.4, 145.1, 142.0, 137.1, 128.5, 126.4, 116.6, 83.8, 61.0, 27.7 ppm. **HRMS (FAB+):** *m/z* calcd for $C_{17}H_{20}O_3N_3$: 314.1505 Da [M], found: 314.1505 Da (Δ = 0.3 ppm). **IR (ATR):** $\tilde{\nu}$ = 3006 (w), 1740 (m), 1637 (vw), 1603 (w), 1581 (w), 1504 (w), 1456 (w), 1415 (w), 1372 (w), 1352 (w), 1279 (w), 1243 (m), 1206 (m), 1138 (m), 843 (m), 785 (w), 747 (w), 670 (m), 638 (w), 585 (vw), 564 (vw), 537 (w), 499 (w), 454 (vw), 388 (vw) cm^{-1}. **UV-Vis (MeCN):** λ_{max} = 255, 375, 536 nm.

(*E*)-4-(Pyridin-3-yldiazenyl)benzene-1,3-diol 100

Procedure cited from master thesis of M. Steinbiß[282]

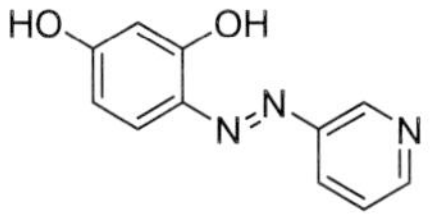

A solution of resorcinol (1,3-dihydroxybenzene) (0.250 g, 2.27 mmol, 1.00 eq.), $NaNO_2$ (0.169 mg, 2.45 mmol, 1.08 eq.) and H_2O (5 mL) was added dropwise to a solution of 3-aminopyridine **69** (0.471 g, 2.73 mmol, 1.20 eq.), conc. HCl (850 µL) and H_2O (5 mL) at 0 °C. The reaction mixture was stirred for 2 h. The mixture was adjusted to pH 6 by adding $NaHCO_3$. The product was extracted with

EtOAc, dried over Na_2SO_4 and the solvent was evaporated. The crystals were purified by silica gel column chromatography (cH:EtOAc 1:1 to EtOAc), resulting in 330 mg (*E*)-4-(pyridin-3-yldiazenyl)benzene-1,3-diol **100** (1.53 mmol, 68%) as orange crystals.

TLC: $R_f = 0.46$ (EtOAc). **^{1}H NMR (300 MHz, DMSO):** δ = 11.74 (s, 1H), 10.64 (s, 1H), 9.06 (d, *J* = 2.1 Hz, 1H), 8.62 (dd, *J* = 4.6, 1.4 Hz, 1H), 8.33 – 8.11 (m, 1H), 7.68 (d, *J* = 8.9 Hz, 1H), 7.54 (dd, *J* = 8.1, 4.7 Hz, 1H), 6.49 (dd, *J* = 8.9, 2.5 Hz, 1H), 6.39 (d, *J* = 2.4 Hz, 1H) ppm. **^{13}C NMR (75 MHz, $DMSO_3$):** δ = 163.7, 157.0, 150.7, 146.7, 145.6, 132.8, 127.8 – 127.3, 126.1, 124.3, 109.3, 102.9 ppm. **MS (ESI+):** *m/z* calcd for $[C_{11}H_{10}N_3O_2]^+$: 216.07 Da $[M+H]^+$, found: 216.09 Da.

(*E*)-1-(2-(*tert*-Butoxy)-2-oxoethyl)-3-((2,4-dihydroxyphenyl)diazenyl)pyridin-1-ium trifluoroacetate 64

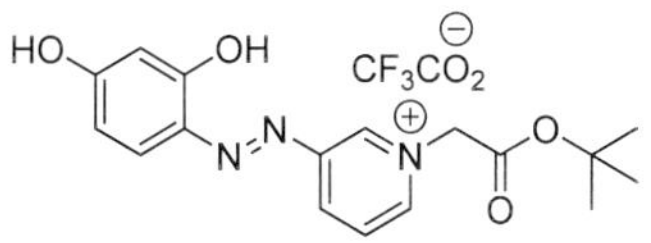

A solution of (*E*)-4-(pyridin-3-yldiazenyl)benzene-1,3-diol **100** (140 mg, 0.651 mmol, 1.00 eq.) and *tert*-butyl 2-bromoacetate (115 µl, 0.781 mmol, 1.20 eq.) in DMF (3.3 mL) was stirred at 50 °C for 3 d. After cooling, the solvent was evaporated under reduced pressure. The mixture dissolved in MeOH (5 mL) and precipitated in diethyl ether (50 mL). The precipitate was filtered of and washed with diethyl ether (3×10 mL). The product (193 mg) was purified by HPLC under standard conditions (retention = 17 min) to obtain 121 mg (*E*)-1-(2-(*tert*-butoxy)-2-oxoethyl)-3-((2,4-dihydroxyphenyl)diazenyl)pyridin-1-ium trifluoroacetate **64** (0.273 mmol, 42%) as a red solid.

^{1}H NMR (400 MHz, DMSO): δ = 11.26 (s, 2H), 9.63 (s, 1H), 9.13 – 9.01 (m, 2H), 8.33 (dd, *J* = 8.5, 6.0 Hz, 1H), 7.69 (d, *J* = 9.1 Hz, 1H), 6.52 (dd, *J* = 9.1, 2.5 Hz, 1H), 6.46 (d, *J* = 2.5 Hz, 1H), 5.64 (d, *J* = 6.0 Hz, 2H), 1.48 (s, 9H) ppm. **^{13}C NMR (101 MHz, DMSO):** δ = 166.5, 165.1, 160.0, 158.2 (q, *J* = 32.3 Hz), 149.7, 145.3, 141.8, 136.8, 133.6, 128.3, 123.2, 110.4, 103.2, 83.9, 61.0, 27.7 ppm. **HRMS (FAB+):** *m/z* calcd for $C_{17}H_{20}O_4N_3$: 330.1454 Da [M], found: 330.1455 Da (Δ = 0.4 ppm). **IR (ATR):** ṽ = 3075 (w), 2981 (w), 1740 (w), 1668 (w), 1614 (m), 1473 (w), 1409 (w), 1315 (w), 1113 (m), 792 (m), 746 (w), 717 (m), 669 (w), 608 (w), 522 (w), 479 (w), 404 (vw) cm^{-1}.

(*E*)-3-((2,4-Dimethoxyphenyl)diazenyl)pyridine 101

A stirred solution of 3-aminopyridine **69** (600 mg, 6.37 mmol, 1.00 eq.) in MeOH (10.8 mL) is placed in an ice-salt bath and treated with 50% aqueous HBF_4 (1.0 mL, 7.65 mmol, 1.20 eq.) and cooled to −5 °C. *tert*-Butyl nitrite (1.0 mL, 7.65 mmol, 1.20 eq.) is added dropwise over 5 min without raising the temperature above 0 °C. Stirring is continued for an additional 5 min and the solution is used directly in the next reaction.[288, 339] The reaction mixture (6.37 mmol, 1.00 eq.) was added dropwise over 5 min to the stirred solution of 1,3-dimethoxybenzene (717 mg, 5.30 mmol, 1.00 eq.), in formic acid (10.6 mL) at 0 °C. The reaction was stirred for 2 h at 0 °C. The mixture was neutralised with sat. sodium bicarbonate (10 mL) and extracted with CH_2Cl_2 (3×10 mL), washed with brine and dried over Na_2SO_4. The solvent was removed under reduced pressure and the crude was purified by column chromatography in EtOAc:cH 1:3 (R_f = 0.25). The 128 mg (*E*)-3-((2,4-dimethoxyphenyl)diazenyl)pyridine **101** (0.526 mmol, 8%) was obtained and used without further purification in the next reaction. To obtain analytic purity the product was further purified by HPLC with a gradient from 5-95% ACN in diH_2O with 0.1% TFA.

^{1}H NMR (400 MHz, DMSO): δ = 9.09 (d, *J* = 2.1 Hz, 1H), 8.74 (dd, *J* = 4.9, 1.5 Hz, 1H), 8.23 (ddd, *J* = 8.3, 2.4, 1.5 Hz, 1H), 7.72 (dd, *J* = 8.2, 4.9 Hz, 1H), 7.68 (d, *J* = 9.0 Hz, 1H), 6.80 (d, *J* = 2.5 Hz, 1H), 6.65 (dd, *J* = 9.1, 2.5 Hz, 1H), 3.99 (s, 3H), 3.89 (s, 3H) ppm. **^{13}C NMR (101 MHz, DMSO):** δ = 164.9, 159.5, 159.2 – 157.8 (m), 148.8, 148.4, 144.1, 136.0, 129.0, 125.4, 117.7, 115.5 (d, *J* = 290.5 Hz), 106.8, 99.1, 56.3, 55.9 ppm. **HRMS (FAB+):** *m/z* calcd for $C_{13}H_{13}O_2N_3$: 243.1008 Da [M], found: 243.1008 Da (Δ = 0.1 ppm). **IR (ATR):** ṽ = 2164 (vw), 1701 (vw), 1600 (w), 1497 (vw), 1474 (vw), 1406 (vw), 1286 (w), 1251 (w), 1180 (w), 1117 (w), 1021 (w), 978 (vw), 907 (vw), 827 (w), 796 (w), 721 (w), 684 (w), 610 (vw), 588 (w), 433 (vw), 383 (vw) cm^{-1}.

1-(4-Phenylpiperazin-1-yl)ethan-1-one 71

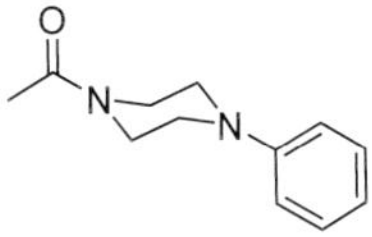

In a round-bottom flask, a mixture of 1-phenylpiperazine (5.00 g, 30.8 mmol, 1.00 eq.) and acetic anhydride (3.49 mL, 37.0 mmol, 1.20 eq.) was stirred at 25 °C for 16 h.[340] After completion of the reaction, as monitored by TLC, the reaction mixture was dissolved in diethyl ether (100 mL) and was allowed to stay in the freezer at −20 °C for 1 h. During this time, crystals of product formed, which were collected by filtration and washed with diethyl ether (3×10 mL). The workup was repeated, and the combined fractions were dried under high vacuum and 5.19 g 1-(4-phenylpiperazin-1-yl)ethan-1-one **71** (25.4 mmol, 83%) was obtained as a white solid.

^{1}H NMR (400 MHz, DMSO): δ = 7.23 (dd, *J* = 8.7, 7.2 Hz, 2H), 6.95 (d, *J* = 7.7 Hz, 2H), 6.81 (t, *J* = 7.3 Hz, 1H), 3.56 (q, *J* = 5.1 Hz, 4H), 3.17 – 3.11 (m, 2H), 3.07 (dd, *J* = 6.2, 4.3 Hz, 2H), 2.04 (s, 3H) ppm. **^{13}C NMR (101 MHz, DMSO):** δ = 168.2, 150.8, 129.0, 119.3, 115.9, 48.7, 48.3, 45.5, 40.7, 21.2 ppm. **HRMS (EI+):** *m/z* calcd for $C_{12}H_{16}N_2O_1$: 204.1257 Da [M+H], found: 204.1259 Da (Δ = 0.9 ppm). **IR (ATR):** ṽ = 3000 (vw), 2816 (w), 1624 (w), 1597 (w), 1494 (w), 1427 (m), 1385 (w), 1340 (w), 1278 (w), 1255 (w), 1225 (m), 1156 (w), 1108 (w), 1046 (w), 1000 (w), 976 (w), 906 (w), 768 (m), 722 (w), 698 (m), 635 (vw), 587 (w), 526 (w), 502 (w), 434 (w) cm^{-1}.

(*E*)-1-(4-(4-(Pyridin-3-yldiazenyl)phenyl)piperazin-1-yl)ethan-1-one 72

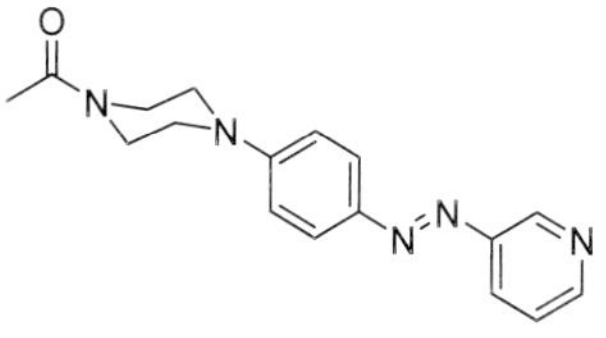

A solution of *p*-sulfonic acid (pTsOH) (7.28 g, 38.3 mmol, 1.20 eq.) in formic acid (29 mL) was slowly added, over a period of 10 min, to a stirred solution of 3-aminopyridine **69** (3.00 g, 31.9 mmol, 1.00 eq.) in formic acid (40 mL), previously cooled to 0 °C in an ice bath. *tert*-butyl nitrite (4.55 mL, 38.3 mmol, 1.20 eq.) was added dropwise at such a rate, that the temperature did not exceed 5 °C (about 10 min overall addition time).[287, 341] White precipitate appeared immediately. After the addition was complete, the cooling bath was removed and the mixture was stirred for further 5 min. Et_2O (500 mL, filtered before use from the remaining particles of KOH as drying agent) was added to complete the precipitation of the diazonium salt. The salt was gathered by filtration on a Büchner funnel and washed with Et_2O (3×20 mL) to remove the remaining formic acid. After drying under reduced pressure, the dry crude 9.01 g pyridine-3-diazonium 4-methylbenzenesulfonate **70** (32.5 mmol, 85%) was obtained and stored at −20 °C under argon. (Note diazonium salts can decompose and safety precautions are necessary, but no decomposition was observed with 4-methylbenzenesulfonate salts.)

HRMS (ESI+): *m/z* calcd for $[C_{12}H_{12}N_3O_4{}^{31}S_1]^+$: 278.0594 Da $[M+H]^+$, found: 278.0594 Da ($\Delta = 0.2$ ppm).

The dry diazonium salt **70** (9.01 g, 32.5 mmol, 1.32 eq.) was added to a solution of N^1-acetyl-N^4-phenylpiperazine **71** (5.03 g, 24.6 mmol, 1.00 eq.) in ACN (123 mL) at 0 °C. The reaction mixture became red within 5 min from the addition time and was stirred for further 2 h at rt. The reaction mixture was quenched with aqueous $NaHCO_3$ and extracted three times with CH_2Cl_2. The organic phase was evaporated under reduced pressure and purified by column chromatography starting with EtOAc and eluting with CH_2Cl_2. After drying under high vacuum, the red solid of 5.84 g (*E*)-1-(4-(4-(pyridin-3-yldiazenyl)phenyl)piperazin-1-yl)ethan-1-one **72** (18.9 mmol, 77%) was obtained.

^{1}H NMR (500 MHz, DMSO): $\delta = 9.01$ (d, $J = 2.3$ Hz, 1H), 8.64 (dd, $J = 4.7$, 1.6 Hz, 1H), 8.09 (ddd, $J = 8.2$, 2.5, 1.6 Hz, 1H), 7.85 (d, $J = 9.1$ Hz, 2H), 7.56 (dd, $J = 8.1$, 4.7 Hz, 1H), 7.11 (d, $J = 9.2$ Hz, 2H), 3.61 (t, $J = 5.2$ Hz, 4H), 3.47 (t, $J = 5.2$ Hz, 2H), 3.40 (t, $J = 5.4$ Hz, 2H), 2.06 (s, 3H) ppm. **^{13}C NMR (126 MHz, DMSO):** $\delta = 168.4$, 153.0, 150.7, 147.7, 145.8, 144.1, 126.4, 124.9, 124.4, 114.0, 46.7, 46.4, 45.0, 40.4, 21.2 ppm. **^{1}H NMR (300 MHz, $CDCl_3$):** $\delta = 9.10$ (d, $J = 1.9$ Hz, 1H), 8.61 (d, $J = 4.6$ Hz, 1H), 8.06 (d, $J = 8.1$ Hz, 1H), 7.88 (d, $J = 8.9$ Hz, 2H), 7.38 (dd, $J = 8.2$, 4.7 Hz, 1H), 6.94 (d, $J = 9.0$ Hz, 2H), 3.86 – 3.70 (m, 2H), 3.70 – 3.55 (m, 2H), 3.36 (dt, $J = 10.3$, 5.2 Hz, 4H), 2.13 (s, 3H) ppm. **^{13}C NMR (75 MHz, $CDCl_3$)**: $\delta = 169.1$, 153.0, 150.7, 148.2, 147.0, 145.7, 126.6, 125.1, 123.9, 114.8, 47.9, 47.7, 45.8, 41.0, 21.4 ppm. **HRMS (FAB+):** *m/z* calcd for $C_{17}H_{20}N_5O_1$: 310.1662 Da [M+H], found: 310.1662 Da ($\Delta = 0.1$ ppm). **IR (ATR):** $\tilde{\nu} = 2855$ (w), 1633 (m), 1597 (m), 1513 (w), 1426 (w), 1383 (w), 1285 (w), 1234 (m), 1151 (m), 1110 (w), 1031 (w), 998 (w), 973 (w), 935 (w), 903 (w), 815 (m), 720 (w), 703 (w), 661 (w), 617 (w), 601 (w), 567 (w), 543 (w), 532 (w), 492 (w), 403 (w) cm^{-1}. **UV-Vis (ACN):** $\lambda_{max} = 258$, 402 nm.

(*E*)-3-((4-(4-Acetylpiperazin-1-yl)phenyl)diazenyl)-1-(2-(*tert*-butoxy)-2-oxoethyl)pyridin-1-ium bromide 60

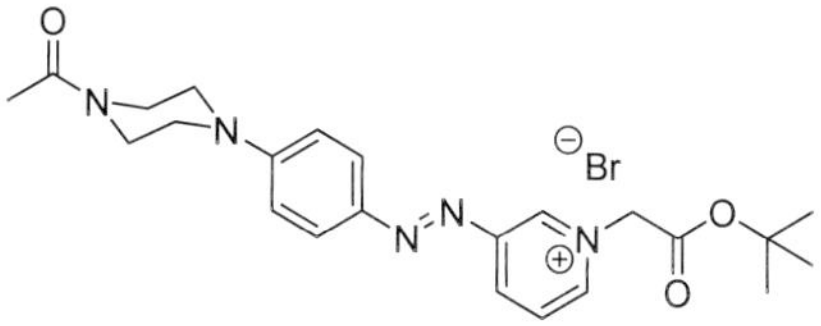

A solution of (*E*)-1-(4-(4-(pyridin-3-yldiazenyl)phenyl)piperazin-1-yl)ethan-1-one **72** (5.23 g, 16.9 mmol, 1.00 eq.) and *tert*-butyl 2-bromoacetate (20.6 g, 20.6 mmol, 1.20 eq.) in ACN:DMF 1:1 (100 mL) was stirred at rt for 4 d.[337] After cooling, the solvent was evaporated under reduced pressure. The mixture was dissolved in MeOH (20 mL) and precipitated with diethyl ether (200 mL). The precipitate was filtered of and washed with diethyl ether (3×20 mL), dried under high vacuum to afford 6.54 g (*E*)-3-((4-(4-acetylpiperazin-1-yl)phenyl)diazenyl)-1-(2-(*tert*-butoxy)-2-oxoethyl)pyridin-1-ium bromide **60** (13.0 mmol, 77%) as a red solid.

^{1}H NMR (400 MHz, DMSO): δ = 9.55 (s, 1H), 9.04 (d, J = 6.0 Hz, 1H), 8.91 (d, J = 9.1 Hz, 1H), 8.33 (dd, J = 8.5, 6.0 Hz, 1H), 7.91 (d, J = 9.2 Hz, 2H), 7.15 (d, J = 9.4 Hz, 2H), 5.66 (s, 2H), 3.74 – 3.43 (m, 8H), 2.06 (s, 3H), 1.48 (s, 9H) ppm. **^{13}C NMR (101 MHz, DMSO):** δ = 168.6, 165.2, 154.1, 150.0, 145.3, 143.4, 141.5, 136.7, 128.4, 126.4, 113.5, 83.8, 61.0, 46.2, 45.93, 44.8, 40.3, 27.7, 21.3 ppm. **HRMS (FAB+):** *m/z* calcd for $C_{23}H_{30}N_5O_3$: 424.2350 Da [M], found: 424.2349 Da (Δ = 0.2 ppm). **IR (ATR):** $\tilde{\nu}$ = 2973 (vw), 1741 (w), 1632 (w), 1599 (m), 1518 (w), 1431 (w), 1388 (m), 1362 (m), 1314 (w), 1238 (m), 1221 (m), 1142 (m), 1109 (m), 999 (w), 908 (w), 832 (w), 749 (w), 667 (w), 584 (w), 568 (w), 535 (w), 505 (w), 440 (vw) cm^{-1}. **UV-Vis (MeCN):** λ_{max} = 278, 480 nm.

(*E*)-3-((4-(4-Acetylpiperazin-1-yl)phenyl)diazenyl)-1-(carboxymethyl)pyridin-1-ium bromide 60OH

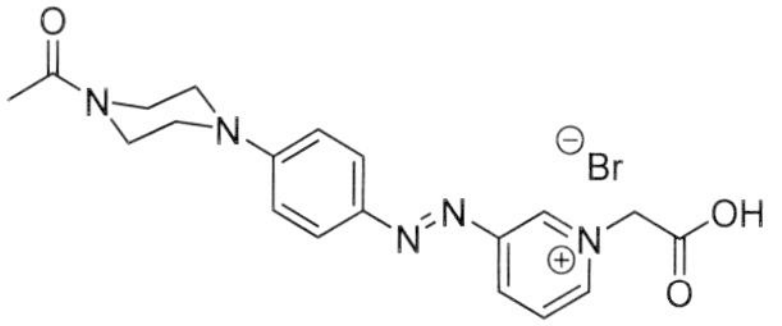

(*E*)-3-((4-(4-acetylpiperazin-1-yl)phenyl)diazenyl)-1-(2-(*tert*-butoxy)-2-oxoethyl)pyridin-1-ium bromide **60** (3.53 g, 7.00 mmol, 1.00 eq.) was dissolved in CH_2Cl_2 (60 mL) yielding an orange solution. Trifluoroacetic acid (60 mL) and 1 vol% Triisopropylsilane (1.2 mL) was added at rt. The mixture was stirred for 1 h at rt. Then, chloroform (50 mL) was added and the mixture was co-evaporated to dryness. The resulting crude was dissolved in a minimal amount of MeOH (20 mL) and precipitated by addition of the solution into stirred diethyl ether (200 mL). The precipitate was filtered off, washed with diethyl ether (3×20 mL) and dried under high vacuum to obtain 3.08 g (*E*)-3-((4-(4-acetylpiperazin-1-yl)phenyl)diazenyl)-1-(carboxymethyl)pyridin-1-ium bromide **60OH** (6.87 mol, 98%) as a red powder.

^{1}H NMR (400 MHz, DMSO): $\delta = 9.55$ (s, 1H), 9.07 (d, $J = 6.0$ Hz, 1H), 8.93 – 8.84 (m, 1H), 8.31 (dd, $J = 8.5$, 6.0 Hz, 1H), 7.89 (d, $J = 9.2$ Hz, 2H), 7.13 (d, $J = 9.3$ Hz, 2H), 5.67 (s, 2H), 3.74 – 3.46 (m, 8H), 2.05 (s, 3H) ppm. **^{13}C NMR (101 MHz, DMSO):** $\delta = 168.6$, 167.5, 154.1, 150.0, 145.3, 143.5, 141.4, 136.4, 128.3, 126.4, 113.6, 61.2, 46.2, 46.0, 44.9, 21.3 ppm. **HRMS (FAB+):** *m/z* calcd for $C_{19}H_{22}O_3N_5$: 368.1717 [M], found: 368.1717 Da ($\Delta = 0.1$ ppm). **IR (ATR):** $\tilde{\nu} = 3380$ (w), 3012 (w), 1731 (w), 1593 (s), 1510 (m), 1441 (m), 1359 (s), 1214 (s), 1136 (s), 994 (m), 902 (m), 824 (m), 795 (m), 718 (w), 696 (m), 597 (w), 565 (m), 537 (w), 501 (w) cm^{-1}.

tert-Butyl (*S*)-2-((((9*H*-fluoren-9-yl)methoxy)carbonyl)amino)-4-oxo-4-(4-phenylpiperazin-1-yl)butanoate 71D

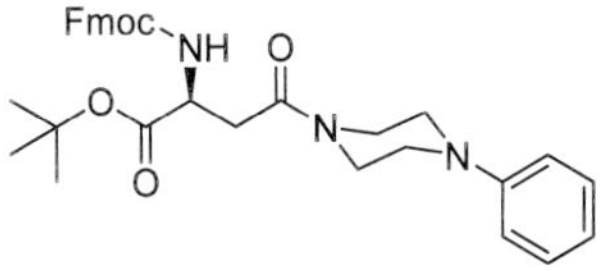

Fmoc-*L*-aspartic acid 4-*tert*-butyl ester (3.00 g, 7.29 mmol, 1.00 eq.), HBTU (2.77 g, 7.29 mmol, 1.00 eq. and DIPEA (3.17 mL, 18.2 mmol, 2.50 eq.) was dissolved in anhydrous CH_2Cl_2:DMF 1:1 (18.2 mL). The reaction was stirred for 10 min at rt under argon. A solution of 1-phenylpiperazine (1.18 g, 7.29 mmol, 1.00 eq.) and DIPEA (3.17 mL, 18.2 mmol, 2.50 eq.) dissolved in CH_2Cl_2 (18.2 mL) was added to the reaction mixture. The reaction was stirred at rt for 2 h and quenched with aqueous ammonium chloride (12 mL), diluted with CH_2Cl_2 (24 mL). The aqueous phase was extracted with CH_2Cl_2 (3×100 mL) and the organic phase was washed with sat. aqueous ammonium chloride (10 mL), brine (10 mL), dried with anhydrous Na_2SO_4 and evaporated under reduced pressure. The crude product was purified by silica gel column chromatography (start 10% EtOAc in cH and eluting with 40% EtOAc in cH $R_f = 0.20$) to yield 4.00 g *tert*-butyl (*S*)-2-((((9*H*-fluoren-9-yl)methoxy)carbonyl)amino)-4-oxo-4-(4-phenylpiperazin-1-yl)butanoate **71D** (7.14 mmol, 98%) as a white solid.

^{1}H NMR (300 MHz, acetonitrile-*d*$_3$): δ = 7.81 (d, *J* = 7.5 Hz, 2H), 7.65 (d, *J* = 7.4 Hz, 2H), 7.32 (ddt, *J* = 31.1, 15.7, 6.9 Hz, 6H), 6.86 (dd, *J* = 18.7, 7.6 Hz, 3H), 6.12 (d, *J* = 8.9 Hz, 1H), 4.41 (dt, *J* = 25.2, 5.6 Hz, 3H), 4.22 (t, *J* = 6.2 Hz, 1H), 3.80 – 3.34 (m, 4H), 3.30 – 2.84 (m, 5H), 2.72 (dd, *J* = 16.6, 3.8 Hz, 1H), 1.41 (s, 9H) ppm. **^{13}C NMR (75 MHz, acetonitrile-*d*$_3$):** δ = 171.3, 169.4, 156.9, 152.1, 145.0, 142.1, 130.0, 128.7, 128.1, 126.1, 120.9, 120.8, 117.2, 82.1, 67.2, 52.2, 50.1, 49.8, 47.9, 45.9, 42.1, 35.9, 28.2 ppm. **HRMS (FAB+):** *m/z* calcd for $C_{33}H_{37}O_5N_3$: 555.2728 Da [M], found: 555.2726 Da (Δ = 0.3 ppm). **IR (ATR):** ṽ = 2975 (vw), 1718 (w), 1638 (w), 1597 (w), 1495 (w), 1446 (w), 1390 (w), 1367 (w), 1227 (w), 1149 (m), 1026 (w), 991 (w), 916 (vw), 845 (w), 755 (w), 733 (m), 692 (w), 619 (vw), 522 (vw), 425 (vw) cm^{-1}.

***tert*-Butyl (*S*,*E*)-2-((((9*H*-fluoren-9-yl)methoxy)carbonyl)amino)-4-oxo-4-(4-(4-(pyridin-3-yldiazenyl)phenyl)piperazin-1-yl)butanoate 72D**

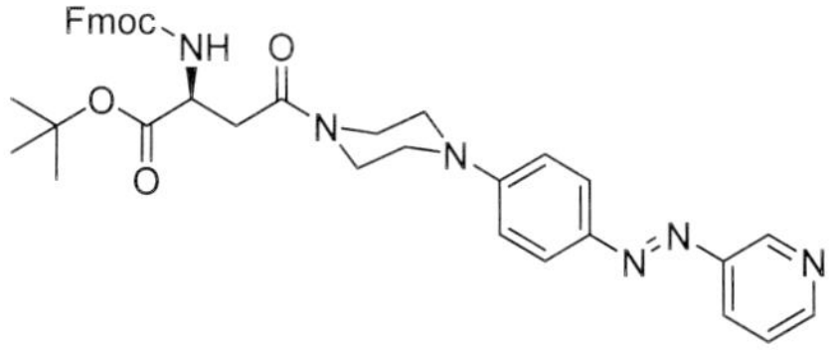

A solution of *p*-TosOH (1.21 g, 6.37 mmol, 1.20 eq.) in formic acid (9.7 mL) was slowly added, over a period of 10 min, to a stirred solution of pyridine-3-amine (0.500 g, 5.31 mmol, 1.00 eq.) in formic acid (9.7 mL), previously cooled to −5 °C in an ice bath. *tert*-butyl nitrite (0.76 mL, 6.37 mmol, 1.20 eq.) was added drop-wise at such a rate, that the temperature did not exceed 5 °C (about 10 min).[287, 341] After the addition was complete, the cooling bath was removed and the mixture was maintained under stirring for a further 5 min. Then, Et_2O (200 mL, filtered over anhydrous Na_2SO_4 and without traces of KOH) was added to complete the precipitation of the diazonium salt, that was gathered by filtration on a Büchner funnel and washed with Et_2O (3×10 mL) to complete the elimination of formic acid. After drying under reduced pressure 1.27 g pyridine-3-diazonium 4-methylbenzenesulfonate **70** (4.58 mmol, 86%) was obtained was obtained and stored at −20 °C under argon. (Note diazonium salts can decompose and safety precautions are necessary, but no decomposition was observed with 4-methylbenzenesulfonate salts.)

HRMS (ESI+): *m/z* calcd for $[C_{12}H_{12}N_3O_4{}^{31}S_1]^+$: 278.0594 Da [M+H]$^+$, found: 278.0594 Da (Δ = 0.2 ppm).

The pyridine-3-diazonium 4-methylbenzenesulfonate **70** (1.05 g, 3.76 mmol, 1.20 eq.) was dissolved in MeOH (15 mL) and added to a solution of *tert*-butyl (*S*)-2-((((9*H*-fluoren-9-yl)methoxy)carbonyl)amino)-4-oxo-4-(4-phenylpiperazin-1-yl)butanoate **71D** (1.75 g, 3.14 mmol, 1.00 eq.) at 0 °C. The reaction turned red within 5 min and was stirred for 3 d at rt. The reaction mixture was quenched with aqueous $NaHCO_3$ and extracted three times with CH_2Cl_2. The organic phase was evaporated under reduced pressure and purified by column chromatography starting from pure CH_2Cl_2 and eluting with a mixture of CH_2Cl_2:EtOAc 1:1 and 1.63 g *tert*-butyl (*S*,*E*)-2-((((9*H*-fluoren-9-yl)methoxy)carbonyl)amino)-4-oxo-4-(4-(4-(pyridin-3-yldiazenyl)phenyl)piperazin-1-yl)butanoate **72D** (2.47 mmol, 66%) was obtained as a red solid.

^{1}H NMR (300 MHz, Acetonitrile-*d*$_3$): δ = 9.00 (d, J = 2.4 Hz, 1H), 8.58 (dd, J = 4.6, 1.4 Hz, 1H), 8.04 (ddd, J = 8.2, 2.5, 1.6 Hz, 1H), 7.93 – 7.71 (m, 4H), 7.62 (d, J = 7.2 Hz, 2H), 7.50 – 7.19 (m, 5H), 6.98 (d, J = 8.8 Hz, 2H), 6.02 (d, J = 9.0 Hz, 1H), 4.55 – 4.25 (m, 3H), 4.20 (t, J = 6.8 Hz, 1H), 3.63 (td, J = 21.1, 18.5, 11.7 Hz, 4H), 3.33 (dq, J = 11.9, 6.5, 5.9 Hz, 4H), 2.84 (ddd, J = 72.5, 16.6, 5.0 Hz, 2H), 1.37 (s, 9H) ppm. **^{13}C NMR (75 MHz, Acetonitrile-*d*$_3$):** δ = 171.3, 169.6, 157.0, 154.2, 151.6, 149.1, 147.2, 145.8, 145.1, 142.1, 128.7, 128.1, 127.3, 126.1, 125.8, 125.1, 121.0, 115.2, 82.1, 67.2, 52.2, 48.0, 48.0, 47.9, 45.4, 41.7, 35.9, 28.2 ppm. **HRMS (FAB+):** *m/z* calcd for $C_{38}H_{40}O_5N_6$: 660.3055 Da [M], found: 660.3053 Da (Δ = 0.2 ppm). **IR (ATR):** $\tilde{\nu}$ = 2973 (vw), 1716 (w), 1641 (w), 1596 (m), 1503 (w), 1443 (w), 1385 (w), 1218 (m), 1143 (m), 1019 (w), 825 (w), 758 (w), 739 (w), 703 (w), 619 (vw), 532 (w), 426 (w) cm^{-1}.

tert-Butyl (*S*,*E*)-2-amino-4-oxo-4-(4-(4-(pyridin-3-yldiazenyl)phenyl)piperazin-1-yl)butanoate trifluoroacetate 72D-NH_2

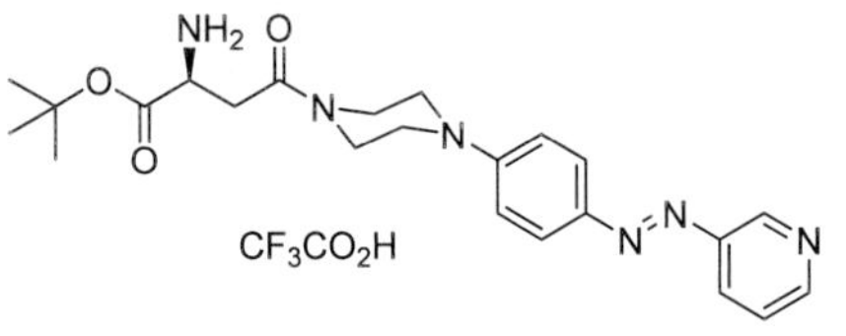

A solution of *tert*-butyl (*S*,*E*)-2-((((9*H*-fluoren-9-yl)methoxy)carbonyl)amino)-4-oxo-4-(4-(4-(pyridin-3-yldiazenyl)phenyl)piperazin-1-yl)butanoate (20 mg, 0.030 mmol, 1.0 eq.) was dissolved in DMF (500 µL). A solution of 2 vol% piperidine in DMF (500 µL) was added and quenched after 20 min with AcOH (500 µL). The mixture was purified by HPLC with the absorption at 330 nm and a gradient 5-95% H_2O to ACN with 0.1% TFA and 12 mg *tert*-butyl (*S*,*E*)-2-amino-

4-oxo-4-(4-(4-(pyridin-3-yldiazenyl)phenyl)piperazin-1-yl)butanoate trifluoro-acetate was isolated (0.022 mmol, 72%).

^{1}H NMR (400 MHz, DMSO): δ = 9.07 (d, *J* = 2.4 Hz, 1H), 8.70 (d, *J* = 4.7 Hz, 1H), 8.37 – 8.11 (m, 4H), 7.87 (d, *J* = 9.1 Hz, 2H), 7.73 – 7.59 (m, 1H), 7.13 (d, *J* = 9.2 Hz, 2H), 4.22 (s, 1H), 3.78 – 3.58 (m, 4H), 3.48 (dddd, *J* = 27.2, 22.7, 7.1, 4.2 Hz, 4H), 3.10 (dd, *J* = 17.5, 5.7 Hz, 1H), 2.96 (dd, *J* = 17.5, 3.7 Hz, 1H), 1.43 (s, 9H) ppm. **^{13}C NMR (101 MHz, DMSO):** δ = 167.8, 167.3, 158.3 (d, *J* = 36.0 Hz), 153.1, 149.3, 148.0, 144.5, 144.2, 127.9, 125.1, 125.0, 117.4, 114.5, 114.0, 82.7, 49.2, 46.6, 46.4, 44.2, 40.9, 33.0, 27.5 ppm. **HRMS (ESI+):** *m/z* calcd for $[C_{23}H_{31}N_6O_3]^+$: 439.2452 Da $[M+H]^+$, found: 439.2438 Da (Δ = 3.2 ppm).

(*S*,*E*)-2-((((*9H*-Fluoren-9-yl)methoxy)carbonyl)amino)-4-oxo-4-(4-(4-(pyridin-3-yldiazenyl)phenyl)piperazin-1-yl)butanoic acid 72D-OH

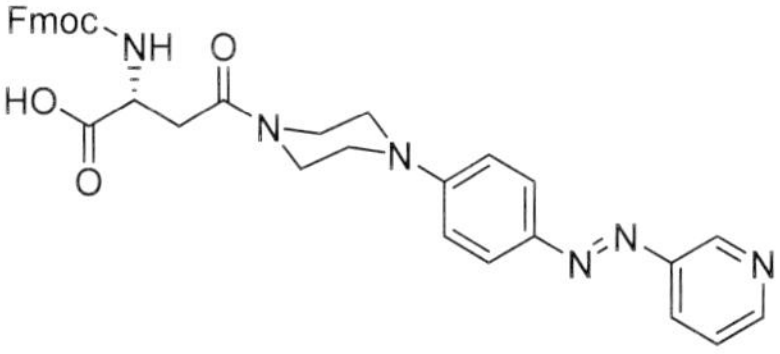

tert-Butyl (*S*,*E*)-2-((((*9H*-fluoren-9-yl)methoxy)carbonyl)amino)-4-oxo-4-(4-(4-(pyridin-3-yldiazenyl)phenyl)piperazin-1-yl)butanoate (1.72 g, 2.60 mmol, 1.00 eq.) was dissolved in CH_2Cl_2 (26 mL) yielding orange solution. Trifluoroacetic acid (TFA, 26 mL) and 1 vol% Triisopropylsilane (0.52 mL) was added at rt. The mixture was stirred for 1 h at rt. Then chloroform (100 mL) was added and the mixture was evaporated to dryness. The resulting crude TFA salt was dissolved in a minimal amount of CH_2Cl_2 (5 mL) and precipitated by addition of the solution into stirred diethyl ether (50 mL). The precipitate was filtrated, washed with diethyl ether (3×10 mL). The TFA salt was dissolved in chloroform and washed with sat. aqueous $NaHCO_3$. Part of the product precipitated and was filtered off, washed with diethyl ether (3×10 mL) and again combined with the red organic phase. The solvent was evaporated under reduced pressure and the product was dried under high vacuum and 1.32 g (*S*,*E*)-2-((((*9H*-fluoren-9-yl)methoxy)carbonyl)amino)-4-oxo-4-(4-(4-(pyridin-3-yldiazenyl)phenyl)piperazin-1-yl)butanoic acid (2.19 mmol, 84%) was obtained as an red powder.

^{1}H NMR (400 MHz, DMSO): δ = 12.71 (s, 1H), 9.05 (d, *J* = 2.3 Hz, 1H), 8.68 (dd, *J* = 4.8, 1.6 Hz, 1H), 8.17 (dt, *J* = 8.3, 1.8 Hz, 1H), 7.87 (dd, *J* = 11.4, 8.3 Hz, 4H), 7.71 (d, *J* = 7.5 Hz, 2H), 7.66 – 7.58 (m, 1H), 7.49 (s, 1H), 7.40 (dt, *J* = 7.8,

3.7 Hz, 2H), 7.31 (ddd, J = 7.4, 6.0, 1.5 Hz, 2H), 7.10 (d, J = 9.3 Hz, 2H), 4.44 (td, J = 7.6, 5.1 Hz, 1H), 4.30 (dd, J = 7.0, 1.9 Hz, 2H), 4.23 (d, J = 6.8 Hz, 1H), 3.63 (d, J = 5.5 Hz, 4H), 3.44 (dd, J = 23.4, 6.5 Hz, 4H), 2.83 (t, J = 6.7 Hz, 2H) ppm. **^{13}C NMR (101 MHz, DMSO):** δ = 173.1, 168.0, 155.8, 153.0, 149.6, 147.9, 144.8, 144.0, 143.7, 140.7, 127.6, 127.4, 127.0, 125.2, 125.0, 124.8, 120.1, 113.8, 65.6, 50.5, 46.6, 46.5, 46.3, 44.2, 40.7, 34.2 ppm. **HRMS (FAB+):** *m/z* calcd for $C_{34}H_{33}O_5N_6$: 605.2512 Da [M+H], found: 605.2511 Da (Δ = 0.3 ppm). **IR (ATR):** $\tilde{\nu}$ = 1716 (w), 1640 (w), 1597 (m), 1509 (w), 1447 (w), 1384 (w), 1230 (w), 1199 (w), 1144 (m), 1025 (w), 825 (w), 761 (w), 741 (w), 678 (vw), 536 (w), 426 (vw) cm^{-1}. **UV-Vis (ACN):** λ_{max} = 265, 288, 300, 404 nm.

(*S*,*E*)-3-((4-(4-(3-((((9*H*-Fluoren-9-yl)methoxy)carbonyl)amino)-3-carboxypropanoyl)piperazin-1-yl)phenyl)diazenyl)-1-(2-(*tert*-butoxy)-2-oxoethyl)pyridin-1-ium bromide 60D

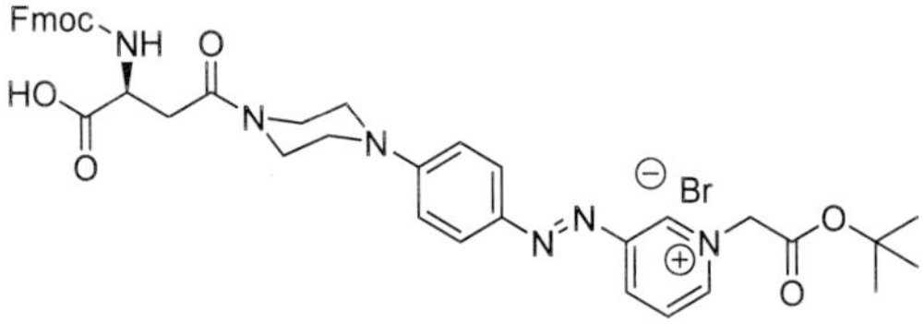

A solution of (*S*,*E*)-2-((((9*H*-fluoren-9-yl)methoxy)carbonyl)amino)-4-oxo-4-(4-(4-(pyridin-3-yldiazenyl)phenyl)piperazin-1-yl)butanoic acid (336 mg, 0.557 mmol, 1.00 eq.) and *tert*-butyl 2-bromoacetate (82 µL, 0.557 mmol, 1.00 eq.) in DMF (2.7 mL) was stirred at rt 72 h in the dark.[342] The mixture was diluted 1:1 with MeOH and precipitated by dropwise addition of the reaction mixture into a stirred solution of diethyl ether (60 mL). The precipitate was filtered of and washed with diethyl ether (3×10 mL), dried under high vacuum to afford the product as a red solid. The crude product was purified by column chromatography (h=7 cm) starting with 2% MeOH and 5% AcOH in CH_2Cl_2 to elute the substrate (orange substrate: R_f = 0.12, red product: R_f = 0.0) and switched to 12.5% MeOH and 5% AcOH in CH_2Cl_2 to elute the red product: R_f = 0.12-0.22. After evaporation of the solvent under reduced pressure, traces of AcOH were removed by lyophilisation in 1:1 H_2O:ACN (50 mL) over night and 337 mg (0.421 mmol, 76%) (*S*,*E*)-3-((4-(4-(3-((((9*H*-fluoren-9-yl)methoxy)carbonyl)amino)-3-carboxypropanoyl)piperazin-1-yl)phenyl)diazenyl)-1-(2-(*tert*-butoxy)-2-oxoethyl)pyridin-1-ium bromide was obtained.

^{1}H NMR (300 MHz, Acetonitrile-d_3): δ = 9.10 (s, 1H), 8.71 (s, 2H), 8.12 (s, 1H), 7.76 (s, 2H), 7.67 (s, 2H), 7.55 (s, 2H), 7.39 – 7.13 (m, 4H), 6.98 – 6.70 (m, 2H),

4.43 (s, 1H), 4.22 (s, 2H), 4.10 (s, 1H), 3.60 (s, 4H), 3.50 – 3.30 (m, 4H), 3.12 – 2.47 (m, 2H), 1.44 (d, J = 3.9 Hz, 9H) 2H peaks overlap with H_2O ppm. **^{13}C NMR (101 MHz, Acetonitrile-d_3):** δ = 175.1, 165.2, 157.1, 154.8, 151.2, 145.2, 144.4, 144.3, 144.2, 141.4, 141.3, 137.6, 129.0, 128.2, 127.6, 127.1, 125.6, 120.4, 113.9, 85.6, 67.1, 52.3, 47.3, 46.5, 46.3, 45.0, 41.5, 35.9, 27.6 ppm. **HRMS (FAB+):** m/z calcd for $C_{46}H_{33}O_4N_5$: 719.2527 Da [M], found: 719.2526 Da (Δ = 0.2 ppm). **IR (ATR):** ṽ = 3390 (vw), 1706 (w), 1633 (w), 1597 (m), 1510 (w), 1447 (w), 1383 (w), 1218 (m), 1143 (m), 1023 (w), 826 (w), 799 (vw), 760 (w), 741 (w), 704 (vw), 671 (vw), 620 (vw), 539 (w), 426 (vw), 385 (vw) cm^{-1}.

(((*9H*-Fluoren-9-yl)methoxy)carbonyl)-*L*-aspartic acid 102

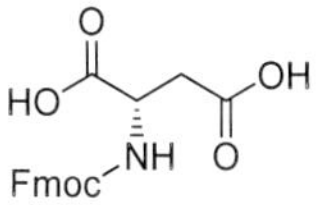

L-aspartic acid (6.06 g, 45.6 mmol, 1.20 eq.) is dissolved in aqueous sodium carbonate solution (108 mL, 138 mmol, 3.60 eq.), in a dry 500 mL round bottom flask. The medium is cooled in an ice bath at 0 °C, then a solution of *N*-(9-fluorenylmethoxycarbonyloxy)succinimide (12.8 g, 38.0 mmol, 1.00 eq.) dissolved in DMF (88 mL) is added with vigorous stirring (a precipitate forms in the reaction medium). The stirring is maintained for 1 h at ambient temperature. The mixture is then diluted in water (1330 mL) and extracted with ether (1×160 mL), then with EtOAc (2×120 mL). The resulting aqueous phase is cooled in an ice bath and acidified to pH 2 with concentrated (6 N) hydrochloric acid. The aqueous phase containing the precipitated product (in the form of an oil) is extracted with EtOAc (6×120 mL). The organic phase derived from the extraction is washed with a sat. aqueous sodium chloride solution (3×70 mL), and then with water (2×70 mL), dried over sodium sulphate, and concentrated under reduced pressure (35 °C.) until a small residual volume is obtained. The crude is precipitated by adding petroleum ether (approximately 10 times the residual volume) with vigorous stirring. After having allowed the mixture to separate by settling out for 12 h at 4 °C, the precipitate is filtered off and then dried for 24 h in a vacuum and 11.2 g (((9*H*-fluoren-9-yl)methoxy)carbonyl)-*L*-aspartic acid **102** (31.6 mmol, 83%) is isolated in the form of a fine white powder.[343]

TLC: R_f = 0.8 (60% AcOH:BuOH 4:6). **^{1}H NMR (300 MHz, DMSO):** δ = 12.63 (s, 2H), 7.89 (d, J = 7.4 Hz, 2H), 7.72 (d, J = 7.6 Hz, 3H), 7.37 (dt, J = 26.5, 7.3 Hz, 4H), 4.30 (td, J = 19.0, 17.1, 6.1 Hz, 4H), 2.75 (d, J = 16.2 Hz, 1H), 2.59 (dd, J = 16.2, 8.1 Hz, 1H) ppm. **^{13}C NMR (75 MHz, DMSO):** δ = 172.8, 171.8,

156.0, 143.9, 140.8, 127.7, 127.2, 125.3, 120.2, 65.82, 50.6, 46.7, 36.1 ppm. **HRMS (EI+):** *m/z* calcd $C_{19}H_{18}NO_6$: 356.1129 Da [M+H], found: 356.1127 Da ($\Delta = 0.5$ ppm). **IR (ATR):** $\tilde{\nu}$ = 3306 (vw), 3038 (vw), 1705 (w), 1682 (w), 1524 (w), 1450 (vw), 1409 (w), 1347 (vw), 1262 (w), 1193 (vw), 1092 (w), 1049 (vw), 1007 (vw), 906 (vw), 781 (vw), 756 (w), 741 (w), 731 (w), 694 (vw), 655 (vw), 642 (w), 621 (w), 589 (vw), 546 (vw), 499 (vw), 456 (vw), 425, (vw), 383 (vw) cm^{-1}.

(*S*)-2-(3-(((9*H*-Fluoren-9-yl)methoxy)carbonyl)-5-oxooxazolidin-4-yl)acetic acid 103

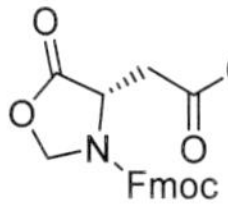

A slurry of (((9*H*-fluoren-9-yl)methoxy)carbonyl)-*L*-aspartic acid **102** (1.78 g, 5.00 mmol, 1.00 eq.), paraformaldehyde (1.00 g, 33.3 mmol, 6.710 eq.) and *p*-toluenesulfonic acid (50 mg, 0.290 mmol, 0.06 eq.) in PhMe (4 mL) was stirred in a vial by hand with a spatula and subjected to microwave irradiation (microwave oven operated at 2450 MHz, 300 W, 105 °C for 12 min, TLC (1% AcOH EtOAc:cH 1:1 $R_f = 0.25$).[344] The reaction was repeated three times with the same conditions and the batches were combined. The reaction mixture was diluted with EtOAc (60 mL), washed with water (2×25 mL), dried over anhydrous Na_2SO_4 and concentrated *in vacuo*. The resulting residue was fractional recrystallized over 3 rounds from acetone:CH_2Cl_2:cH 2:1:3 and 4.32 g (*S*)-2-(3-(((9*H*-fluoren-9-yl)methoxy)carbonyl)-5-oxooxazolidin-4-yl)acetic acid **103** (11.8 mmol, 78%).

^{1}H NMR (300 MHz, DMSO): δ = 12.78 (s, 1H), 7.89 (d, *J* = 7.5 Hz, 2H), 7.65 (t, *J* = 6.3 Hz, 2H), 7.38 (dt, *J* = 25.1, 7.4 Hz, 4H), 5.38 (s, 1H), 5.15 (d, *J* = 29.4 Hz, 1H), 4.71 – 4.01 (m, 4H), 3.52 – 2.55 (m, 2H) ppm. **^{13}C NMR (75 MHz, DMSO):** δ = 171.5, 143.6, 140.8, 127.8, 127.2, 125.1, 120.2, 78.0, 67.1, 51.5, 46.5, 34.2 ppm. **HRMS (EI+):** *m/z* calcd for $C_{20}H_{18}O_6N_1$: 368.1129 Da [M+H], found: 368.1129 Da ($\Delta = 0.2$ ppm). **IR (ATR):** $\tilde{\nu}$ = 3302 (w), 1771 (m), 1728 (m), 1689 (s), 1497 (vw), 1430 (m), 1409 (m), 1358 (m), 1301 (w), 1261 (m), 1170 (m), 1134 (m), 1102 (w), 1079 (w), 1052 (m), 999 (m), 926 (w), 818 (w), 749 (m), 738 (m), 686 (w), 620 (w), 583 (w), 545 (w), 509 (m), 473 (w), 425 (w), 383 (w) cm^{-1}

(9*H*-Fluoren-9-yl)methyl (*S*)-5-oxo-4-(2-oxo-2-(4-phenylpiperazin-1-yl)ethyl)oxazolidine-3-carboxylate 104

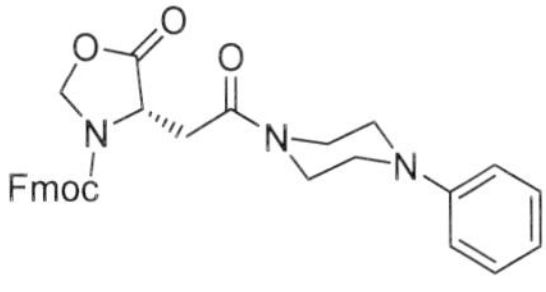

To a mixture of 1-phenylpiperazine (1.30 g, 8.01 mmol, 1.00 eq.) in CH_2Cl_2 (9.3 mL) at 0 °C was added (*S*)-2-(3-(((9*H*-fluoren-9-yl)methoxy)carbonyl)-5-oxooxazolidin-4-yl)acetic acid **103** (2.94, 8.01 mmol, 1.00 eq.) and EDC·HCl (1.54 g, 8.01 mmol, 1.00 eq.) DMAP (196 mg, 1.60 mmol, 0.200 eq.) was added in one portion.[345] The mixture gradually warmed to rt, and was quenched by addition of water after 16 h. The mixture was diluted with CH_2Cl_2 (50 mL) and washed with diluted aqueous ammonium chloride (2×20 mL). The organic phase was dried over $MgSO_4$, filtered through paper, and evaporated under reduced pressure. The crude product was purified by silica gel column chromatography: Start in pure EtOAc:cH 2:8 $R_f = 0.0$, elution with EtOAc:cH 1:1 ($R_f = 0.25$). Evaporation of combined fractions and drying under high vacuum overnight resulted in 2.77 g (9*H*-fluoren-9-yl)methyl (*S*)-5-oxo-4-(2-oxo-2-(4-phenylpiperazin-1-yl)ethyl)oxazolidine-3-carboxylate **104** (5.42 mmol, 68%).

^{1}H NMR (300 MHz, Acetonitrile-*d*$_3$): δ = 7.78 (s, 2H), 7.54 (s, 2H), 7.45 – 7.24 (m, 4H), 7.19 (s, 2H), 7.03 – 6.53 (m, 3H), 5.29 (s, 1H), 5.10 (d, *J* = 40.0 Hz, 1H), 4.58 (s, 1H), 4.28 (d, *J* = 31.9 Hz, 2H), 3.79 – 3.08 (m, 5H), 3.08 – 2.88 (m, 5H), 2.45 (d, *J* = 21.8 Hz, 1H) ppm. **^{13}C NMR (75 MHz, Acetonitrile-*d*$_3$):** δ = 173.7, 168.3, 153.7, 152.1, 144.8, 142.1, 130.0, 128.7, 128.2, 125.8, 121.0, 120.8, 117.2, 79.1, 67.6, 52.7, 49.8, 49.6, 47.8, 45.8, 42.0, 34.2 ppm. **HRMS (EI+):** *m/z* calcd for $C_{30}H_{30}O_5N_3$: 512.2180 Da [M+H], found: 512.2179 Da (Δ = 0.2 ppm). **IR (ATR):** ṽ = 2916 (vw), 1793 (w), 1708 (w), 1640 (w), 1597 (w), 1494 (w), 1446 (w), 1407 (w), 1353 (w), 1229 (w), 1130 (w), 1043 (w), 990 (w), 940 (w), 910 (w), 884 (vw), 756 (m), 739 (m), 692 (w), 620 (vw), 563 (vw), 518 (w), 481 (vw), 427 (vw) cm^{-1}.

(*S*,*E*)-5-Oxo-4-(2-oxo-2-(4-(4-(pyridin-3-yldiazenyl)phenyl)piperazin-1-yl)ethyl)oxazolidine-3-carboxylate 105

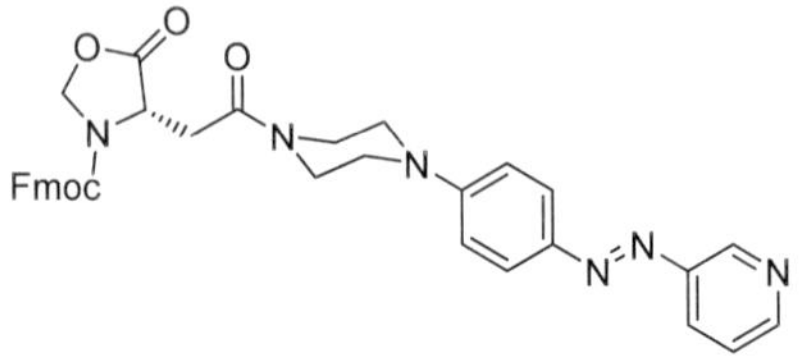

A mixture of $NaNO_2$ (0.122 g, 1.76 mmol, 1.20 eq.,) in H_2O (5 mL) combined with (*S*)-5-oxo-4-(2-oxo-2-(4-phenylpiperazin-1-yl)ethyl)oxazolidine-3-carboxylate **104** (752 mg, 1.47 mmol, 1.00 eq.) dissolved in 2-butanol:THF 2:1(3 mL) was added by a syringe pump within 30 min to a stirred solution of 3-aminopyridine **69** (166 mg, 1.76 mmol, 1.20 eq.) in aqueous HCl (0.85 mL, 10.1 mmol, 6.90 eq.) and H_2O (5 mL) at 0 °C. The reaction mixture was stirred for 16 h at 0 °C to rt. The mixture was neutralised with aqueous $NaHCO_3$ and the aqueous phase was extracted CH_2Cl_2 (3×50 mL) and the organic phase was washed with brine (10 mL), dried with anhydrous Na_2SO_4 and evaporated under reduced pressure. The crude product was purified by silica gel column chromatography with EtOAc:CH_2Cl_2 1:1 (R_f = 0.20). Evaporation of combined fractions and drying under high vacuum overnight resulted in 117 mg (*9H*-fluoren-9-yl)methyl (*S*,*E*)-5-oxo-4-(2-oxo-2-(4-(4-(pyridin-3-yldiazenyl)phenyl)piperazin-1-yl)ethyl)oxazolidine-3-carboxylate **105** (1.90 mmol, 13%).

^{1}H NMR (300 MHz, Acetonitrile-*d*$_3$): δ = 9.07 (s, 1H), 8.68 (d, *J* = 5.2 Hz, 1H), 8.36 (d, *J* = 8.3 Hz, 1H), 7.79 (d, *J* = 17.7 Hz, 5H), 7.58 (s, 2H), 7.48 – 7.11 (m, 4H), 6.95 (s, 2H), 6.45 (s, 2H), 5.32 (s, 1H), 5.12 (d, *J* = 37.4 Hz, 1H), 4.62 (s, 1H), 4.33 (d, *J* = 28.9 Hz, 2H), 3.63 (d, *J* = 47.8 Hz, 3H), 3.34 (d, *J* = 18.1 Hz, 5H), 2.49 (d, *J* = 23.1 Hz, 1H) ppm. **^{13}C NMR (75 MHz, Acetonitrile-*d*$_3$):** δ = 173.7, 168.7, 154.6, 150.2, 146.6, 145.5, 144.8 (low s/n), 142.4, 142.1, 132.3, 128.7, 128.2, 127.0, 126.5, 125.8, 121.0, 114.9, 79.1, 67.6, 52.7, 47.8, 47.5, 47.4, 45.2, 41.5, 34.2 ppm. **HRMS (EI+):** *m/z* calcd for $C_{35}H_{33}O_5N_6$: 617.2507 [M+H], found: 617.2506 Da (Δ = 0.1 ppm).

(*E*)-3,5-Dimethoxy-2-(pyridin-3-yldiazenyl)aniline 106

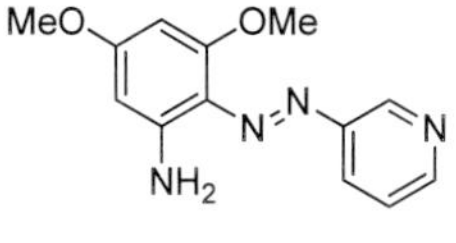

A solution of *p*-TosOH·H_2O (2.91 g, 15.3 mmol, 1.20 eq.) in formic acid:MeOH 10:1 (12.8 mL) was added slowly to a solution of 3-aminopyridine **69** (1.20 g, 12.8 mmol, 1.00 eq.) in formic acid:MeOH 10:1 (12.8 mL), placed in an ice-salt bath and cooled to −5 °C. *tert*-Butyl nitrite (2.02 mL, 15.3 mmol, 1.20 eq.) is added, so that the temperature does not rise above 0 °C. Stirring is continued for an additional 10 min.[288] Then the reaction mixture (12.8 mmol, 1.00 eq.) was added to the stirred solution of 3,5-dimethoxyaniline (1.95 g, 12.8 mmol, 1.00 eq.), in formic acid:MeOH 10:1 (12.8 mL) at 0 °C. The reaction was stirred for 1 h at 0 °C and 24 h at rt. The mixture diluted with CH_2Cl_2 (40 mL) was filtered through a silica pad (4 cm, Ø=5 cm), washed with 30% MeOH in CH_2Cl_2 (3×50 mL) and evaporated under reduced pressure. The crude was basified with Et_3N (2 mL) and purified by silica gel column chromatography in 1% Et_3N, in cH:EtOAc 1:1 ($R_f = 0.20$) and 317 mg (*E*)-3,5-dimethoxy-2-(pyridin-3-yldiazenyl)aniline **106** (1.23 mmol, 10%) was obtained as an orange solid.

^{1}H NMR (400 MHz, DMSO): δ = 9.17 (d, *J* = 2.2 Hz, 1H), 8.71 (dd, *J* = 5.3, 1.4 Hz, 1H), 8.50 (ddd, *J* = 8.4, 2.3, 1.4 Hz, 1H), 7.92 (dd, *J* = 8.4, 5.3 Hz, 1H), 5.99 (d, *J* = 2.5 Hz, 1H), 5.89 (d, *J* = 2.5 Hz, 1H), 3.82 (d, *J* = 21.8 Hz, 6H) ppm. **^{13}C NMR (101 MHz, DMSO):** δ = 165.6, 160.88, 158.7 (q, *J* = 35.9 Hz), 150.4, 146.0, 141.9, 139.1, 131.8, 126.75, 123.1, 115.9 (q, *J* = 291.9 Hz), 90.5, 89.2, 56.0, 55.5 ppm. **HRMS (FAB+):** *m/z* calcd for $C_{13}H_{14}O_2N_4$: 258.1117 Da [M], found: 258.1115 Da (Δ = 0.6 ppm). **IR (ATR):** ṽ = 3350 (w), 3076 (vw), 2945 (vw), 2469 (vw), 2121 (vw), 1781 (vw), 1689 (w), 1595 (w), 1457 (w), 1398 (w), 1363 (w), 1320 (w), 1295 (w), 1120 (m), 995 (w), 938 (w), 834 (w), 802 (w), 720 (w), 676 (w), 607 (w), 573 (w), 509 (w), 458 (w), 388 (vw) cm^{-1}.

(*E*)-*N*-(3,5-Dimethoxy-2-(pyridin-3-yldiazenyl)phenyl)acetamide trifluoro-acetate 107

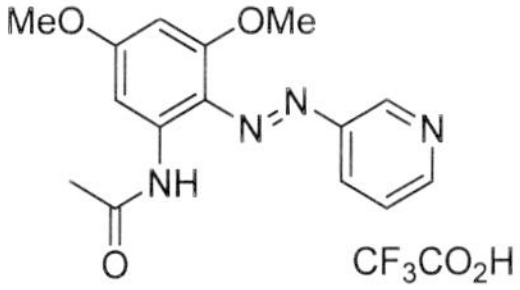

MeO OMe N N N NH O CF_3CO_2H

A solution of (*E*)-3,5-dimethoxy-2-(pyridin-3-yldiazenyl)aniline **106** (297 mg, 1.15 mmol, 1.00 eq.) and acetyl chloride (108 mg, 1.38 mmol, 1.20 eq.) in Ac_2O (12 mL) was stirred at rt for 48 h. The solvent was removed under reduced pressure and the product (258 mg) was purified by HPLC under standard conditions to obtain 258 mg (*E*)-*N*-(3,5-dimethoxy-2-(pyridin-3-yldiazenyl)phenyl)acetamide trifluoroacetate **107** (0.623 mmol, 54%) as a red solid.

^{1}H NMR (400 MHz, DMSO): δ = 12.26 (s, 1H), 9.08 (d, *J* = 2.3 Hz, 1H), 8.71 (dd, *J* = 4.9, 1.5 Hz, 1H), 8.21 (ddd, *J* = 8.3, 2.4, 1.5 Hz, 1H), 7.83 (d, *J* = 2.6 Hz, 1H), 7.71 (ddd, *J* = 8.3, 4.9, 0.7 Hz, 1H), 6.47 (d, *J* = 2.6 Hz, 1H), 3.92 (s, 3H), 3.86 (s, 3H), 2.19 (s, 3H) ppm. **^{13}C NMR (101 MHz, DMSO):** δ = 170.0, 165.3, 160.2, 158.4 (q, *J* = 36.9 Hz), 148.7, 147.9, 143.9, 134.8, 128.1, 125.3, 123.3, 115.6 (d, *J* = 290.7 Hz), 96.3, 93.8, 56.5, 55.8, 25.5 ppm. **HRMS (FAB+):** *m/z* calcd for $C_{15}H_{16}O_3N_4$: 300.1222 Da [M], found: 300.1224 Da (Δ = 0.5 ppm). **IR (ATR):** ṽ = 3570 (vw), 3047 (vw), 2947 (vw), 1696 (w), 1613 (w), 1576 (m), 1460 (m), 1421 (m), 1372 (w), 1336 (w), 1285 (m), 1230 (m), 1156 (m), 1114 (w), 984 (w), 811 (m), 795 (m), 719 (m), 683 (w), 607 (w), 568 (w), 513 (w), 440 (w) cm^{-1}. **UV-Vis (MeCN):** λ_{max} = 231, 260, 382 nm.

((*E*)-3-((2-Acetamido-4,6-dimethoxyphenyl)diazenyl)-1-(2-(*tert*-butoxy)-2-oxoethyl)pyridin-1-ium trifluoroacetate 108

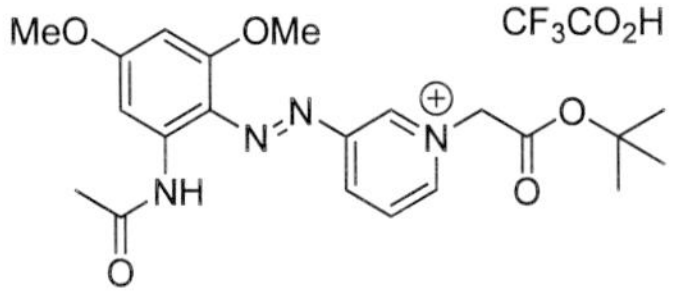

A solution of (*E*)-*N*-(3,5-dimethoxy-2-(pyridin-3-yldiazenyl)phenyl)acetamide **107** (70 mg, 0.233 mmol, 1.00 eq.) and *tert*-butyl 2-bromoacetate (68 mg, 0.35 mmol, 1.5 eq.) in 2 mL DMF:DMSO 4:1 was stirred at 60 °C 1 d. After cooling, the solvent was evaporated under reduced pressure. The residue was dissolved in 5 mL MeOH and precipitated with 50 mL diethyl ether. The precipitate was filtered of and washed with diethyl ether (3×10 mL) dried under high vacuum to afford a crude. The product was purified by HPLC under standard conditions (retention=22 min) to obtain 52 mg ((*E*)-3-((2-acetamido-4,6-dimethoxyphenyl)diazenyl)-1-(2-(*tert*-butoxy)-2-oxoethyl)pyridin-1-ium trifluoroacetate **108** (0.98 mmol, 42%) as a red solid.

^{1}H NMR (400 MHz, DMSO): δ = 11.75 (s, 1H), 9.52 (t, *J* = 1.6 Hz, 1H), 9.06 (d, *J* = 6.0 Hz, 1H), 8.96 – 8.82 (m, 1H), 8.35 (dd, *J* = 8.5, 6.0 Hz, 1H), 7.87 (d, *J* = 2.5 Hz, 1H), 6.56 (d, *J* = 2.6 Hz, 1H), 5.68 (s, 2H), 3.97 (s, 3H), 3.92 (s, 3H), 2.24 (s, 3H), 1.49 (s, 9H) ppm. **^{13}C NMR (101 MHz, DMSO):** δ = 170.0, 167.1, 165.1, 159.8, 158.8 – 156.7 (m), 150.1, 145.5, 141.6, 136.6, 136.1, 128.3, 124.0,

116.3 (d, J = 295.5 Hz), 96.9, 94.0, 83.9, 61.0, 56.6, 56.1, 27.6, 25.5 ppm. **HRMS (EI+):** m/z calcd for $C_{21}H_{27}O_5N_4$: 415.1981 Da [M], found: 415.1981 Da (Δ = 0.1 ppm). **IR (ATR):** $\tilde{\nu}$ = 3077 (vw), 2946 (vw), 2985 (vw), 1741 (w), 1689 (w), 1617 (w), 1571 (w), 1453 (w), 1420 (w), 1397 (w), 1369 (w), 1296 (w), 1233 (w), 1148 (m), 841 (w), 794 (w), 747 (w), 704 (m), 671 (w), 588 (vw), 567 (w), 514 (vw), 428 (vw) cm^{-1}. **UV-Vis (PBS):** λ_{max} = 394 nm.

(*E*)-3,5-Dichloro-4-(pyridin-3-yldiazenyl)aniline trifluoroacetate 109

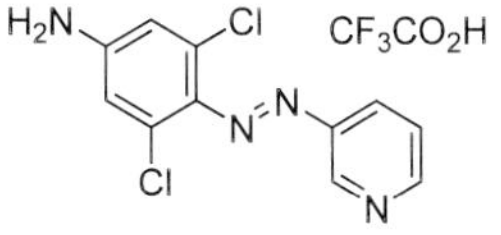

The pyridine-3-diazonium 4-methylbenzenesulfonate **70** (860 mg, 3.10 mmol, 1.00 eq.) was added to a solution of 3,5-dichloroaniline (502 mg, 3.10 mmol, 1.00 eq.) in AcOH (6 mL). The reaction was stirred for 2 h at rt and was quenched with sat. aqueous $NaHCO_3$ (20 mL) and extracted three times with CH_2Cl_2 (3×50 mL). The organic phase was removed under reduced pressure and purified by HPLC under standard conditions. Three different fractions were observed (retention = 18, 20 and 22 min) and the first fraction with 22 mg (*E*)-3,5-dichloro-4-(pyridin-3-yldiazenyl)aniline trifluoroacetate **109** (0.082 mmol, 2%) was isolated as a red solid.

^{1}H NMR (400 MHz, DMSO): δ = 9.08 (s, 1H), 8.77 (s, 1H), 8.20 (d, J = 8.1 Hz, 1H), 7.73 (dd, J = 8.6, 4.5 Hz, 1H), 6.77 (s, 2H) ppm. **^{13}C NMR (101 MHz, DMSO):** δ = 158.5 (d, J = 36.7 Hz), 152.2, 149.5, 144.1, 134.0, 131.4, 128.5, 125.6, 113.9 ppm. **HRMS (FAB+):** m/z calcd for $C_{11}H_8N_4Cl_2$: 266.0126 Da [M], found: 266.0125 Da (Δ = 0.4 ppm). **IR (ATR):** $\tilde{\nu}$ = 3369 (w), 3179 (m), 2470 (w), 2112 (w), 1676 (m), 1634 (m), 1594 (s), 1547 (m), 1471 (w), 1449 (w), 1405 (m), 1311 (m), 1202 (m), 1165 (m), 1119 (s), 1068 (m), 1011 (w), 954 (w), 906 (m), 886 (w), 843 (m), 816 (m), 803 (m), 788 (s), 725 (m), 670 (m), 642 (m), 601 (m), 512 (w), 464 (w), 399 (w) cm^{-1}.

(*E*)-*N*^{1},*N*^{1},*N*^{3},*N*^{3}-Tetramethyl-4-(pyridin-3-yldiazenyl)benzene-1,3-diamine 110

Procedure cited from master thesis of M. Steinbiß[282]

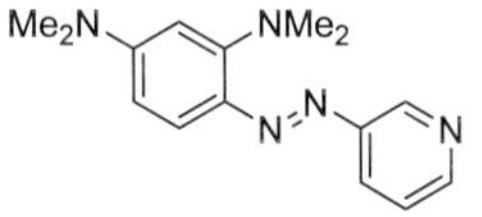

A mixture of *N*^{1},*N*^{1},*N*^{3},*N*^{3}-tetramethyl-1,3-phenylenediamine (0.410 mL, 0.400 g, 2.44 mmol, 1.00 eq.), $NaNO_2$ (181 mg, 2.63 mmol, 1.08 eq.) and H_2O (5 mL) was added dropwise to a solution of 3-aminopyridine **69** (0.275 g, 2.92 mmol, 1.20 eq.), conc. HCl (850 µL) and H_2O (5 mL) at 0 °C. The reaction mixture was stirred over 2 h. The mixture was adjusted to pH 6 by adding $NaHCO_3$.The crude product was extracted with EtOAc, dried over anhydrous Na_2SO_4 and the solvent was evaporated. The crystals were purified by silica gel column chromatography (gradient cH:EtOAc 9:1 to 1:1), resulting in 635 mg (*E*)-*N*^{1},*N*^{1},*N*^{3},*N*^{3}-tetramethyl-4-(pyridin-3-yldiazenyl)benzene-1,3-diamine **110** (2.36 mmol, 97%) of red crystals.

TLC: R_f = 0.38 (EtOAc). **^{1}H NMR ($CDCl_3$, 300 MHz):** δ = 8.98 (d, *J* = 1.9 Hz, 1H), 8.43 (dd, *J* = 4.6, 1.0 Hz, 1H), 7.95 (d, *J* = 8.2 Hz, 1H), 7.77 (d, *J* = 9.3 Hz, 1H), 7.26 (dd, *J* = 8.1, 4.7 Hz, 1H), 7.26 (dd, *J* = 8.1, 4.7 Hz, 1H), 6.21 (dd, *J* = 9.3, 2.5 Hz, 1H, $C_{Ar}H$), 5.98 (d, *J* = 2.3 Hz, 1H), 3.06 (s, 6H), 2.99 (s, 6H) ppm. **^{13}C NMR ($CDCl_3$, 75 MHz):** δ = 153.8, 153.7, 149.0, 148.9, 146.0, 135.1, 126.7, 123.8, 118.7, 105.2, 98.7, 45.1, 40.2. ppm. **MS (ESI+):** *m/z* calcd for $[C_{15}H_{20}N_5]^+$: 270.1713 Da $[M]^+$, found: 270.1692 Da.

(*E*)-3-((2,4-Bis(dimethylamino)phenyl)diazenyl)-1-(2-(*tert*-butoxy)-2-oxoethyl)pyridin-1-ium bromide 62

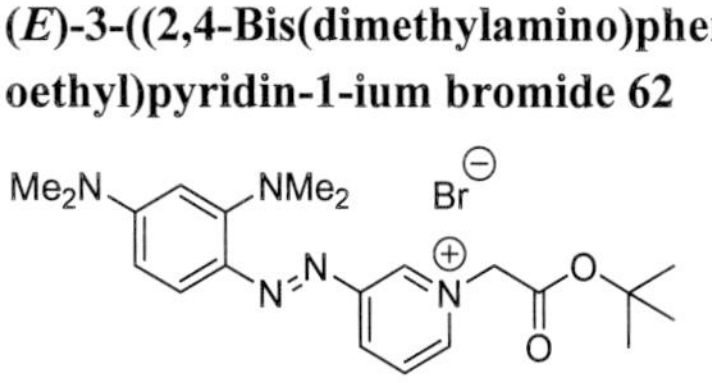

A solution of (*E*)-*N*^{1},*N*^{1},*N*^{3},*N*^{3}-tetramethyl-4-(pyridin-3-yldiazenyl)benzene-1,3-diamine **110** (129 mg, 0.479 mmol, 1.00 eq.) and *tert*-butyl 2-bromoacetate (85 µl, 0.575 mmol, 1.20 eq.) in DMF (2.4 mL) was stirred at 50 °C for 3 d. After cooling, the solvent was evaporated under reduced pressure. The mixture dissolved in MeOH (5 mL) and precipitated in diethyl ether (50 mL). The precipitate was filtered of and washed with diethyl ether (3×10 mL), dried under high vacuum to afford 205 mg (0.441 mmol, 92%) (*E*)-3-((2,4-bis(dimethylamino)phenyl)diazenyl)-1-(2-(*tert*-butoxy)-2-oxoethyl)pyridin-1-ium bromide as a red solid.

^{1}H NMR (300 MHz, DMSO): δ = 9.18 (s, 1H), 8.75 (d, *J* = 5.9 Hz, 1H), 8.60 (d, *J* = 8.8 Hz, 1H), 8.15 (dd, *J* = 8.6, 5.9 Hz, 1H), 7.83 (d, *J* = 9.7 Hz, 1H), 6.56 (dd, *J* = 9.8, 2.5 Hz, 1H), 5.92 (d, *J* = 2.5 Hz, 1H), 5.59 (s, 2H), 3.27 (s, 6H), 3.16 (s, 6H), 1.48 (s, 9H) ppm. **^{13}C NMR (75 MHz, DMSO):** δ = 165.3, 155.1, 154.3, 151.6, 142.1, 134.4, 119.4, 119.1, 106.6, 95.9, 95.8, 83.6, 60.8, 44.8, 44.6, 27.6 ppm.[282] **HRMS (FAB+):** *m/z* calcd for $C_{21}H_{30}O_2N_5$: 384.2400 Da [M], found: 384.2399 Da (Δ = 0.2 ppm). **IR (ATR):** ṽ = 3089 (w), 2990 (w), 2945 (vw), 1782 (m), 1734 (m), 1672 (m), 1625 (m), 1594 (m), 1542 (m), 1495 (m), 1460 (m), 1428 (m), 1360 (m), 1334 (m), 1295 (w), 1254 (m), 1186 (s), 1129 (s), 931 (m), 840 (m), 813 (s), 716 (m), 665 (s), 583 (m), 424 (m) cm^{-1}.

(*E*)-4-(Pyridin-3'-yldiazenyl)aniline 111

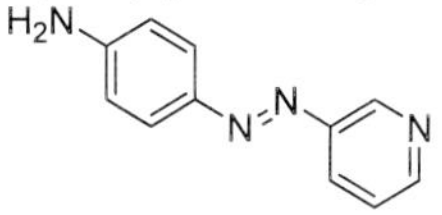

A solution of *p*-TosOH·H_2O (5.82 g, 30.6 mmol, 1.20 eq.) in formic acid (25.5 mL) was added slowly to a solution of 3-aminopyridine **69** (2.40 g, 25.5 mmol, 1.00 eq.) in formic acid (25.5 mL), placed in an ice-salt bath and cooled to 0 °C. *tert*-Butyl nitrite (4.1 mL, 30.6 mmol, 1.20 eq.) is added dropwise over 10 min, so that the temperature does not rise above 0 °C.[288, 339] Then the reaction mixture (25.5 mmol, 1.00 eq.) was added dropwise over 5 min to the stirred solution of aniline (2.38 g, 25.5 mmol, 1.00 eq.), in formic acid (25.5 mL) at 0 °C. The reaction was stirred for 1 h at 0 °C and 1 h at rt. The mixture diluted with CH_2Cl_2 (80 mL) was filtered through a silica pad (h=4 cm, Ø=5 cm), washed with 30% MeOH in CH_2Cl_2 (3×50 mL). The combined organic fractions were quenched with slow addition of sat. aqueous $NaHCO_3$ and extracted with EtOAc (3×100 mL) and the organic fractions were evaporated under reduced pressure. The remaining red aqueous fraction was discarded. The crude was basified with 2 mL Et_3N and purified by column chromatography in 1% Et_3N, in cH:EtOAc 1:1 (R_f = 0.20) and 945 mg *(E)*-4-(pyridin-3'-yldiazenyl)aniline **111** (4.77 mol, 19%). To obtain analytic purity it was necessary to further purify the product by HPLC under standard conditions.

^{1}H NMR (400 MHz, DMSO): δ = 9.03 (d, *J* = 2.3 Hz, 1H), 8.67 (dd, *J* = 4.9, 1.5 Hz, 1H), 8.24 (dt, *J* = 8.4, 1.8 Hz, 1H), 7.71 (dd, *J* = 8.6, 4.4 Hz, 3H), 6.75 – 6.65 (m, 2H) ppm. **^{13}C NMR (101 MHz, DMSO):** δ = 153.9, 148.5, 147.5,

143.2, 143.0, 128.8, 126.0, 125.3, 113.6. **HRMS (FAB+):** *m/z* calcd for $C_{11}H_{10}N_4$: 198.0905 Da [M], found: 198.0904 Da (Δ = 0.8 ppm). **IR (ATR):** $\tilde{\nu}$ =2424 (w), 1732 (m), 1606 (w), 1569 (s), 1477 (s), 1418 (w), 1372 (m), 1310 (m), 1278 (s), 1240 (s), 1180 (m), 1151 (m), 1125 (w), 1104 (m), 1048 (m), 1020 (m), 894 (m), 838 (m), 798 (s), 700 (s), 638 (s), 549 (m), 530 (m), 494 (w), 438 (m), 414 (w) cm^{-1}. **UV-Vis (MeCN):** λ_{max} = 215, 244, 277, 350 nm.

(*E*)-*N*-(4-(Pyridin-3-yldiazenyl)phenyl)acetamide 112

Procedure cited from master thesis of Michael Steinbiß.[282]

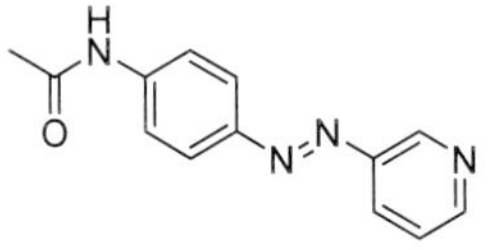

(E)-4-(pyridin-3'-yldiazenyl)aniline **111** (50 mg, 0.25 mmol, 1.00 eq.) was dissolved in Ac_2O (4 mL) and stirred at 50 °C until complete conversion was observed *via* TLC. The reaction mixture was cooled to rt., added to H_2O (10 mL), adjusted to pH 6 and extracted with EtOAc. The organic layer was washed with brine, dried over Na_2SO_4 and solvents were removed under reduced pressure. The crude product was purified by silica gel column chromatography (cH:EtOAc 1:1 → EtOAc), resulting in 31 mg (*E*)-*N*-(4-(pyridin-3-yldiazenyl)phenyl)acetamide **112** (1.3 mmol, 51%) of orange crystals.

TLC: R_f = 0.33 (EtOAc). **^{1}H NMR (MeOD, 300 MHz):** δ = 8.90 (s, 1H), 8.49 (d, *J* = 4.7 Hz, 1H), 8.08 (d, *J* = 7.6 Hz, 1H), 7.77 (d, *J* = 8.8 Hz, 2H), 7.64 (d, *J* = 8.8 Hz, 2H), 7.45 (dd, *J* = 8.2, 4.8 Hz, 1H), 4.78 (s, 2H), 2.06 (s, 3H) ppm. **^{13}C NMR (MeOD, 75 MHz):** δ = 170.3, 150.1, 148.3, 148.2, 145.3, 142.3, 127.7, 124.5, 123.7, 119.4, 22.6 ppm. **^{1}H NMR (300 MHz, DMSO):** δ = 10.38 (s, 1H), 9.14 (s, 1H), 8.77 (s, 1H), 8.23 (d, *J* = 8.2 Hz, 1H), 7.93 (d, *J* = 8.9 Hz, 2H), 7.83 (d, *J* = 9.0 Hz, 2H), 7.68 (dd, *J* = 8.4, 4.4 Hz, 1H), 2.11 (s, 3H) ppm. **^{13}C NMR (101 MHz, DMSO):** δ = 168.9, 150.7, 147.7, 147.4, 145.2, 143.2, 127.7, 125.0, 124.1, 119.1, 24.2 ppm. **HRMS (FAB+):** *m/z* calcd for $C_{13}H_{12}O_1N_4$: 240.1011 Da [M], found: 240.1012 Da (Δ = 0.5 ppm). **IR (ATR):** $\tilde{\nu}$ =3261 (vw), 3066 (vw), 1689 (w), 1593 (w), 1538 (w), 1500 (w), 1435 (w), 1409 (w), 1370 (w), 1302 (w), 1262 (w), 1175 (w), 1120 (w), 989 (w), 915 (w), 844 (w), 795 (w), 720 (w), 680 (w), 633 (vw), 619 (vw), 589 (w), 555 (w), 526 (w), 499 (w), 448 (vw) cm^{-1}.

(*E*)-3-((4-Acetamidophenyl)diazenyl)-1-(2-(*tert*-butoxy)-2-oxoethyl)pyridin-1-ium bromide 65

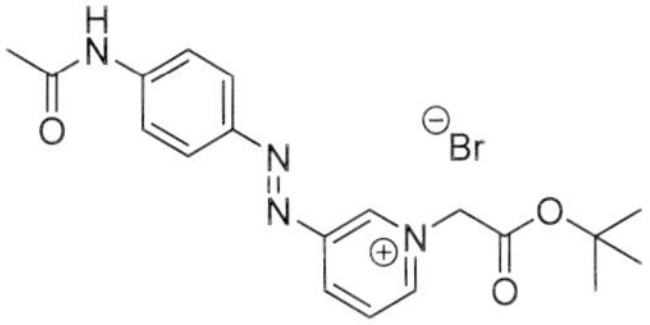

(*E*)-*N*-(4-(pyridin-3-yldiazenyl)phenyl)acetamide (31 mg, 0.129 mmol, 1.00 eq.) was dissolved in DMF (5 mL). *tert*-Butyl bromoacetate (40 μL, 50 mg, 0.258 mmol, 2.00 eq.) was added under stirring at rt. The reaction mixture was stirred at 60 °C for 60 h. DMF was evaporated. The remaining salt was dissolved in 5 mL MeOH and precipitated in 300 mL Et_2O, resulting in 55 mg (*E*)-3-((4-acetamidophenyl)diazenyl)-1-(2-(*tert*-butoxy)-2-oxoethyl)pyridin-1-ium bromide **65** (13 mmol, 98%) of orange crystals.

^{1}H NMR (300 MHz, MeOD): δ = 9.58 (s, 1H), 9.07 (s, 1H), 9.05 (s, 1H), 8.40 – 8.24 (m, 1H), 8.07 (d, *J* = 9.0 Hz, 2H), 7.88 (d, *J* = 9.0 Hz, 2H), 5.66 (s, 2H), 2.21 (s, 3H), 1.57 (s, 9H) ppm. **^{13}C NMR (75 MHz, MeOD):** δ = 172.1, 166.2, 151.8, 149.5, 147.9, 145.7, 143.6, 139.0, 129.3, 126.3, 120.8, 86.2, 62.8, 28.3, 24.2 ppm. **MS (ESI+):** *m/z* calcd for $[C_{19}H_{23}N_4O_3]^+$: 355.1765 Da $[M]^+$, found: 355.1696 Da.

5.9.1 Solid-phase synthesis 3-arylazopyridinium derivatives

The solid-phase synthesis of the 3-arylazopyridinium derivatives was performed by Tobias Bantle.[281]

(*E*)-2-(4-(Phenyldiazenyl)phenyl)acetic acid 112

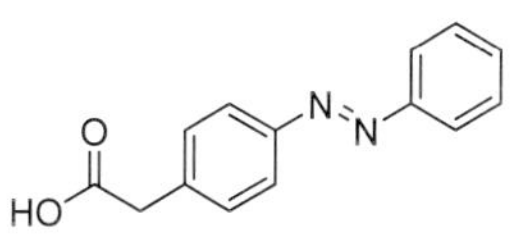

4-aminophenylacetic acid (706 mg, 4.67 mmol, 1.00 eq.) was dissolved in AcOH (20 mL). Nitrosobenzene (750 mg, 7.00 mmol, 1.50 eq.) was added and the reaction mixture was stirred at rt overnight. The mixture was then diluted with water (100 mL) and the aqueous phase was extracted with CH_2Cl_2 (3×100 mL). The combined organic layers were washed with water, dried over Na_2SO_4 and the solvents were evaporated under reduced pressure. The crude was purified by column chromatography (CH_2Cl_2:MeOH 1:0 to 95:5) resulting in 505 mg (*E*)-2-(4-(phenyldiazenyl)phenyl)acetic acid **112** (2.10 mmol, 45% yield) as orange solid.

^{1}H NMR (400 MHz, DMSO-d_6,): δ = 12.48 (s, 1H), 7.89 (dd, J = 8.1, 1.7 Hz, 2H), 7.86 (d, J = 8.4 Hz, 2H), 7.59 (d, J = 7.5 Hz, 3H), 7.49 (d, J = 8.4 Hz, 2H), 3.71 (s, 2H) ppm. **^{13}C NMR (101 MHz, DMSO):** δ = 172.3, 151.9, 150.7, 138.9, 131.5, 130.6, 129.5, 122.5, 122.5, 40.5 ppm. **HRMS (ESI+):** *m/z* calcd for $C_{14}H_{13}N_2O_2^+$: 241.0972 Da $[M+H]^+$, found: 241.0971 Da (Δ = 0.4 ppm). **Analytical HPLC:** water+0.1% TFA:acetonitrile+0.1% TFA 95:5 to 5:95 in 30 min, flow rate: 1.00 mL/min, detection: 330 nm, t_{Ret} = 16.3 min.

Fmoc-Phe-Rink resin 113

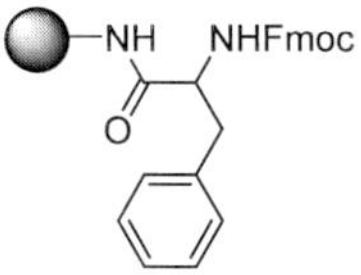

NH NHFmoc
O

Fmoc-protected Rink resin (1.00 g, 200 mesh, 0.61 mmol/g) was swollen for 1 h in CH_2Cl_2 (10 mL), filtered off and treated 5 min with 20% piperidine in DMF (10 mL) to remove the protecting group. After extensive washing with DMF and CH_2Cl_2 the deprotected resin was shook overnight with a coupling mixture composed of Fmoc-*L*-phenylalanine (77.5 mg, 200 µmol, 1.00 eq.), HBTU (75.9 mg, 200 µmol, 1.00 eq.), DIPEA (65 µL, 200 µmol 1.00 eq.) and 2,6-lutidine (65 µL, 300 µmol, 1.50 eq.) in NMP (*N*-methylpyrrolidinone, 5 mL), and washed extensively with CH_2Cl_2. The remaining unreacted amine groups were acetylated ('capped') with 5 mL of the capping mixture: the solution of acetic anhydride (0.92 mL) and 2,6-lutidine (1.30 mL) in DMF (18 mL). The overall result was decrease of the resin loading from 0.61 mmol/g to 0.20 mmol/g. The resulting Fmoc-Phe-Rink resin **113** was extensively washed with DMF and CH_2Cl_2, dried with Et_2O and stored in the fridge for further use.

Boc-Ala-Ala-Aha-Phe-Rink resin 114

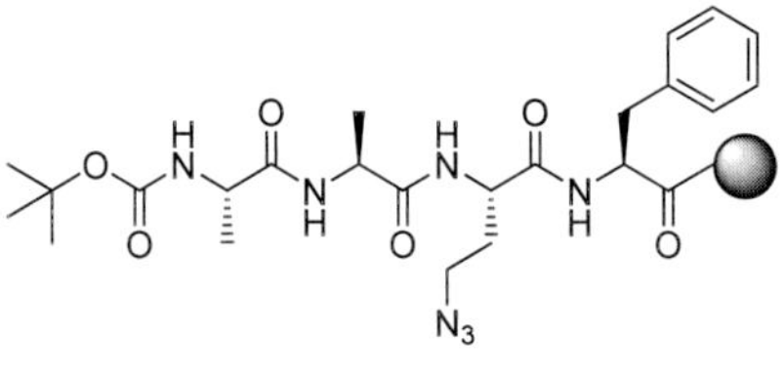

H O H O
O N N
N N
O H O H O
N_3

Fmoc-Phe-Rink resin **113** (200 mg, 0.04 mmol, 1.00 eq.) was swollen with CH_2Cl_2 for 1 h, filtered off and treated with 20% piperidine in DMF (5 mL) to remove the N-terminal Fmoc protecting group. After extensive washing with DMF and CH_2Cl_2, the resin was shook with a coupling mixture comprised of *(S)*-*N*-Fmoc-2-amino-4-azidobutanoic acid (73.3 mg, 0.20 mmol, 5.00 eq.), HBTU (75.9 mg, 0.20 μmol, 5.00 eq.), DIPEA (65 μL, 0.20 μmol, 5.00 eq.) and 2,6-lutidine (65 μL, 0.30 μmol, 7.50 eq.) in NMP (5 mL) for 2 h. The resin was washed with CH_2Cl_2 and treated with the 'capping mixture' mentioned above (previous procedure) to acetylate the potentially unreacted amino groups. The same procedure (*N*-terminal Fmoc group deprotection, coupling and acetylation) was further performed with Fmoc-Ala-OH (62.3 mg, 0.20 mmol, 5.00 eq.) and Boc-Ala-OH (73.3 mg, 0.20 mmol, 5.00 eq.) (and identical amounts of HBTU, DIPEA and 2,6-lutidine). In order to check the outcome of the synthesis and purity of the product on the solid support, some resin beads were treated with 100 μL of 20% *m*-cresol solution in TFA for 1 h, in order to cleave the oligopeptide from the solid support and concomitantly deprotect the N-terminal Boc-group. The polymeric beads were filtered off and the remaining solution was mixed with Et_2O (1 mL) in an Eppendorf tube (1.5 mL), spun down with a centrifuge (5 min, 14.000 rpm) and the resulting pellet was washed with fresh Et_2O (1 mL). It was dissolved in 100 μL of MeCN:H_2O 1:1 and used for MALDI-MS as well as HPLC analysis, which both confirmed the purity and the expected composition of the oligopeptide **114**.

MS (MALDI-TOF, CHCA matrix): *m/z* calcd for $C_{19}H_{28}N_8O_4$: 433.2234 Da [M+H], found: 433.2401 Da. **Analytical HPLC:** water+0.1% TFA:acetonitrile+0.1%TFA 95:5 to 5:95 in 30 min, flow rate: 1.00 mL/min, t_{Ret} = 11.6 min.

H-Ala-Ala-APy-Phe-NH_2 67

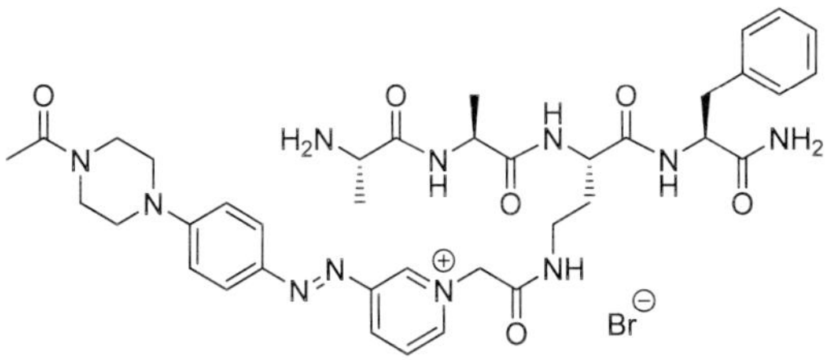

Boc-Ala-Ala-Aha-Phe-Rink resin **114** (120 mg, 0.02 mmol, 1.00 eq.) was treated with 17% solution of tri-*n*-butylphosphine in a mixture of NMP:THF:water (10:9:1) (3.0 mL) for 4 h to reduce the azide to an amino group. After extensive washing with THF and CH_2Cl_2, the resin was shook with a coupling mixture of the

(*E*)-3-((4-(4-acetylpiperazin-1-yl)phenyl)diazenyl)-1-(carboxymethyl)pyridin-1-ium bromide **60OH** (173 mg, 0.400 mmol, 20.0 eq.), HOBt (216 mg, 1.60 mmol, 80.0 eq.) and DIC (253 mg, 2.00 µmol, 100 eq.) in NMP (5 mL) overnight and then extensively washed with DMF and CH_2Cl_2. In order to cleave the product from resin and remove the Boc protecting group, the beads were treated with 500 µL of 20% *m*-cresol in TFA for 2 h. The resulting solution was filtered to separate the remaining resin and mixed with 10 mL Et_2O to precipitate the product. The crude product was isolated by spinning the mixture in a 50-mL Falcon tube (20 min centrifugation, 8.000 rpm) and washing the pellet with fresh Et_2O (20 mL). The product **67** was purified *via* preparative HPLC.

HRMS (ESI+): *m/z* calcd for $C_{38}H_{50}N_{11}O_6^+$: 756.3940 Da $[M]^+$, 757.3974 Da [41%], found: 756.3976 Da ($\Delta = 4.8$ ppm), 757.3951 Da ($\Delta = 3.0$ ppm). **Analytical HPLC:** water+0.1% TFA/acetonitrile+0.1%TFA 95:5 to 5:95 in 30 min, flow rate: 1.00 mL/min, $t_{Ret} = 5.6$ min.

H-Ala-Ala-Azo-Phe-NH_2 68

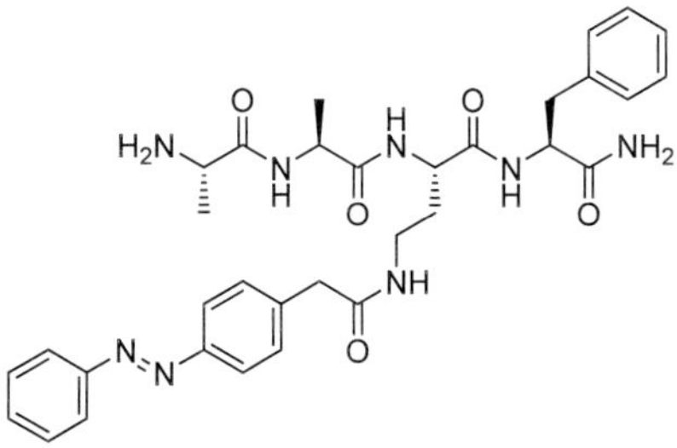

Boc-Ala-Ala-Aha-Phe-Rink resin **114** (120 mg, 0.0200 mmol, 1.00 eq.) was treated with 17% solution of tri-*n*-butylphosphine in a mixture of NMP:THF:water (10:9:1) (3.0 mL) for 4 h to reduce the azide to an amino group. After extensive washing with THF and CH_2Cl_2 the resin was shook with a coupling mixture of the (*E*)-2-(4-(phenyldiazenyl)phenyl)acetic acid **112** (97.0 mg, 0.400 mmol, 20.0 eq.), HOBt (216 mg, 1.60 mmol, 80.0 eq.) and DIC (253 mg, 2.00 µmol, 100 eq.) in NMP (5 mL) overnight, and then extensively washed with DMF and CH_2Cl_2. In order to cleave the product from resin and remove the Boc protecting group, the beads were treated with 500 µL of 20% *m*-cresol in TFA for 2 h. The resulting solution was filtered to separate the remaining resin and mixed with 10 mL Et_2O to precipitate the product. The crude product was isolated by spinning the mixture in a 50-mL Falcon tube (20 min centrifugation, 8.000 rpm) and washing the pellet with fresh Et_2O (20 mL). The product **68** was purified *via* preparative HPLC.

HRMS (ESI+): *m/z* calcd for $C_{33}H_{41}N_8O_5^+$: 629.3194 Da $[M+H]^+$, found: 629.3196 Da ($\Delta = 0.3$ ppm). **Analytical HPLC:** water+0.1% TFA:acetonitrile+0.1%TFA 95:5 to 5:95 in 30 min, flow rate: 1.00 mL/min, $t_{Ret} = 6.82$ min (*cis*-conformation); $t_{Ret} = 7.3$ min (*trans*-conformation).

5.9.2 Photophysical properties of 3-arylazopyridinium derivatives

The analytics and photophysical properties of 3-arylazopyridinium derivatives were measured by Jens Lackner and Mathilde A. Bichelberger.[281]

Determination of the photophysical properties[281]

Azo compound solutions were prepared by dissolving the solid compounds in phosphate buffered saline, pH 7.4 (PBS, Dulbecco's PBS without Ca^{2+} and Mg^{2+}, PAA Labs, Cölbe, Germany). Those compounds with absorbance maxima above 450 nm in aqueous solutions at neutral pH were selected for further characterization. Molecules **62** and **63** were also dissolved in 10 mM citric acid-sodium phosphate buffer, pH 2.7 – 7.5 (citric acid monohydrate, Merck, Darmstadt, Germany; disodium hydrogen phosphate dehydrate, Carl Roth, Karlsruhe, Germany) to determine the pK_a values.

The kinetics of *cis-trans* isomerization were measured by using a homebuilt nanosecond flash photolysis system described earlier in detail.[346] Briefly, the azo compounds were isomerized by a 6 ns (full width at half maximum) pulse from a frequency doubled Nd:YAG laser (Surelite II-10, Continuum, Santa Clara, CA). Absorbance changes were monitored with light from a tungsten source that was passed through a monochromator (which was set to 480 nm unless noted otherwise). The transmitted light was detected with a photomultiplier tube (R5600U, Hamamatsu Corp., Middlesex, NJ). After amplification, the voltage output was recorded by two analogue-to-digital converter cards, NIPCI 5114 and NI-PCI 6221 (National Instruments, Munich, Germany), and digitized in the time ranges 4 ns – 3 ms and > 4 μs, respectively, to yield a rebinding trace. 50 traces were averaged and fitted with a mono-exponential decay function to determine the lifetimes of the *cis*-isomers.

Dependence of the physical properties on the pH value

For compounds **56** and **60**, we have determined the pH-dependence of the physical properties. Therefore, we measured absorption spectra of the stable *trans*-isomer and *cis-trans* isomerization kinetics at different pH values. The absorption spectra of compound **5** (Figure 119) are similar over the whole pH range from 2.7 to 10.2, with only minor shifts of the absorption maximum between 478 and 483 nm. In contrast, the *cis-trans* relaxation kinetics are strongly dependent on pH (Figure 121). The pH-dependent variation can be modelled with a simple protonation reaction following a Henderson-Hasselbalch relation (Figure 122), with $pK_a \approx 7.4 \pm 0.1$. With increasing pH, the relaxation back to the stable *trans*-isomer becomes slower (Table 4). Possibly, isomerization is affected by the protonation state of a functional group of the *Z*-isomer. The same trend is visible for compound **56**, however, the pH dependence is less clear (Figure 122 and Figure 123).

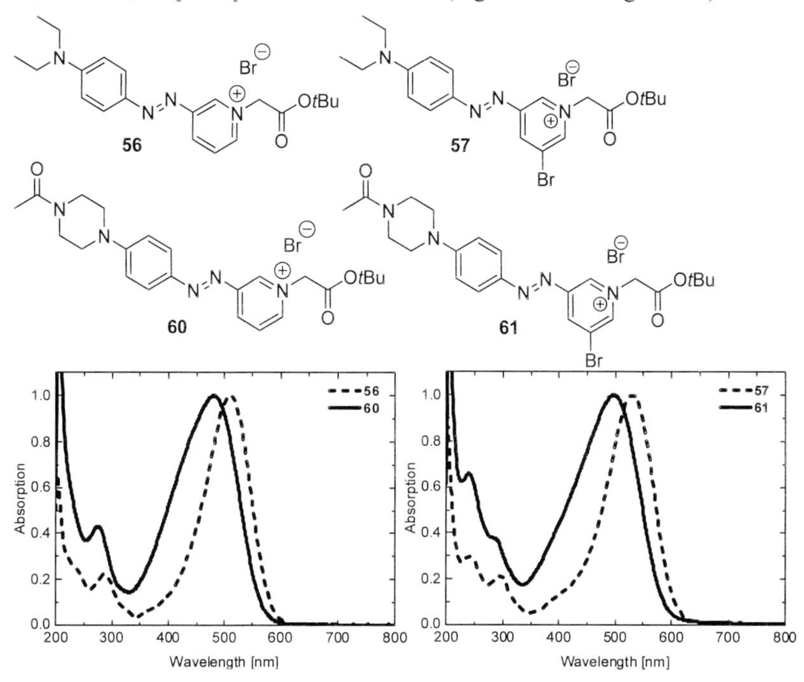

Figure 117. Structures and absorption spectra of the *trans*-forms of 56 vs. 60 (left) and 57 vs. 61 (right).

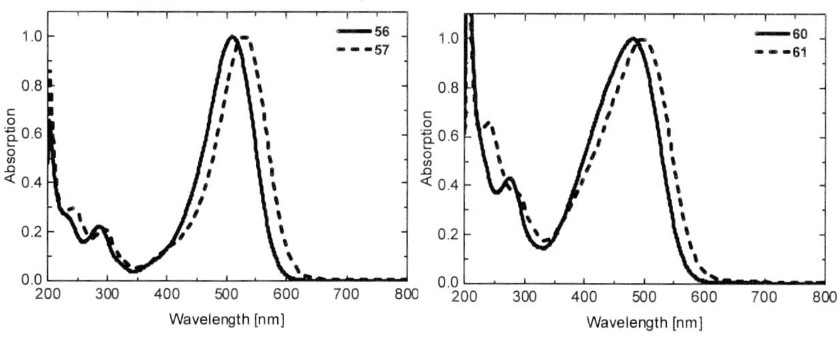

Figure 118. Absorption spectra of the *trans*-forms of molecules 56 vs. 57 and 60 vs. 61 in PBS.

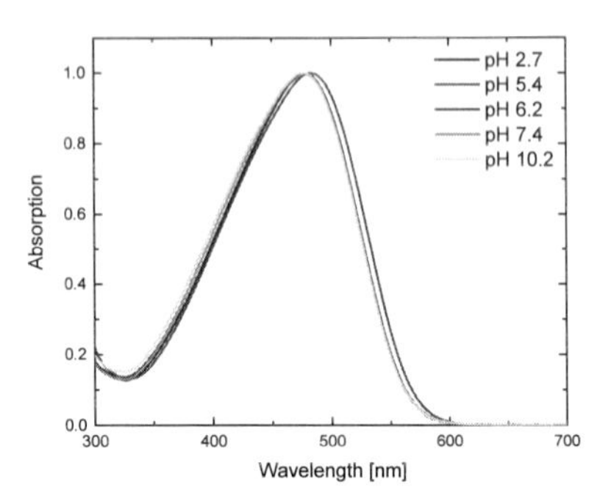

Figure 119. Absorption spectra of compound 60 in a pH range between 2.7 and 10.2. All spectra are normalized to the same peak amplitude.

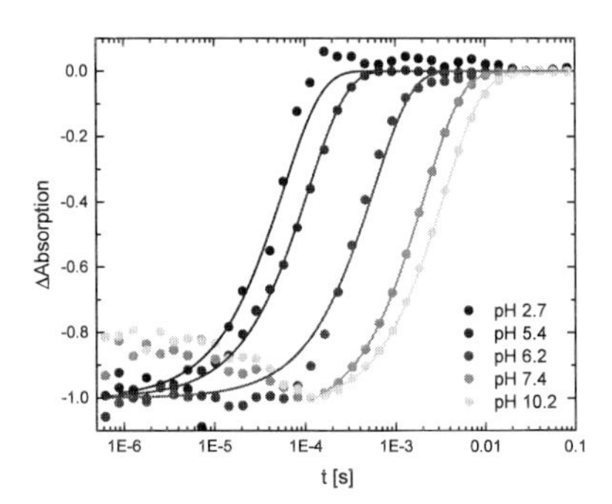

Figure 120. *Cis-trans* relaxation kinetics of compound 60 at different pH values. Lines represent the mono-exponential fits to the data. All traces are scaled between -1 and 0 for better comparison.

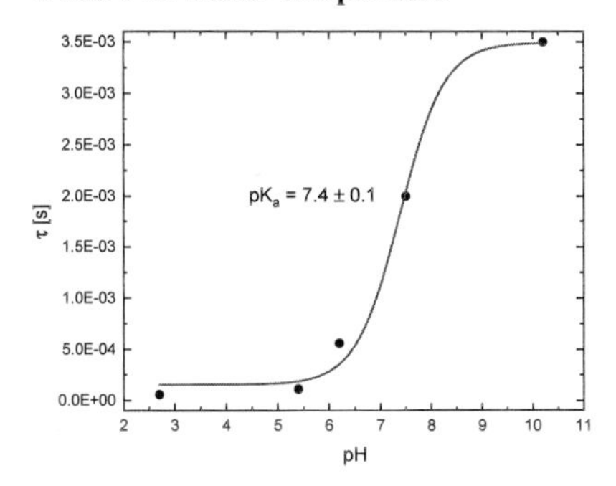

Figure 121. Dependence of *Z*-lifetime of compound 60 on the pH value. The line represents the Henderson-Hasselbalch fit to the data.

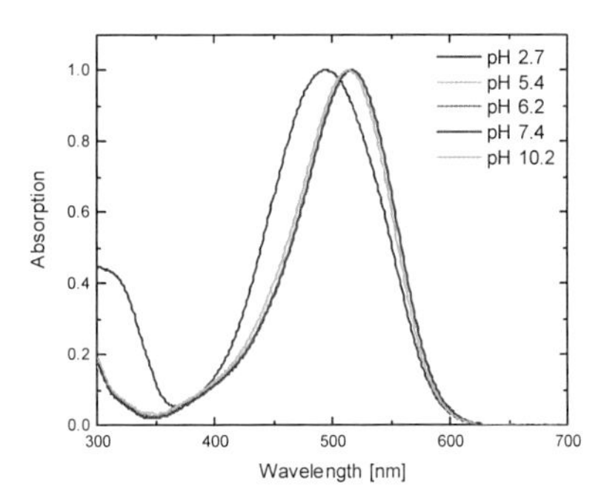

Figure 122. Absorption spectra of compound 56 in a pH range between 2.7 and 10.2. All spectra are normalized to the same peak amplitude.

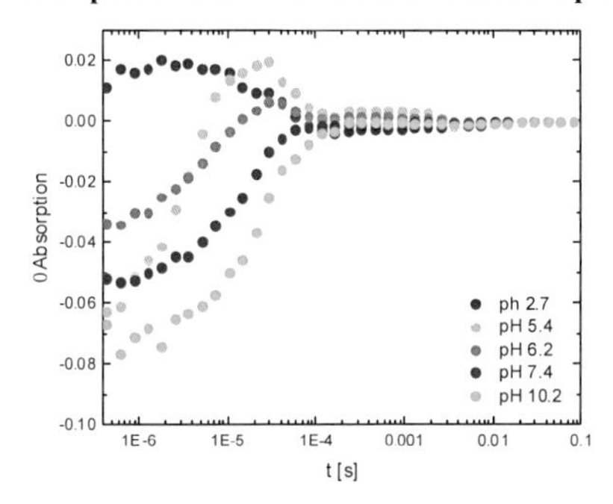

Figure 123. *Cis-trans* relaxation kinetics of compound 56 at different pH values.

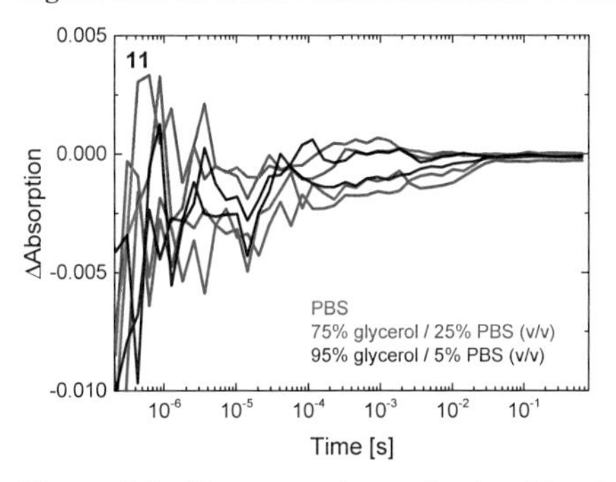

Figure 124. *Cis* to *trans* isomerization kinetics of molecule 4AFM2-11 in PBS and glycerol/PBS mixtures.

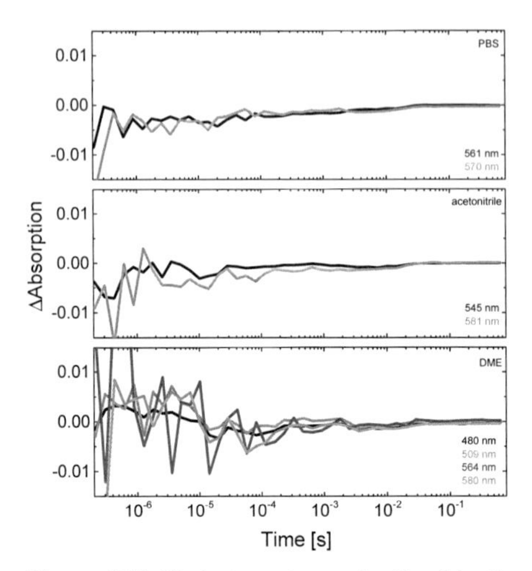

Figure 125. ***Cis*** **to** ***trans*** **isomerization kinetics of molecule 4AFM2-11 in PBS, acetonitrile and dimethoxyethane (DME). Monitoring wavelengths are given in each panel.**

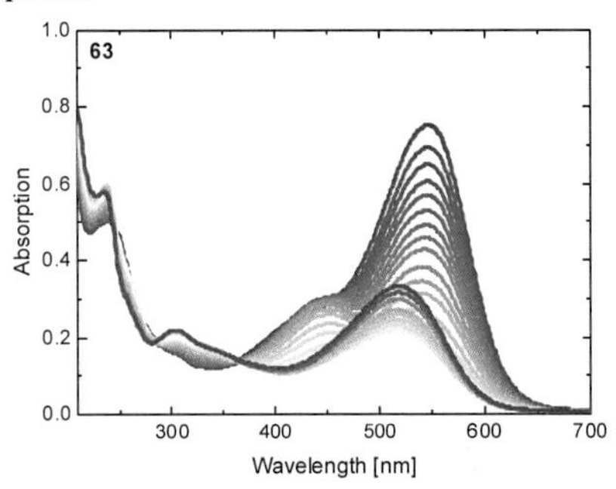

Figure 126. Absorption spectra of 63 upon continuous illumination with 523 nm light (from blue to red).

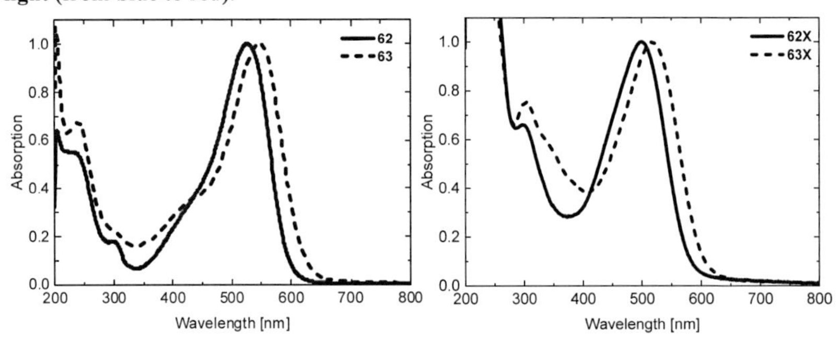

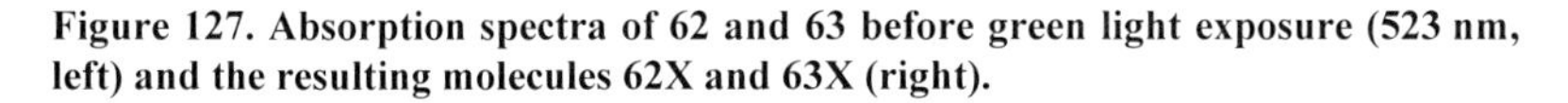

Figure 127. Absorption spectra of 62 and 63 before green light exposure (523 nm, left) and the resulting molecules 62X and 63X (right).

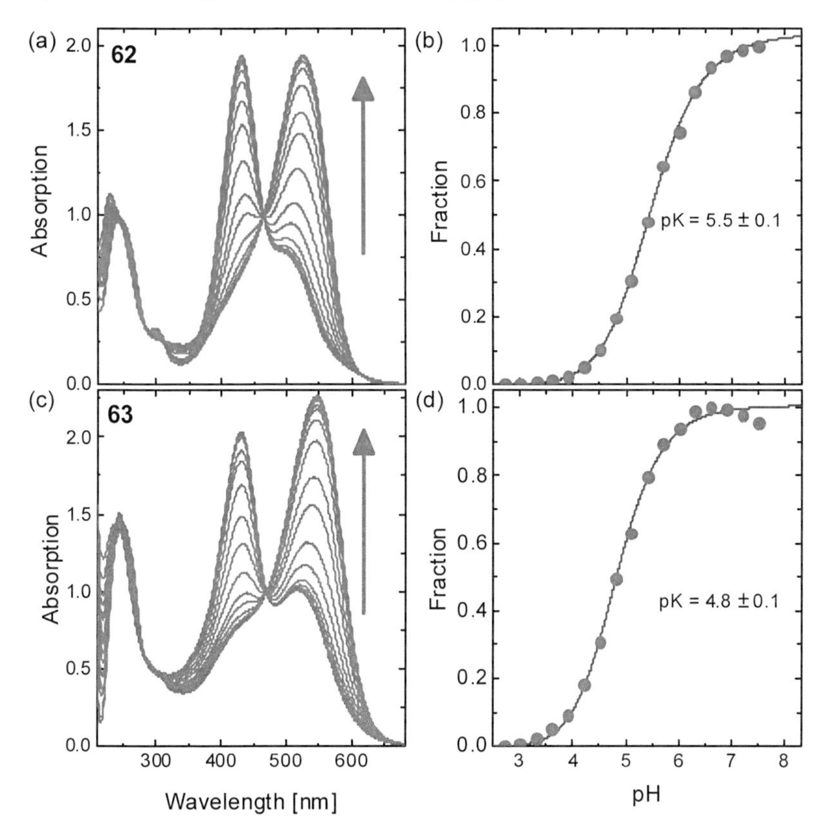

Figure 128. Optical properties of 62 and 63. (a) Absorption spectra of 62 at pH 2.7 – 7.5; (b) Fraction of deprotonated species 62 as a function of pH; (c) Absorption spectra of 63 at pH 2.7 – 7.5; (d) Fraction of deprotonated species 63 as a function of pH. Arrows point toward increasing pH.

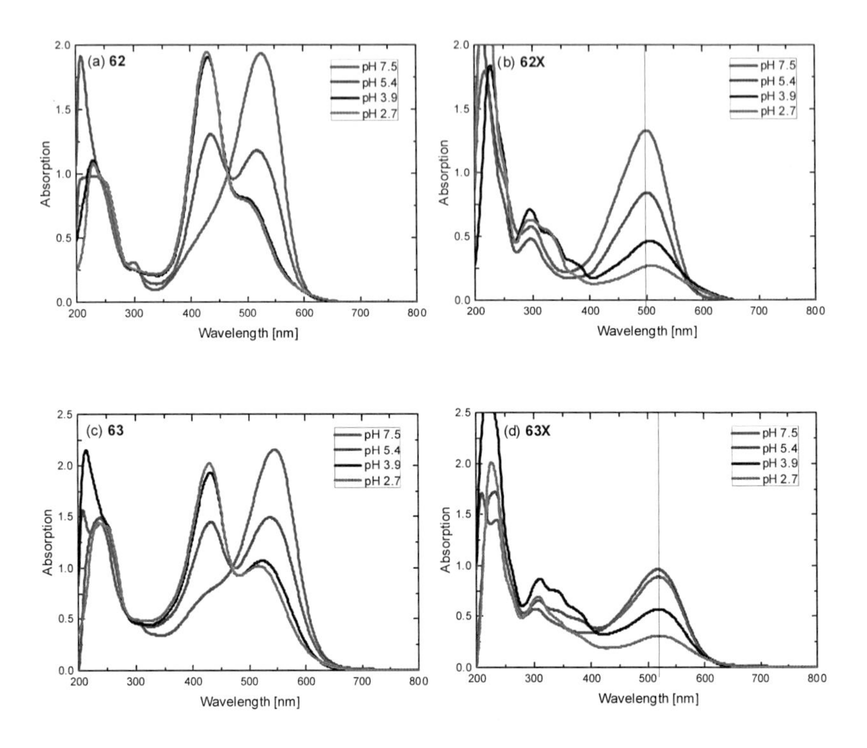

Figure 129. Absorption spectra of (a) 62 and (c) 63 at different pH values in the dark (a, c) and upon 523 nm light exposure (as 62X and 63X, respectively) (b, d).

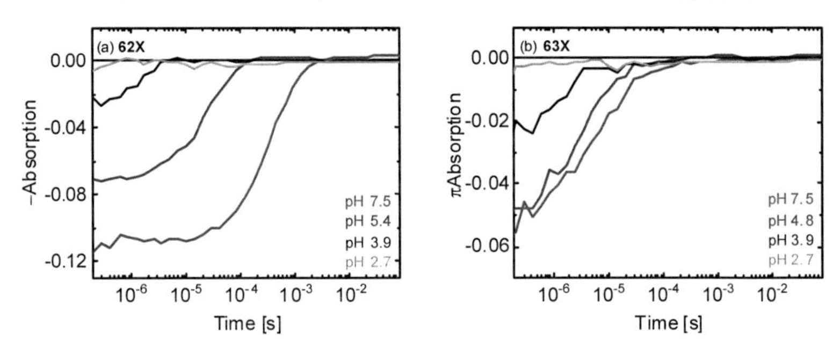

Figure 130. pH-dependent isomerization kinetics of the 62X and 63X molecules.

Table 25. Lifetimes of the *cis*-forms of 62X and 63X at different pH values.

pH citric acid/sodium phosphate	τ [s] 63X	 62X
7.5	$1.05 \pm 0.05 \cdot 10^{-5}$	$4.33 \pm 0.08 \cdot 10^{-4}$
4.8	$5.3 \pm 0.2 \cdot 10^{-6}$	-
5.4	-	$2.98 \pm 0.07 \cdot 10^{-5}$
3.9	$2.3 \pm 0.2 \cdot 10^{-6}$	$1.3 \pm 0.1 \cdot 10^{-6}$
2.7	-	-

5.9.2.1 Photooxidation of molecule 62

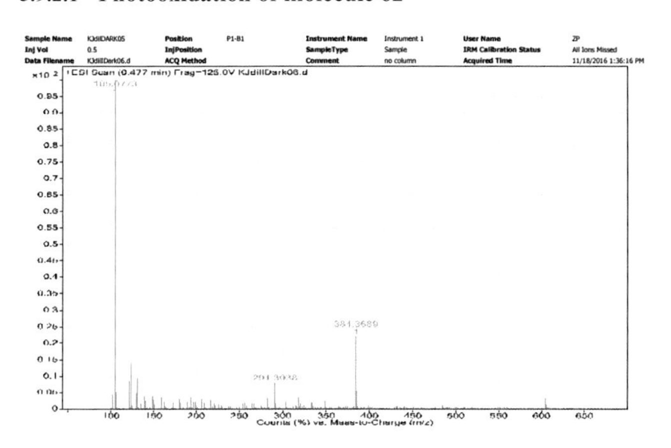

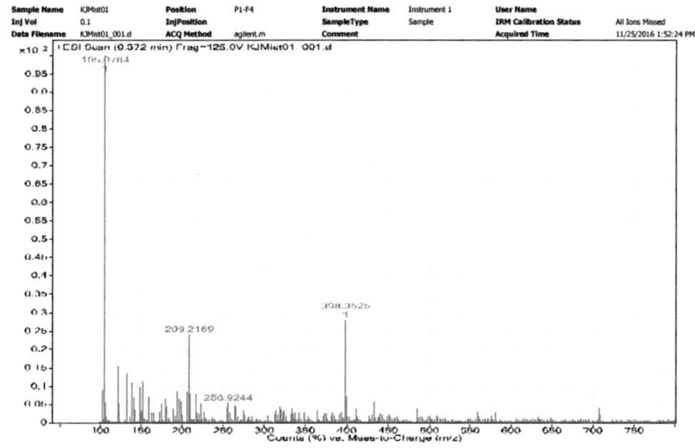

Figure 131. Mass spectrometry of molecule 62 and the photooxidation product 62X.

The photooxidation process was followed with ^{1}H NMR. In the first experiment, irradiation of a 5 mM solution of **62** in PBS, pH 7.4 with a 523 nm 10 W LED for 30 min did not result in significant conversion. In the second experiment, a more diluted sample (31.8 μM of **62** in PBS, pH 7.4) was irradiated under the same light source and intensity for 2 h. The substrate **62** was no longer detected with LCMS. Instead, a new product **62X** with the mass bigger by 14 Da was observed. That is also reflected by the resulting NMR spectrum. In both cases, the irradiated samples (and a non-irradiated sample of **62** as a reference) were lyophilized and re-dissolved in deuterated acetonitrile (CD_3CN) prior to the NMR analysis.[281]

5.10 Dicarboxylic acid linker

Dimethyl 2',5'-dimethyl-[1,1':4',1''-terphenyl]-4,4''-dicarboxylate 115

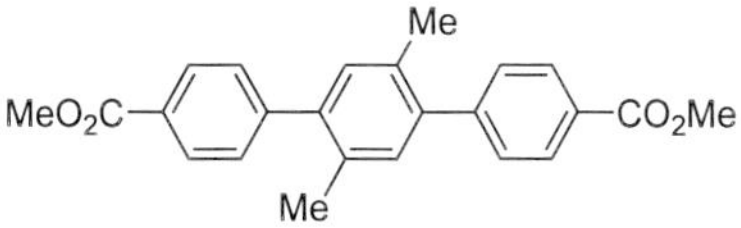

Ethylene glycol dimethyl ether (250 mL) was bubbling with nitrogen for around 1 h before introducing into argon-protected solid mixture of 1,4-dibromo-2,5-dimethylbenzene (4.00 g, 15.2 mmol, 1.00 eq.), 4-methoxyl carbonylphenylboronic acid (6.55 g, 36.4 mmol, 2.40 eq.), potassium carbonate (12.6 g, 91.0 mmol, 6.00 eq.) and tetrakis (triphenylphosphine) palladium (0.490 g, 0.420 mmol, 0.0300 eq.). The mixture was allowed to reflux (95 °C) for 4 d under nitrogen protection. After cooling to rt., the solvent was evaporated to dryness. The residue was washed with a large amount of water followed by acetone. After removing the solvent, the residue was purified with column chromatography (silica gel, CH_2Cl_2) to give the 3.49 g dimethyl 2',5'-dimethyl-[1,1':4',1''-terphenyl]-4,4''-dicarboxylate **115** (9.32 mmol, 62%) as a white solid.[347]

^{1}H NMR (300 MHz, $CDCl_3$): δ = 8.11 (d, J = 8.1 Hz, 4H), 7.45 (d, J = 8.1 Hz, 4H), 7.16 (s, 2H), 3.95 (s, 6H), 2.28 (s, 6H) ppm. **^{13}C NMR (75 MHz, $CDCl_3$):** δ = 167.1, 146.4, 140.5, 132.8, 131.8, 129.6, 129.4, 128.8, 52.3, 20.0 ppm. **HRMS (EI+):** *m/z* calcd for $C_{24}H_{22}O_4$: 374.1513 Da [M], found: 374.1514 Da (Δ = 0.4 ppm). **IR (ATR):** $\tilde{\nu}$ = 2944 (vw), 1713 (m), 1608 (w), 1519 (w), 1492 (vw), 1466 (w), 1436 (w), 1411 (w), 1385 (w), 1278 (m), 1192 (m), 1106 (m), 1044 (w), 1015 (w), 990 (w), 970 (w), 922 (w), 856 (m), 828 (w), 784 (w), 758 (m), 702 (m), 652 (w), 507 (w), 466 (w) cm^{-1}.

2',5'-Dimethyl-[1,1':4',1''-terphenyl]-4,4''-dicarboxylic acid 116

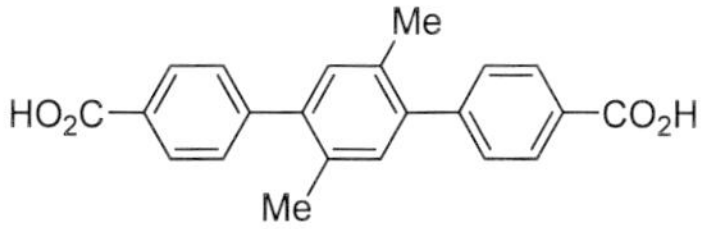

The dimethyl 2',5'-dimethyl-[1,1':4',1''-terphenyl]-4,4''-dicarboxylate **115** (1.00 g, 2.67 mmol) was stirred in THF (40 mL) and MeOH (40 mL) mixed solvent, to which a solution of KOH (4.50 g, 80.1 mmol, 30.0 eq.) in H_2O (40 mL) was introduced. This mixture was refluxed for 12 h. After cooling down to rt, THF and MeOH were evaporated. Additional water was added to the resulting water phase and the mixture was slightly heated until the solid was fully dissolved, then the homogeneous solution was acidified with diluted HCl until no further precipitate was detected (pH ≈ 2-3). The solid 823 mg 2',5'-dimethyl-[1,1':4',1''-terphenyl]-4,4''-dicarboxylic acid **116** (2.38 mmol, 89%) was collected by filtration, washed with water (3×20 mL), Et_2O (3×20 mL) and dried *in vacuo*.[347]

^{1}H NMR (300 MHz, DMSO): δ = 13.00 (s, 1H), 8.02 (d, *J* = 8.2 Hz, 2H), 7.50 (d, *J* = 8.2 Hz, 2H), 7.19 (s, 1H), 2.23 (s, 3H) ppm. **^{13}C NMR (75 MHz, DMSO):** δ = 167.2, 145.4, 139.8, 132.3, 131.6, 129.4, 129.3, 129.2, 19.6 ppm. **HRMS (EI+):** *m/z* calcd for $C_{22}H_{18}O_4$: 346.1200 Da [M], found: 346.1201 Da (Δ = 0.3 ppm). **IR (ATR):** ṽ = 2984 (w), 2669 (w), 1684 (s), 1604 (m), 1564 (w), 1489 (vw), 1421 (m), 1385 (w), 1311 (m), 1282 (m), 1172 (w), 1125 (w), 1100 (w), 1036 (vw), 1016 (w), 895 (w), 854 (m), 809 (w), 786 (w), 757 (m), 707 (m), 546 (m), 505 (w), 461 (w) cm^{-1}.

Dimethyl 2',5'-bis(bromomethyl)-[1,1':4',1''-terphenyl]-4,4''-dicarboxylate 117

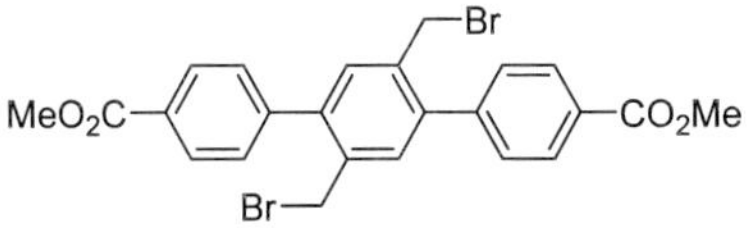

The dimethyl 2',5'-dimethyl-[1,1':4',1''-terphenyl]-4,4''-dicarboxylate **116** (1.79 g, 4.78 mmol, 1.00 eq.), *N*-bromosuccinimide (1.79 g, 11.2 mmol, 2.10 eq.), and benzoyl peroxide (107 mg, 0.440 mmol, 0.09 eq.) were stirred in carbon tetrachloride (45 mL) under nitrogen protection to reflux for 18 h. After cooling to rt., the solvent was removed. The residue was washed with Et_2O (1×100 mL), MeOH (2×10 mL) to give 2.01 g dimethyl 2',5'-bis(bromomethyl)-[1,1':4',1''-terphenyl]-4,4''-dicarboxylate **117** (3.79 mmol, 79%) as a white solid.[347]

^{1}H NMR (300 MHz, DMSO): δ = 8.09 (t, *J* = 8.2 Hz, 5H), 7.91 (d, *J* = 8.0 Hz, 2H), 7.74 – 7.41 (m, 2H), 4.66 (s, 4H), 3.91 (s, 6H) ppm. **^{1}H NMR (300 MHz, $CDCl_3$):** δ = 8.34 – 8.07 (m, 4H), 7.69 (d, *J* = 8.4 Hz, 1H), 7.58 (d, *J* = 8.3 Hz, 3H), 7.44 (s, 2H), 4.41 (s, 4H), 3.97 (s, 6H) ppm. **^{13}C NMR (75 MHz, $CDCl_3$):** δ = 166.9, 143.8, 141.3, 135.8, 132.9, 130.3, 129.9, 129.2, 127.4, 52.4, 30.8 ppm. **HRMS (EI+):** *m/z* calcd for $C_{24}H_{20}{}^{79}Br_2O_4$: 529.9723 Da [M], found: 529.9722 Da (Δ = 0.1 ppm); *m/z* calcd for $C_{24}H_{20}{}^{79}Br_1{}^{81}Br_1O_4$: 531.9702 Da [M], found: 531.9702 Da (Δ = 0.1 ppm). **IR (ATR):** ṽ = 2950 (vw), 1715 (m), 1609 (w), 1432 (w), 1278 (m), 1221 (w), 1191 (w), 1109 (m), 1016 (w), 962 (vw), 915 (vw), 861 (w), 800 (w), 772 (w), 728 (w), 712 (m), 657 (vw), 638 (vw), 574 (w), 513 (w), 466 (vw) cm^{-1}.

2,2-Bis((propioloyloxy)methyl)propane-1,3-diyl dipropiolate 118

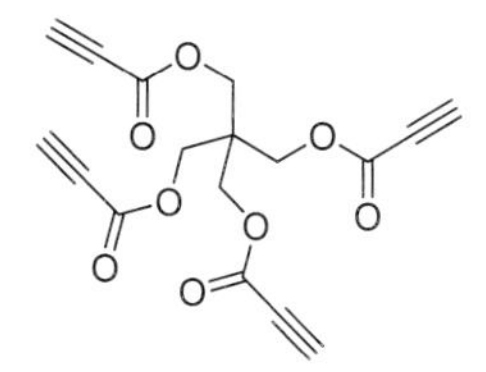

In a dry 250-mL round-bottom flask equipped with a Dean-Stark trap and magnetic stirring bar was placed trimethylolethane (1.45 g, 10.7 mmol, 1.00 eq.) and *p*-Toluenesulfonic acid monohydrate (203 mg, 1.07 mmol, 0.10 eq.) dissolved in dry 90 mL PhMe. Propiolic acid (3.30 g, 47.1 mmol, 4.40 eq.) was added dropwise and then the mixture was heated to reflux (130 °C). After 2 h the light brown solution was cooled to rt, 90 mL of EtOAc was added and washed twice with 5% sodium hydroxide solution (2×25 mL), then with brine solution (20 mL). Thereafter the product was dried over anhydrous magnesium sulfate, filtered and concentrated *in vacuo*. The crude was purified by silica gel column chromatography (R_f = 0.23 3:7 EtOAc:cH, R_f = 0.4 4:6 EtOAc:cH) yielding 0.126 g 2,2-bis((propioloyloxy)methyl)propane-1,3-diyl dipropiolate **118** (0.366 mmol, 3%) as a slightly yellow oil.[348]

^{1}H NMR (300 MHz, $CDCl_3$): δ = 4.29 (s, 8H), 2.98 (s, 4H) ppm. **^{13}C NMR (75 MHz, $CDCl_3$):** δ = 151.9, 76.6, 73.8, 63.5, 41.9 ppm. **^{1}H NMR (300 MHz, DMSO):** δ = 4.66 (s, 4H), 4.26 (s, 8H) ppm. **HRMS (EI+):** *m/z* calcd for $C_{17}H_{13}O_8$: 345.0605 Da [M], found: 345.0606 Da (Δ = 0.3 ppm). **IR (ATR):** $\tilde{\nu}$ = 3285 (m), 2922 (vw), 2119 (m), 1712 (s), 1465 (w), 1383 (w), 1210 (s), 993 (m), 924 (m), 896 (w), 747 (s), 703 (m), 672 (s); 607 (w), 580 (m), 522 (w), 438 (w) cm^{-1}.

Tetrakis(2-propynyloxymethyl)methane 119

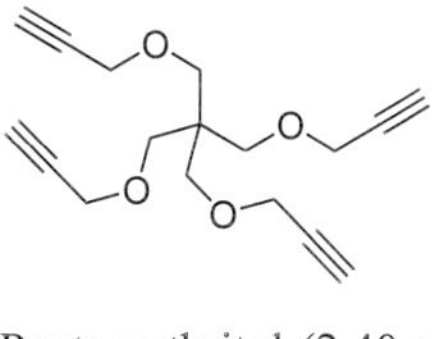

Pentaerythritol (2.40 g, 17.6 mmol, 1.00 eq.) was added into a solution of KOH (14.8 g, 264 mmol, 15.0 eq.) in anhydrous DMF (30 mL). After stirring at 0 °C for 30 min, propargyl bromide (21.0 g, 176.3 mmol, 10.0 eq.) was slowly added over a period of 20 min. The colour of the solution turned brown and the reaction mixture was stirring at 40 °C overnight. The reaction mixture was quenched with sat. aqueous NH_4Cl (20 mL) and extracted with ethyl ether (3×100 mL). The organic layers were combined, washed with water, then with brine, and dried over Na_2SO_4. After removal of the ethyl ether under reduced pressure, the product was further purified by silica gel column chromatography, using mixed 15% EtOAc in cH (R_f = 0.32, detected with aqueous MnO_4) as the eluent and 0.819 g tetrakis(2-propynyloxymethyl)methane **119** (2.84 mmol, 16%) were obtained.[349]

^{1}H NMR (300 MHz, $CDCl_3$): δ = 4.11 (d, J = 2.4 Hz, 8H), 3.52 (s, 8H), 2.40 (s, 4H) ppm. **^{13}C NMR (75 MHz, $CDCl_3$):** δ = 80.2, 74.2, 69.1, 58.8, 44.9 ppm. **HRMS (EI+):** *m/z* calcd for $C_{17}H_{21}O_4$: 289.1434 Da [M], found: 289.1433 Da (Δ = 0.3 ppm). **IR (ATR):** ṽ = 3288 (w), 2875 (w), 2115 (vw), 1474 (w), 1441 (w), 1358 (w), 1263 (w), 1194 (vw), 1170 (w), 1083 (s), 1018 (m), 959 (w), 903 (w), 632 (m), 535 (w) cm^{-1}.

1,1,1-Trimethylolethane tripropiolate 120

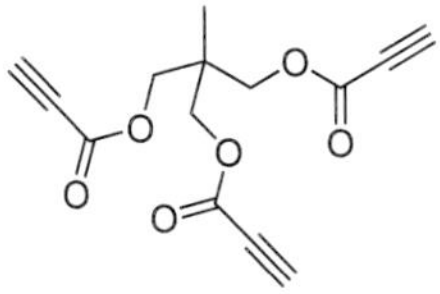

In a dry 250-mL round-bottom flask equipped with a Dean-Stark trap and magnetic stirring bar was placed trimethylolethane (1.00 g, 10.3 mmol, 1.00 eq.) and *p*-Toluenesulfonic acid monohydrate (195 mg, 1.03 mmol, 0.100 eq.) dissolved in dry PhMe (87 mL). Propiolic acid (2.37 g, 33.9 mmol, 3.30 eq.) was added dropwise, and then the mixture was heated to reflux (oil bath 100 °C). After 16 h the solution was concentrated *in vacuo*.[348a] The crude product was concentrated *in vacuo* and purified by silica gel column chromatography (R_f = 0.28 3:7 EtOAc:cH) to yield 0.271 g 1,1,1-trimethylolethane tripropiolate **120** (0.981 mmol, 10%).

^{1}H NMR (300 MHz, DMSO): δ = 4.62 (s, 3H), 4.13 (s, 6H), 0.98 (s, 3H) ppm. **^{13}C NMR (75 MHz, DMSO):** δ = 152.0, 79.7, 74.3, 66.8, 38.0, 16.3 ppm. **HRMS (EI+):** *m/z* calcd for $C_{14}H_{13}O_6$: 277.0707 Da [M+H], found: 277.0705 Da (Δ = 0.6 ppm). **IR (ATR):** ṽ = 3285 (w), 3249 (w), 2976 (vw), 2116 (m), 1719 (m), 1702 (m), 1476 (w), 1376 (w), 1287 (w), 1218 (s), 1000 (w), 980 (m), 909 (w), 875 (w), 749 (m), 696 (m), 667 (m), 604 (w), 548 (w), 522 (w), 475 (vw) cm^{-1}.

5.11 [4,4'-Bipyridine]-3,5-dicarbonitrile

The synthesis of the asymmetric [4,4'-bipyridine]-3,5-dicarbonitrile **47** is inspired by the analogue synthetic route towards 4-aryldihydropyridines with enamines by the group of IBRAHIM et al. (Scheme 15).[350]

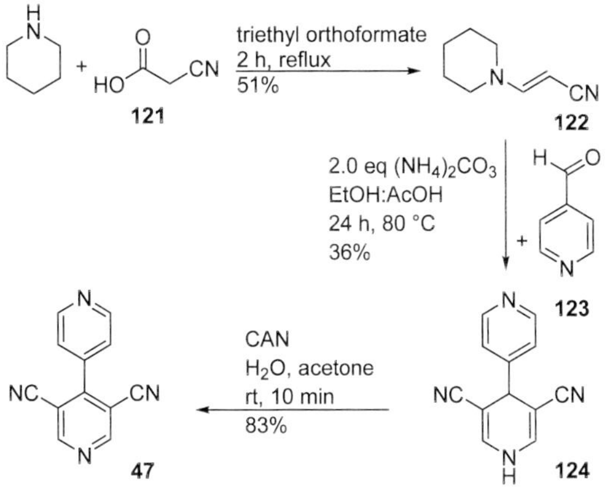

Scheme 15. Synthesis of asymmetric [4,4'-bipyridine]-3,5-dicarbonitrile 47.

(*E*)-3-(Piperidin-1-yl)acrylonitrile 122

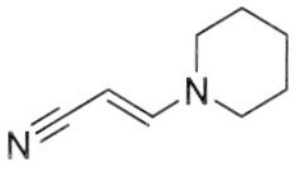

A mixture of cyanoacetic acid **121** (20 g, 235 mmol, 1.00 eq.), triethyl orthoformate (34.6 g, 235 mmol, 1.00 eq.), and piperidine (20.0 g, 235 mmol, 1.00 eq.) is heated under reflux during 2 h (oil bath 130 °C). The mixture is evaporated *in vacuo* and the crude residue is diluted with CH_2Cl_2 (110 mL), washed with 1 molar sodium hydrogen carbonate solution (36 mL) and water (36 mL). After drying with anhydrous Na_2SO_4 the solution is evaporated *in vacuo*. The residue recrystallized from diethyl ether and 12.7 g (*E*)-3-(piperidin-1-yl)acrylonitrile **122** (93.3 mmol, 51%) was obtained as pale crystals.[351]

^{1}H NMR (300 MHz, DMSO): δ = 7.11 (d, *J* = 13.7 Hz, 1H), 4.04 (d, *J* = 13.7 Hz, 1H), 3.14 (s, 2H), 1.71 – 1.26 (m, 6H) ppm. **^{13}C NMR (75 MHz, DMSO):** δ = 153.9, 122.8, 58.7, 24.9, 23.5 ppm. **HRMS (ESI+):** *m/z* calcd for $C_8H_{12}N_2^+$: 137.1073 Da [M]$^+$, found: 137.1072 Da (Δ = 0.7 ppm). **IR (ATR):** ṽ = 3064 (w), 2942 (m), 2859 (w), 2185 (m), 1612 (s), 1466 (w), 1426 (m), 1346 (m), 1287 (m),

1233 (m), 1200 (m), 1130 (w), 1110 (m), 1020 (m), 992 (m), 976 (m), 951 (m), 847 (w), 729 (m), 583 (w), 545 (w), 520 (w), 433 (w), 400 (w) cm^{-1}.

1,4-Dihydro-[4,4'-bipyridine]-3,5-dicarbonitrile 124

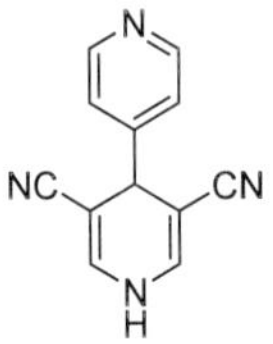

(*E*)-3-(piperidin-1-yl)acrylonitrile **122** (1.37 g, 10.1 mmol, 2.00 eq.) was dissolved in EtOH:AcOH 1:1 (5.0 mL) and 4-pyridinecarboxaldehyde **123** (540 mg, 5.04 mmol, 1.00 eq.), ammonium carbonate (969 mg, 10.1 mmol, 2.00 eq.) was added to an open vial.[350] The vial was closed, the syringe was punched with a needle and stirred for 24 h at 80 °C. The mixture was evaporated under reduced pressure and purified by column chromatography (1% MeOH 1% AcOH in EtOAc, R_f = 0.20), eluting with 10% MeOH in EtOAc (side product R_f = 0.25 1% AcOH, 20% MeOH) and 379 mg 1,4-dihydro-[4,4'-bipyridine]-3,5-dicarbonitrile **124** (1.82 mmol, 36%) was isolated.

^{1}H NMR (300 MHz, Methanol-*d*$_4$): δ = 8.62 (d, J = 4.7 Hz, 2H), 7.48 (d, J = 4.6 Hz, 2H), 7.10 (s, 2H), 4.62 (s, 1H) ppm. **^{1}H NMR (300 MHz, DMSO):** δ = 9.71 (br, 1H), 8.65 (d, J = 6.1 Hz, 2H), 7.36 (d, J = 6.1 Hz, 2H), 7.27 (s, 2H), 4.63 (s, 1H) ppm. **^{13}C NMR (75 MHz, DMSO):** δ = 151.1, 150.4, 138.5, 122.9, 118.9, 84.0, 38.8 ppm. **HRMS (EI+):** *m/z* calcd for $C_{12}H_8N_4$: 208.0743 Da [M], found: 208.0741 Da (Δ = 1.0 ppm). **IR (ATR):** $\tilde{\nu}$ = 2748 (vw), 2198 (vw), 1664 (vw), 1602 (vw), 1523 (vw), 1421 (vw), 1364 (vw), 1269 (vw), 1205 (vw), 1131 (vw), 1085 (vw), 1068 (vw), 1006 (vw), 847 (vw), 804 (vw), 747 (vw), 702 (vw), 660 (vw), 514 (vw), 478 (vw), 432 (vw) cm^{-1}.

[4,4'-Bipyridine]-3,5-dicarbonitrile 47

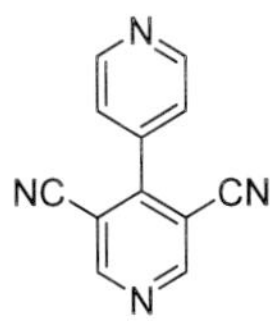

To a suspension of the 1,4-dihydro-[4,4'-bipyridine]-3,5-dicarbonitrile **124** (197 mg, 0.946 mmol, 1.00 eq.) in acetone (7.1 mL) is added a solution of ceric ammonium nitrate (1.04 g, 1.89 mmol, 2.00 eq.) in H_2O (1.7 mL) fairly rapidly dropwise from a hand-held pipette at rt. The orange colour of the reagent disappears immediately on addition of each drop. After stirring for 10 min, the resulting solution is concentrated to a small volume under reduced pressure. H_2O (20 mL) is added and the mixture is extracted with CH_2Cl_2 (2×30 mL). The organic phase is washed with brine (20 mL), dried ($MgSO_4$), and evaporated under reduced pressure.[352] The crude was purified by silica gel column chromatography cH:EtOAc 4:6 (R_f = 0.23) and 161 mg [4,4'-bipyridine]-3,5-dicarbonitrile **47** (7.81 mmol, 83%) was obtained.

^{1}H NMR (300 MHz, DMSO): δ = 9.47 (s, 2H), 8.90 (d, *J* = 6.1 Hz, 2H), 7.73 (d, *J* = 6.1 Hz, 2H) ppm. **^{13}C NMR (75 MHz, DMSO):** δ = 156.4, 152.3, 150.4, 140.3, 123.2, 114.8, 109.6 ppm. **HRMS (EI+):** *m/z* calcd for $C_{12}H_6N_4$: 206.0587 Da [M], found: 206.0586 Da (Δ = 0.3 ppm). **IR (ATR):** ṽ = 3044 (vw), 2919 (vw), 2237 (w), 1730 (vw), 1595 (vw), 1558 (w), 1530 (w), 1451 (w), 1403 (w), 1217 (vw), 1149 (vw), 1076 (vw), 1032 (w), 915 (w), 836 (m), 764 (w), 737 (w), 654 (w), 628 (w), 578 (w), 527 (w), 497 (vw), 466 (w) cm^{-1}.

6 Abbreviation

6.1 Abbreviation index

Ac	acetyl
ACN	acetonitrile
Arg	arginine
Az	azobenzene
Bn	benzyl
Bz	benzoyl
calcd	calculated
cH	cyclohexane
cm^{-1}	wavenumber
conc.	concentrated
config.	configuration
conv.	conversion
DE	diarylethene
DIPEA	*N,N*-diisopropylethylamine
DMF	*N,N*-dimethylformamide
DMSO	dimethyl sulfoxide
DNA	deoxyribonucleic acid
DTE	dithienylethene
EA	elemental analysis
ee	enantiomeric excess
eq.	equivalents
Et	ethyl
EtOAc	ethyl acetate
eV	electronvolt
FAB	fast atom bombardment
Fmoc	fluorenylmethoxycarbonyl
g	grams
GC	gas chromatography
HOAt	1-hydroxy-7-azabenzo-triazole
HPLC	high pressure liquid chromatography
HTS	high throughput screening
Hz	hertz
iPr	isopropyl
IR	infrared
L-Ala	*L*-alanine

Lys	lysine
m/z	mass to charge ratio
Me	methyl
MeOH	methanol
MHz	megahertz
mL	millilitres
mm	millimetre
MS	mass spectrometry
*n*BuLi	*n*-butyllithium
NMO	*N*-methylmorpholine *N*-oxide
NMP	*N*-methylpyrrolidone
NMR	nuclear magnetic resonance spectroscopy
Nu	nucleophile
Ph	phenyl
ppm	parts per million
pTsOH	*para*-toluenesulfonic acid
quant.	quantitative
R_f	retention factor
rpm	revolutions per minute
rt	room temperature
sat.	saturated
SEC	size-exclusion chromatography
SP	spiropyran
SPO	spirooxazines
tBu	*tert*-butyl
TFA	trifluoroacetic acid
THF	tetrahydrofuran
TIPS	triisopropylsilyl
TLC	thin-layer chromatography
$T_{g\text{-}s}$	gel-to-sol transition temperature
T_m	melting point
Ts	tosyl
TSTU	*N*,*N*,*N'*,*N'*-Tetramethyl-*O*-(*N*-succinimidyl)uronium tetrafluoroborate
UV	ultraviolet
UV-Vis	ultraviolet and visible

7 Literature

7.1 Literature index

[1] A. C. Clarke, *The Promise of Space*, Berkley Publishing Group, **1985**.
[2] C. Knie, M. Utecht, F. Zhao, H. Kulla, S. Kovalenko, A. M. Brouwer, P. Saalfrank, S. Hecht, D. Bleger, *Chem. Eur. J.* **2014**, *20*, 16492-16501.
[3] A. R. Hirst, B. Escuder, J. F. Miravet, D. K. Smith, *Angew. Chem. Int. Ed.* **2008**, *47*, 8002-8018.
[4] *Supramolecular Chemistry: From Molecules to Nanomaterials*, John Wiley & Sons, Ltd., **2012**.
[5] J. Fu, M. Liu, Y. Liu, N. W. Woodbury, H. Yan, *J. Am. Chem. Soc.* **2012**, *134*, 5516-5519.
[6] L. Cong, F. A. Ran, D. Cox, S. Lin, R. Barretto, N. Habib, P. D. Hsu, X. Wu, W. Jiang, L. A. Marraffini, F. Zhang, *Science* **2013**, *339*, 819-823.
[7] J. Fredens, K. Wang, D. de la Torre, L. F. H. Funke, W. E. Robertson, Y. Christova, T. Chia, W. H. Schmied, D. L. Dunkelmann, V. Beranek, C. Uttamapinant, A. G. Llamazares, T. S. Elliott, J. W. Chin, *Nature* **2019**.
[8] K. R. Zahs, K. H. Ashe, *Trends Neurosci.* **2010**, *33*, 381-389.
[9] B. I. Lee, Y. S. Suh, Y. J. Chung, K. Yu, C. B. Park, *Sci. Rep.* **2017**, *7*, 7523.
[10] M. J. Webber, E. A. Appel, E. W. Meijer, R. Langer, *Nat. Mater.* **2016**, *15*, 13-26.
[11] Z. Mao, H. Xu, D. Wang, *Adv. Funct. Mater.* **2010**, *20*, 1053-1074.
[12] R. Bucci, P. Das, F. Iannuzzi, M. Feligioni, R. Gandolfi, M. L. Gelmi, M. Reches, S. Pellegrino, *Org. Biomol. Chem.* **2017**, *15*, 6773-6779.
[13] G. V. Oshovsky, D. N. Reinhoudt, W. Verboom, *Angew. Chem. Int. Ed.* **2007**, *46*, 2366-2393.
[14] X. Zhang, W. Meining, M. Fischer, A. Bacher, R. Ladenstein, *J. Mol. Biol.* **2001**, *306*, 1099-1114.
[15] E. Sasaki, D. Bohringer, M. van de Waterbeemd, M. Leibundgut, R. Zschoche, A. J. Heck, N. Ban, D. Hilvert, *Nat. Commun.* **2017**, *8*, 14663.
[16] E. Y. Kim, D. Tullman-Ercek, *Curr. Opin. Biotechnol.* **2013**, *24*, 627-632.
[17] P. C. Jordan, D. P. Patterson, K. N. Saboda, E. J. Edwards, H. M. Miettinen, G. Basu, M. C. Thielges, T. Douglas, *Nat. Chem.* **2016**, *8*, 179-185.
[18] A. J. Simon, Y. Zhou, V. Ramasubramani, J. Glaser, A. Pothukuchy, J. Gollihar, J. C. Gerberich, J. C. Leggere, B. R. Morrow, C. Jung, S. C. Glotzer, D. W. Taylor, A. D. Ellington, *Nat. Chem.* **2019**, *11*, 204-212.
[19] S. Li, Q. Zou, Y. Li, C. Yuan, R. Xing, X. Yan, *J. Am. Chem. Soc.* **2018**, *140*, 10794-10802.
[20] S. Goskulwad, D. D. La, M. A. Kobaisi, S. V. Bhosale, V. Bansal, A. Vinu, K. Ariga, S. V. Bhosale, *Sci. Rep.* **2018**, *8*, 11220.
[21] M. Kathan, S. Hecht, *Chem. Soc. Rev.* **2017**, *46*, 5536-5550.
[22] Z. L. Pianowski, *Chem. Eur. J.* **2019**, *25*, 5128-5144.
[23] P. Stacko, J. C. M. Kistemaker, T. van Leeuwen, M. C. Chang, E. Otten, B. L. Feringa, *Science* **2017**, *356*, 964-968.

[24] a) L. Wu, K. Koumoto, N. Sugimoto, *Chem. Commun.* **2009**, 1915-1917; b) L. Wu, Y. Wu, H. Jin, L. Zhang, Y. He, X. Tang, *MedChemComm* **2015**, *6*, 461-468.
[25] A. S. Lubbe, Q. Liu, S. J. Smith, J. W. de Vries, J. C. M. Kistemaker, A. H. de Vries, I. Faustino, Z. Meng, W. Szymanski, A. Herrmann, B. L. Feringa, *J. Am. Chem. Soc.* **2018**, *140*, 5069-5076.
[26] D. Zhao, T. van Leeuwen, J. Cheng, B. L. Feringa, *Nat. Chem.* **2016**, *9*, 250-256.
[27] M. Irie, *Chem. Rev.* **2000**, *100*, 1683-1684.
[28] A. A. Beharry, G. A. Woolley, *Chem. Soc. Rev.* **2011**, *40*, 4422-4437.
[29] J. Garcia-Amoros, D. Velasco, *Beilstein J. Org. Chem.* **2012**, *8*, 1003-1017.
[30] a) D. Bleger, J. Schwarz, A. M. Brouwer, S. Hecht, *J. Am. Chem. Soc.* **2012**, *134*, 20597-20600; b) M. Dong, A. Babalhavaeji, S. Samanta, A. A. Beharry, G. A. Woolley, *Acc. Chem. Res.* **2015**, *48*, 2662-2670; c) S. Samanta, A. A. Beharry, O. Sadovski, T. M. McCormick, A. Babalhavaeji, V. Tropepe, G. A. Woolley, *J. Am. Chem. Soc.* **2013**, *135*, 9777-9784; d) Y. Yang, R. P. Hughes, I. Aprahamian, *J. Am. Chem. Soc.* **2014**, *136*, 13190-13193.
[31] a) S. Castellanos, A. Goulet-Hanssens, F. Zhao, A. Dikhtiarenko, A. Pustovarenko, S. Hecht, J. Gascon, F. Kapteijn, D. Bleger, *Chem. Eur. J.* **2016**, *22*, 746-752; b) K. Muller, A. Knebel, F. Zhao, D. Bleger, J. Caro, L. Heinke, *Chem. Eur. J.* **2017**, *23*, 5434-5438.
[32] W. Szymanski, J. M. Beierle, H. A. V. Kistemaker, W. A. Velema, B. L. Feringa, *Chem. Rev.* **2013**, *113*, 6114-6178.
[33] a) M. Banghart, K. Borges, E. Isacoff, D. Trauner, R. H. Kramer, *Nat. Neurosci.* **2004**, *7*, 1381-1386; b) M. A. Kienzler, A. Reiner, E. Trautman, S. Yoo, D. Trauner, E. Y. Isacoff, *J. Am. Chem. Soc.* **2013**, *135*, 17683-17686; c) M. Volgraf, P. Gorostiza, R. Numano, R. H. Kramer, E. Y. Isacoff, D. Trauner, *Nat. Chem. Biol.* **2006**, *2*, 47-52.
[34] J. Garcia-Amoros, M. C. Castro, P. Coelho, M. M. Raposo, D. Velasco, *Chem. Commun.* **2016**, *52*, 5132-5135.
[35] A. A. Beharry, O. Sadovski, G. A. Woolley, *J. Am. Chem. Soc.* **2011**, *133*, 19684-19687.
[36] D. B. Konrad, J. A. Frank, D. Trauner, *Chem. Eur. J.* **2016**, *22*, 4364-4368.
[37] M. J. Hansen, M. M. Lerch, W. Szymanski, B. L. Feringa, *Angew. Chem. Int. Ed.* **2016**, *55*, 13514-13518.
[38] K. Hull, J. Morstein, D. Trauner, *Chem. Rev.* **2018**, *118*, 10710-10747.
[39] J. Broichhagen, J. A. Frank, D. Trauner, *Acc. Chem. Res.* **2015**, *48*, 1947-1960.
[40] J. B. Trads, J. Burgstaller, L. Laprell, D. B. Konrad, L. de la Osa de la Rosa, C. D. Weaver, H. Baier, D. Trauner, D. M. Barber, *Org. Biomol. Chem.* **2016**, *15*, 76-81.
[41] A. Rullo, A. Reiner, A. Reiter, D. Trauner, E. Y. Isacoff, G. A. Woolley, *Chem. Commun.* **2014**, *50*, 14613-14615.

[42] W. A. Velema, M. J. Hansen, M. M. Lerch, A. J. Driessen, W. Szymanski, B. L. Feringa, *Bioconjugate Chem.* **2015**, *26*, 2592-2597.
[43] T. Stafforst, D. Hilvert, *Chem. Commun.* **2009**, 287-288.
[44] R. Klajn, *Chem. Soc. Rev.* **2014**, *43*, 148-184.
[45] V. I. Minkin, *Chem. Rev.* **2004**, *104*, 2751-2776.
[46] J. T. Wojtyk, A. Wasey, N. N. Xiao, P. M. Kazmaier, S. Hoz, C. Yu, R. P. Lemieux, E. Buncel, *J. Phys. Chem. A* **2007**, *111*, 2511-2516.
[47] X. Meng, G. Qi, X. Li, Z. Wang, K. Wang, B. Zou, Y. Ma, *J. Mater. Chem. C* **2016**, *4*, 7584-7588.
[48] D. Samanta, D. Galaktionova, J. Gemen, L. J. W. Shimon, Y. Diskin-Posner, L. Avram, P. Kral, R. Klajn, *Nat. Commun.* **2018**, *9*, 641.
[49] D. A. Parthenopoulos, P. M. Rentzepis, *Science* **1989**, *245*, 843-845.
[50] C. Liu, D. Yang, Q. Jin, L. Zhang, M. Liu, *Adv. Mater.* **2016**, *28*, 1644-1649.
[51] C. Liu, Q. Jin, K. Lv, L. Zhang, M. Liu, *Chem. Commun.* **2014**, *50*, 3702-3705.
[52] D. Yang, C. Liu, L. Zhang, M. Liu, *Chem. Commun.* **2014**, *50*, 12688-12690.
[53] W. Miao, S. Wang, M. Liu, *Adv. Funct. Mater.* **2017**, *27*.
[54] T. van Leeuwen, T. C. Pijper, J. Areephong, B. L. Feringa, W. R. Browne, N. Katsonis, *J. Mater. Chem.* **2011**, *21*.
[55] M. Hanazawa, R. Sumiya, Y. Horikawa, M. Irie, *Chem. Commun.* **1992**.
[56] W. R. B. Ben L. Feringa, *Molecular switches*, 2. ed., Wiley-VCH, Weinheim, **2011**.
[57] M. Irie, *Chem. Rev.* **2000**, *100*, 1685-1716.
[58] G. M. Tsivgoulis, J.-M. Lehn, *Adv. Mater.* **1997**, *9*, 627-630.
[59] W. Li, X. Li, Y. Xie, Y. Wu, M. Li, X. Y. Wu, W. H. Zhu, H. Tian, *Sci. Rep.* **2015**, *5*, 9186.
[60] D. J. van Dijken, J. M. Beierle, M. C. Stuart, W. Szymanski, W. R. Browne, B. L. Feringa, *Angew. Chem. Int. Ed.* **2014**, *53*, 5073-5077.
[61] M. M. Lerch, W. Szymanski, B. L. Feringa, *Chem. Soc. Rev.* **2018**, *47*, 1910-1937.
[62] M. M. Lerch, M. J. Hansen, W. A. Velema, W. Szymanski, B. L. Feringa, *Nat. Commun.* **2016**, *7*, 12054.
[63] D. J. van Dijken, P. Kovaricek, S. P. Ihrig, S. Hecht, *J. Am. Chem. Soc.* **2015**, *137*, 14982-14991.
[64] Z. Kokan, M. J. Chmielewski, *J. Am. Chem. Soc.* **2018**, *140*, 16010-16014.
[65] J. M. Lehn, *Chem. Eur. J.* **2006**, *12*, 5910-5915.
[66] L. Greb, A. Eichhofer, J. M. Lehn, *Angew. Chem. Int. Ed.* **2015**, *54*, 14345-14348.
[67] C. Petermayer, H. Dube, *Acc. Chem. Res.* **2018**, *51*, 1153-1163.
[68] C. Petermayer, S. Thumser, F. Kink, P. Mayer, H. Dube, *J. Am. Chem. Soc.* **2017**, *139*, 15060-15067.
[69] M. Guentner, M. Schildhauer, S. Thumser, P. Mayer, D. Stephenson, P. J. Mayer, H. Dube, *Nat. Commun.* **2015**, *6*, 8406.

[70] A. Gerwien, P. Mayer, H. Dube, *J. Am. Chem. Soc.* **2018**, *140*, 16442-16445.
[71] D. J. van Dijken, J. Chen, M. C. Stuart, L. Hou, B. L. Feringa, *J. Am. Chem. Soc.* **2016**, *138*, 660-669.
[72] N. Basílio, L. García-Río, *Curr. Opin. Colloid Interface Sci.* **2017**, *32*, 29-38.
[73] B. C. Buddingh, J. C. M. van Hest, *Acc. Chem. Res.* **2017**, *50*, 769-777.
[74] Y. Tu, F. Peng, A. Adawy, Y. Men, L. K. Abdelmohsen, D. A. Wilson, *Chem. Rev.* **2016**, *116*, 2023-2078.
[75] C. S. Chen, X. D. Xu, S. Y. Li, R. X. Zhuo, X. Z. Zhang, *Nanoscale* **2013**, *5*, 6270-6274.
[76] H. Yotsuji, K. Higashiguchi, R. Sato, Y. Shigeta, K. Matsuda, *Chem. Eur. J.* **2017**, *23*, 15059-15066.
[77] T. Li, X. Li, J. Wang, H. Ågren, X. Ma, H. Tian, *Adv. Opt. Mater.* **2016**, *4*, 840-847.
[78] M. B. Baker, L. Albertazzi, I. K. Voets, C. M. Leenders, A. R. Palmans, G. M. Pavan, E. W. Meijer, *Nat. Commun.* **2015**, *6*, 6234.
[79] P. Wei, T. R. Cook, X. Yan, F. Huang, P. J. Stang, *J. Am. Chem. Soc.* **2014**, *136*, 15497-15500.
[80] K. Kreger, P. Wolfer, H. Audorff, L. Kador, N. Stingelin-Stutzmann, P. Smith, H. W. Schmidt, *J. Am. Chem. Soc.* **2010**, *132*, 509-516.
[81] Z. Shen, T. Wang, M. Liu, *Angew. Chem. Int. Ed.* **2014**, *53*, 13424-13428.
[82] J. Lee, S. Oh, J. Pyo, J. M. Kim, J. H. Je, *Nanoscale* **2015**, *7*, 6457-6461.
[83] R. O. Al-Kaysi, C. J. Bardeen, *Adv. Mater.* **2007**, *19*, 1276-1280.
[84] J. Li, X. J. Loh, *Adv. Drug Deliv. Rev.* **2008**, *60*, 1000-1017.
[85] a) S. Tamesue, Y. Takashima, H. Yamaguchi, S. Shinkai, A. Harada, *Angew. Chem. Int. Ed.* **2010**, *49*, 7461-7464; b) P. Zheng, X. Hu, X. Zhao, L. Li, K. C. Tam, L. H. Gan, *Macromol. Rapid Commun.* **2004**, *25*, 678-682.
[86] Y. Chen, Y. Liu, *Chem. Soc. Rev.* **2010**, *39*, 495-505.
[87] Y. M. Zhang, Y. H. Liu, Y. Liu, *Adv. Mater.* **2019**, e1806158.
[88] Y. M. Zhang, N. Y. Zhang, K. Xiao, Q. Yu, Y. Liu, *Angew. Chem. Int. Ed.* **2018**, *57*, 8649-8653.
[89] H. Xiao, P. Verdier-Pinard, N. Fernandez-Fuentes, B. Burd, R. Angeletti, A. Fiser, S. B. Horwitz, G. A. Orr, *Proc. Natl. Acad. Sci. U.S.A.* **2006**, *103*, 10166-10173.
[90] A. Akhmanova, M. O. Steinmetz, *Nat. Rev. Mol. Cell Biol.* **2015**, *16*, 711-726.
[91] M. Borowiak, W. Nahaboo, M. Reynders, K. Nekolla, P. Jalinot, J. Hasserodt, M. Rehberg, M. Delattre, S. Zahler, A. Vollmar, D. Trauner, O. Thorn-Seshold, *Cell* **2015**, *162*, 403-411.
[92] a) M. C. Jiménez, C. Dietrich-Buchecker, J.-P. Sauvage, *Angew. Chem. Int. Ed.* **2000**, *39*, 3284-3287; b) M. C. Jimenez-Molero, C. Dietrich-Buchecker, J.-P. Sauvage, *Chem. Commun.* **2003**.

[93] a) J.-M. Lehn, *Angew. Chem. Int. Ed.* **1990**, *29*, 1304-1319; b) T. Aida, E. W. Meijer, S. I. Stupp, *Science* **2012**, *335*, 813-817; c) B. Xue, M. Qin, T. Wang, J. Wu, D. Luo, Q. Jiang, Y. Li, Y. Cao, W. Wang, *Adv. Funct. Mater.* **2016**, *26*, 9053-9062.
[94] C. J. Bruns, J. F. Stoddart, *Acc. Chem. Res.* **2014**, *47*, 2186-2199.
[95] J.-P. Collin, C. Dietrich-Buchecker, P. Gaviña, M. C. Jimenez-Molero, J.-P. Sauvage, *Acc. Chem. Res.* **2001**, *34*, 477-487.
[96] J. Chen, F. K. Leung, M. C. A. Stuart, T. Kajitani, T. Fukushima, E. van der Giessen, B. L. Feringa, *Nat. Chem.* **2018**, *10*, 132-138.
[97] S. Zhang, M. A. Greenfield, A. Mata, L. C. Palmer, R. Bitton, J. R. Mantei, C. Aparicio, M. O. de la Cruz, S. I. Stupp, *Nat. Mater.* **2010**, *9*, 594-601.
[98] a) P. Terech, R. G. Weiss, *Chem. Rev.* **1997**, *97*, 3133-3160; b) D. J. Abdallah, R. G. Weiss, *Adv. Mater.* **2000**, *12*, 1237-1247.
[99] A. R. Hirst, S. Roy, M. Arora, A. K. Das, N. Hodson, P. Murray, S. Marshall, N. Javid, J. Sefcik, J. Boekhoven, J. H. van Esch, S. Santabarbara, N. T. Hunt, R. V. Ulijn, *Nat. Chem.* **2010**, *2*, 1089-1094.
[100] G. O. Lloyd, J. W. Steed, *Nat. Chem.* **2009**, *1*, 437-442.
[101] X. Du, J. Zhou, J. Shi, B. Xu, *Chem. Rev.* **2015**, *115*, 13165-13307.
[102] M. Halperin-Sternfeld, M. Ghosh, R. Sevostianov, I. Grigoriants, L. Adler-Abramovich, *Chem. Commun.* **2017**, *53*, 9586-9589.
[103] D. Wang, J. Hao, *Colloid Polym. Sci.* **2013**, *291*, 2935-2946.
[104] X.-Q. Dou, C.-L. Zhao, N. Mehwish, P. Li, C.-L. Feng, H. Schönherr, *Chin. J. Polym. Sci.* **2019**, *37*, 437-443.
[105] L. Ji, G. Ouyang, M. Liu, *Langmuir* **2017**, *33*, 12419-12426.
[106] G. F. Liu, W. Ji, W. L. Wang, C. L. Feng, *ACS Appl. Mater. Interfaces* **2015**, *7*, 301-307.
[107] X. Li, J. Fei, Y. Xu, D. Li, T. Yuan, G. Li, C. Wang, J. Li, *Angew. Chem. Int. Ed.* **2018**, *57*, 1903-1907.
[108] N. Sun, A. Wu, Y. Yu, X. Gao, L. Zheng, *Chem. Eur. J.* **2019**, *25*, 6203-6211.
[109] H. Hu, Y. Qiu, J. Wang, D. Zhao, H. Wang, Q. Wang, Y. Liao, H. Peng, X. Xie, *Macromol. Rapid Commun.* **2019**, *40*, e1800629.
[110] Y. Wang, N. Ma, Z. Wang, X. Zhang, *Angew. Chem. Int. Ed.* **2007**, *46*, 2823-2826.
[111] K. L. Liu, Z. Zhang, J. Li, *Soft Matter* **2011**, *7*.
[112] T. Kakuta, Y. Takashima, M. Nakahata, M. Otsubo, H. Yamaguchi, A. Harada, *Adv. Mater.* **2013**, *25*, 2849-2853.
[113] a) Y. Guan, H.-B. Zhao, L.-X. Yu, S.-C. Chen, Y.-Z. Wang, *RSC Adv.* **2014**, *4*; b) Y. Lu, H. Zou, H. Yuan, S. Gu, W. Yuan, M. Li, *Eur. Polym. J.* **2017**, *91*, 396-407.
[114] H. Yamaguchi, Y. Kobayashi, R. Kobayashi, Y. Takashima, A. Hashidzume, A. Harada, *Nat. Commun.* **2012**, *3*, 603.
[115] F. Xie, G. Ouyang, L. Qin, M. Liu, *Chem. Eur. J.* **2016**, *22*, 18208-18214.
[116] D. Wang, M. Wagner, H. J. Butt, S. Wu, *Soft Matter* **2015**, *11*, 7656-7662.

[117] S. Lee, S. Oh, J. Lee, Y. Malpani, Y. S. Jung, B. Kang, J. Y. Lee, K. Ozasa, T. Isoshima, S. Y. Lee, M. Hara, D. Hashizume, J. M. Kim, *Langmuir* **2013**, *29*, 5869-5877.
[118] P. Fatas, J. Bachl, S. Oehm, A. I. Jimenez, C. Cativiela, D. Diaz Diaz, *Chem. Eur. J.* **2013**, *19*, 8861-8874.
[119] Y. Lin, Y. Qiao, P. Tang, Z. Li, J. Huang, *Soft Matter* **2011**, *7*.
[120] Y. Matsuzawa, N. Tamaoki, *J. Phys. Chem. B* **2010**, *114*, 1586-1590.
[121] W. A. Velema, M. C. Stuart, W. Szymanski, B. L. Feringa, *Chem. Commun.* **2013**, *49*, 5001-5003.
[122] J. Li, I. Cvrtila, M. Colomb-Delsuc, E. Otten, S. Otto, *Chem. Eur. J.* **2014**, *20*, 15709-15714.
[123] R. Yang, S. Peng, W. Wan, T. C. Hughes, *J. Mater. Chem. C* **2014**, *2*, 9122-9131.
[124] C. Lin, S. Maisonneuve, R. Metivier, J. Xie, *Chem. Eur. J.* **2017**, *23*, 14996-15001.
[125] S. J. Wezenberg, C. M. Croisetu, M. C. A. Stuart, B. L. Feringa, *Chem. Sci.* **2016**, *7*, 4341-4346.
[126] G. Liu, Y.-M. Zhang, X. Xu, L. Zhang, Y. Liu, *Adv. Opt. Mater.* **2017**, *5*.
[127] X. Yao, T. Li, J. Wang, X. Ma, H. Tian, *Adv. Opt. Mater.* **2016**, *4*, 1322-1349.
[128] Q. Zou, K. Liu, M. Abbas, X. Yan, *Adv. Mater.* **2016**, *28*, 1031-1043.
[129] J. Zhou, J. Li, X. Du, B. Xu, *Biomaterials* **2017**, *129*, 1-27.
[130] A. S. Lubbe, W. Szymanski, B. L. Feringa, *Chem. Soc. Rev.* **2017**, *46*, 1052-1079.
[131] S. M. Thomas, B. Sahu, S. Rapireddy, R. Bahal, S. E. Wheeler, E. M. Procopio, J. Kim, S. C. Joyce, S. Contrucci, Y. Wang, S. I. Chiosea, K. L. Lathrop, S. Watkins, J. R. Grandis, B. A. Armitage, D. H. Ly, *ACS Chem. Biol.* **2013**, *8*, 345-352.
[132] T. Stafforst, D. Hilvert, *Angew. Chem. Int. Ed.* **2010**, *49*, 9998-10001.
[133] A. L. Le Ny, C. T. Lee, Jr., *J. Am. Chem. Soc.* **2006**, *128*, 6400-6408.
[134] a) Y. Zakrevskyy, P. Cywinski, M. Cywinska, J. Paasche, N. Lomadze, O. Reich, H. G. Lohmannsroben, S. Santer, *J. Chem. Phys.* **2014**, *140*, 044907; b) Y. Zakrevskyy, E. Titov, N. Lomadze, S. Santer, *J. Chem. Phys.* **2014**, *141*, 164904; c) Y. Zakrevskyy, A. Kopyshev, N. Lomadze, E. Morozova, L. Lysyakova, N. Kasyanenko, S. Santer, *Phys Rev E Stat Nonlin Soft Matter Phys* **2011**, *84*, 021909; d) Y. Zakrevskyy, J. Roxlau, G. Brezesinski, N. Lomadze, S. Santer, *J. Chem. Phys.* **2014**, *140*, 044906.
[135] M. Geoffroy, D. Faure, R. Oda, D. M. Bassani, D. Baigl, *ChemBioChem* **2008**, *9*, 2382-2385.
[136] Y. C. Liu, A. L. Le Ny, J. Schmidt, Y. Talmon, B. F. Chmelka, C. T. Lee, Jr., *Langmuir* **2009**, *25*, 5713-5724.
[137] Y. Li, J. Yang, L. Sun, W. Wang, W. Liu, *J. Mater. Chem.* **2014**, *2*.
[138] A. Venancio-Marques, A. Bergen, C. Rossi-Gendron, S. Rudiuk, D. Baigl, *ACS Nano* **2014**, *8*, 3654-3663.

[139] M. Deiana, Z. Pokladek, J. Olesiak-Banska, P. Mlynarz, M. Samoc, K. Matczyszyn, *Sci. Rep.* **2016**, *6*, 28605.
[140] A. Bergen, S. Rudiuk, M. Morel, T. Le Saux, H. Ihmels, D. Baigl, *Nano Lett.* **2016**, *16*, 773-780.
[141] A. A. Zinchenko, M. Tanahashi, S. Murata, *ChemBioChem* **2012**, *13*, 105-111.
[142] J. Andersson, S. Li, P. Lincoln, J. Andreasson, *J. Am. Chem. Soc.* **2008**, *130*, 11836-11837.
[143] M. Hammarson, J. R. Nilsson, S. Li, P. Lincoln, J. Andreasson, *Chem. Eur. J.* **2014**, *20*, 15855-15862.
[144] M. Hammarson, J. Andersson, S. Li, P. Lincoln, J. Andreasson, *Chem. Commun.* **2010**, *46*, 7130-7132.
[145] M. Balter, M. Hammarson, P. Remon, S. Li, N. Gale, T. Brown, J. Andreasson, *J. Am. Chem. Soc.* **2015**, *137*, 2444-2447.
[146] A. Mammana, G. T. Carroll, J. Areephong, B. L. Feringa, *J. Phys. Chem. B* **2011**, *115*, 11581-11587.
[147] T. C. Pace, V. Muller, S. Li, P. Lincoln, J. Andreasson, *Angew. Chem. Int. Ed.* **2013**, *52*, 4393-4396.
[148] K. Liu, Y. Wen, T. Shi, Y. Li, F. Li, Y. L. Zhao, C. Huang, T. Yi, *Chem. Commun.* **2014**, *50*, 9141-9144.
[149] C. Dohno, S. N. Uno, K. Nakatani, *J. Am. Chem. Soc.* **2007**, *129*, 11898-11899.
[150] J. Rubio-Magnieto, T. A. Phan, M. Fossepre, V. Matot, J. Knoops, T. Jarrosson, P. Dumy, F. Serein-Spirau, C. Niebel, S. Ulrich, M. Surin, *Chem. Eur. J.* **2018**, *24*, 706-714.
[151] D. Paolantoni, J. Rubio-Magnieto, S. Cantel, J. Martinez, P. Dumy, M. Surin, S. Ulrich, *Chem. Commun.* **2014**, *50*, 14257-14260.
[152] J.-M. Lehn, *Angew. Chem.* **2015**, *127*, 3326-3340.
[153] R. J. Mart, R. K. Allemann, *Chem. Commun.* **2016**, *52*, 12262-12277.
[154] M. Lindgren, M. Hällbrink, A. Prochiantz, Ü. Langel, *Trends Pharmacol. Sci.* **2000**, *21*, 99-103.
[155] A. D. Frankel, C. O. Pabo, *Cell* **1988**, *55*, 1189-1193.
[156] J. P. Richard, K. Melikov, E. Vives, C. Ramos, B. Verbeure, M. J. Gait, L. V. Chernomordik, B. Lebleu, *J. Biol. Chem.* **2003**, *278*, 585-590.
[157] C. Madhu, C. Voshavar, K. Rajasekhar, T. Govindaraju, *Org. Biomol. Chem.* **2017**, *15*, 3170-3174.
[158] J. Farrera-Sinfreu, E. Giralt, S. Castel, F. Albericio, M. Royo, *J. Am. Chem. Soc.* **2005**, *127*, 9459-9468.
[159] a) T. Endoh, T. Ohtsuki, *Adv. Drug Deliv. Rev.* **2009**, *61*, 704-709; b) D. Schaffert, E. Wagner, *Gene Ther.* **2008**, *15*, 1131-1138.
[160] P. Kumar, H. Wu, J. L. McBride, K. E. Jung, M. H. Kim, B. L. Davidson, S. K. Lee, P. Shankar, N. Manjunath, *Nature* **2007**, *448*, 39-43.
[161] N. L. Benner, R. L. McClellan, C. R. Turlington, O. A. W. Haabeth, R. M. Waymouth, P. A. Wender, *J. Am. Chem. Soc.* **2019**, *141*, 8416-8421.
[162] A. Barhoumi, Q. Liu, D. S. Kohane, *J. Control. Release* **2015**, *219*, 31-42.

[163] A. Prestel, H. M. Moller, *Chem. Commun.* **2016**, *52*, 701-704.
[164] A. A. Beharry, L. Wong, V. Tropepe, G. A. Woolley, *Angew. Chem. Int. Ed.* **2011**, *50*, 1325-1327.
[165] a) G. A. Woolley, *Acc. Chem. Res.* **2005**, *38*, 486-493; b) S. L. Dong, M. Loweneck, T. E. Schrader, W. J. Schreier, W. Zinth, L. Moroder, C. Renner, *Chem. Eur. J.* **2006**, *12*, 1114-1120; c) S. Samanta, C. Qin, A. J. Lough, G. A. Woolley, *Angew. Chem. Int. Ed.* **2012**, *51*, 6452-6455; d) D. G. Flint, J. R. Kumita, O. S. Smart, G. A. Woolley, *Chem. Biol.* **2002**, *9*, 391-397; e) C. Hoppmann, V. K. Lacey, G. V. Louie, J. Wei, J. P. Noel, L. Wang, *Angew. Chem. Int. Ed.* **2014**, *53*, 3932-3936; f) G. C. Kim, J. H. Ahn, J. H. Oh, S. Nam, S. Hyun, J. Yu, Y. Lee, *Biomacromolecules* **2018**, *19*, 2863-2869.
[166] L. Nevola, A. Martin-Quiros, K. Eckelt, N. Camarero, S. Tosi, A. Llobet, E. Giralt, P. Gorostiza, *Angew. Chem. Int. Ed.* **2013**, *52*, 7704-7708.
[167] C. Hoppmann, I. Maslennikov, S. Choe, L. Wang, *J. Am. Chem. Soc.* **2015**, *137*, 11218-11221.
[168] Y. Q. Yeoh, J. Yu, S. W. Polyak, J. R. Horsley, A. D. Abell, *ChemBioChem* **2018**, *19*, 2591-2597.
[169] A. Pantos, I. Tsogas, C. M. Paleos, *Biochim. Biophys. Acta* **2008**, *1778*, 811-823.
[170] W. A. Velema, W. Szymanski, B. L. Feringa, *J. Am. Chem. Soc.* **2014**, *136*, 2178-2191.
[171] A. J. Kleinsmann, B. J. Nachtsheim, *Chem. Commun.* **2013**, *49*, 7818-7820.
[172] a) Z. Xie, A. Zhang, L. Ye, Z.-g. Feng, *Soft Matter* **2009**, *5*, 1474-1482; b) Z. Xie, A. Zhang, L. Ye, X. Wang, Z.-g. Feng, *J. Mater. Chem. C* **2009**, *19*, 6100.
[173] A. D. Borthwick, *Chem. Rev.* **2012**, *112*, 3641-3716.
[174] N. Kameta, A. Tanaka, H. Akiyama, H. Minamikawa, M. Masuda, T. Shimizu, *Chem. Eur. J.* **2011**, *17*, 5251-5255.
[175] a) M. Bose, D. Groff, J. Xie, E. Brustad, P. G. Schultz, *J. Am. Chem. Soc.* **2006**, *128*, 388-389; b) W. Li, I. S. Park, S. K. Kang, M. Lee, *Chem. Commun.* **2012**, *48*, 8796-8798; c) J. Raeburn, T. O. McDonald, D. J. Adams, *Chem. Commun.* **2012**, *48*, 9355-9357; d) X. Cao, X. Peng, L. Zhong, R. Sun, *J. Agric. Food Chem.* **2014**, *62*, 10000-10007; e) Z. Yan, (University Nanjing), WO2012092718A1, **2013**.
[176] M. Jaspers, A. E. Rowan, P. H. J. Kouwer, *Adv. Funct. Mater.* **2015**, *25*, 6503-6510.
[177] J. Jeon, M. S. Shell, *J. Phys. Chem. B* **2014**, *118*, 6644-6652.
[178] K. Shimaoka, D. A. Schoenfeld, W. D. Dewys, R. H. Creech, R. Deconti, *Cancer* **1985**, *56*, 2155-2160.
[179] V. G. Box, *J. Mol. Struct.* **2007**, *26*, 14-19.
[180] C. Tanford, *Science* **1978**, *200*, 1012-1018.
[181] E. Gazit, *FASEB J.* **2002**, *16*, 77-83.
[182] Y. Xu, J. Shen, X. Luo, W. Zhu, K. Chen, J. Ma, H. Jiang, *Proc. Natl. Acad. Sci. U.S.A.* **2005**, *102*, 5403-5407.

[183] S. E. Feller, *Curr. Opin. Colloid Interface Sci.* **2000**, *5*, 217-223.
[184] a) S. Manchineella, T. Govindaraju, *RSC Adv.* **2012**, *2*, 5539-5542; b) T. Govindaraju, *Supramolecular Chemistry* **2011**, *23*, 759-767.
[185] N. Singh, M. Kumar, J. F. Miravet, R. V. Ulijn, B. Escuder, *Chem. Eur. J.* **2017**, *23*, 981-993.
[186] G. S. Hartley, *Nature* **1937**, *140*, 281-281.
[187] E. M. Ahmed, *J. Adv. Res.* **2015**, *6*, 105-121.
[188] J. M. Lehn, *Chem. Soc. Rev.* **2007**, *36*, 151-160.
[189] a) R. M. Putri, J. W. Fredy, J. J. Cornelissen, M. S. Koay, N. Katsonis, *ChemPhysChem* **2016**, *17*, 1815-1818; b) I. Vlassiouk, C. D. Park, S. A. Vail, D. Gust, S. Smirnov, *Nano Lett.* **2006**, *6*, 1013-1017.
[190] S. Yagi, K. Maeda, H. Nakazumi, *J. Mater. Chem. C* **1999**, *9*, 2991-2997.
[191] a) Z. Qiu, H. Yu, J. Li, Y. Wang, Y. Zhang, *Chem. Commun.* **2009**, 3342-3344; b) W. Francis, A. Dunne, C. Delaney, L. Florea, D. Diamond, *Sens. Actuator B-Chem.* **2017**, *250*, 608-616.
[192] C. Maity, W. E. Hendriksen, J. H. van Esch, R. Eelkema, *Angew. Chem. Int. Ed.* **2015**, *54*, 998-1001.
[193] J. E. Trosko, E. H. Y. Chu, W. L. Carrier, *Radiat. Res.* **1965**, *24*.
[194] R. Cai, Y. Kubota, T. Shuin, H. Sakai, K. Hashimoto, A. Fujishima, *Cancer Res.* **1992**, *52*, 2346-2348.
[195] J. W. Karcher, unpublished Vertiefer thesis, The Karlsruhe Institute of Technology (Karlsruhe), **2015**.
[196] J. W. Snow, T. M. Hooker, *J. Am. Chem. Soc.* **1975**, *97*, 3506-3511.
[197] S. Palacin, D. N. Chin, E. E. Simanek, J. C. MacDonald, G. M. Whitesides, M. T. McBride, G. T. R. Palmore, *J. Am. Chem. Soc.* **1997**, *119*, 11807-11816.
[198] a) M. Wegener, M. J. Hansen, A. J. M. Driessen, W. Szymanski, B. L. Feringa, *J. Am. Chem. Soc.* **2017**, *139*, 17979-17986; b) W. A. Velema, J. P. van der Berg, M. J. Hansen, W. Szymanski, A. J. Driessen, B. L. Feringa, *Nat. Chem.* **2013**, *5*, 924-928.
[199] a) C. A. Lipinski, F. Lombardo, B. W. Dominy, P. J. Feeney, *Adv. Drug Deliv. Rev.* **1997**, *23*, 3-25; b) C. A. Lipinski, *Drug Discov. Today* **2004**, *1*, 337-341.
[200] K. D. Eichenbaum, A. A. Thomas, G. M. Eichenbaum, B. R. Gibney, D. Needham, P. F. Kiser, *Macromolecules* **2005**, *38*, 10757-10762.
[201] a) Y. Brudno, D. J. Mooney, *J. Control. Release* **2015**, *219*, 8-17; b) N. Fomina, J. Sankaranarayanan, A. Almutairi, *Adv. Drug Deliv. Rev.* **2012**, *64*, 1005-1020.
[202] F. Silva, O. Lourenco, J. A. Queiroz, F. C. Domingues, *J. Antibiot.* **2011**, *64*, 321-325.
[203] A. J. Kleinsmann, N. M. Weckenmann, B. J. Nachtsheim, *Chem. Eur. J.* **2014**, *20*, 9753-9761.
[204] a) N. Galie, B. H. Brundage, H. A. Ghofrani, R. J. Oudiz, G. Simonneau, Z. Safdar, S. Shapiro, R. J. White, M. Chan, A. Beardsworth, L. Frumkin, R. J. Barst, H. Pulmonary Arterial, G. Response to Tadalafil Study,

Circulation **2009**, *119*, 2894-2903; b) G. N. Maw, C. M. N. Allerton, E. Gbekor, W. A. Million, *Bioorg. Med. Chem. Lett.* **2003**, *13*, 1425-1428.

[205] J. Liddle, M. J. Allen, A. D. Borthwick, D. P. Brooks, D. E. Davies, R. M. Edwards, A. M. Exall, C. Hamlett, W. R. Irving, A. M. Mason, G. P. McCafferty, F. Nerozzi, S. Peace, J. Philp, D. Pollard, M. A. Pullen, S. S. Shabbir, S. L. Sollis, T. D. Westfall, P. M. Woollard, C. Wu, D. M. Hickey, *Bioorg. Med. Chem. Lett.* **2008**, *18*, 90-94.

[206] Y. Yamazaki, M. Sumikura, Y. Masuda, Y. Hayashi, H. Yasui, Y. Kiso, T. Chinen, T. Usui, F. Yakushiji, B. Potts, S. Neuteboom, M. Palladino, G. K. Lloyd, Y. Hayashi, *Bioorg. Med. Chem.* **2012**, *20*, 4279-4289.

[207] M. Hasanpourghadi, C. Karthikeyan, A. K. Pandurangan, C. Y. Looi, P. Trivedi, K. Kobayashi, K. Tanaka, W. F. Wong, M. R. Mustafa, *J. Exp. Clin. Cancer Res.* **2016**, *35*, 58.

[208] A. J. Ross, H. L. Lang, R. F. Jackson, *J. Org. Chem.* **2010**, *75*, 245-248.

[209] S. Kirchner, unpublished master thesis, Heidelberg University (Heidelberg), **2018**.

[210] J. Luo, S. Samanta, M. Convertino, N. V. Dokholyan, A. Deiters, *ChemBioChem* **2018**, *19*, 2178-2185.

[211] S. Cohen, E. Lobel, A. Trevgoda, Y. Peled, *J. Control. Release* **1997**, *44*, 201-208.

[212] a) B. Li, J. He, D. Gevans, X. Duan, *Appl. Clay Sci.* **2004**, *27*, 199-207; b) S.-J. Xia, Z.-M. Ni, Q. Xu, B.-X. Hu, J. Hu, *J. Solid State Chem.* **2008**, *181*, 2610-2619; c) B. Li, J. He, D. G. Evans, X. Duan, *Int. J. Pharm.* **2004**, *287*, 89-95; d) B. D. Kevadiya, G. V. Joshi, H. M. Mody, H. C. Bajaj, *Appl. Clay Sci.* **2011**, *52*, 364-367.

[213] K. Muguruma, T. Shirasaka, D. Akiyama, K. Fukumoto, A. Taguchi, K. Takayama, A. Taniguchi, Y. Hayashi, *Angew. Chem. Int. Ed.* **2018**, *130*, 2192-2195.

[214] a) F. Yakushiji, H. Tanaka, K. Muguruma, T. Iwahashi, Y. Yamazaki, Y. Hayashi, *Chem. Eur. J.* **2011**, *17*, 12587-12590; b) S. Liao, X. Qin, D. Li, Z. Tu, J. Li, X. Zhou, J. Wang, B. Yang, X. Lin, J. Liu, X. Yang, Y. Liu, *Eur. J. Med. Chem.* **2014**, *83*, 236-244; c) F. Yakushiji, H. Tanaka, K. Muguruma, T. Iwahashi, Y. Yamazaki, Y. Hayashi, *Chem. Pharm. Bull.* **2012**, *60*, 877-881; d) K. Muguruma, F. Yakushiji, R. Kawamata, D. Akiyama, R. Arima, T. Shirasaka, Y. Kikkawa, A. Taguchi, K. Takayama, T. Fukuhara, T. Watabe, Y. Ito, Y. Hayashi, *Bioconjugate Chem.* **2016**, *27*, 1606-1613; e) Y. Hayashi, Y. Yamazaki-Nakamura, F. Yakushiji, *Chem. Pharm. Bull.* **2013**, *61*, 889-901; f) F. Yakushiji, K. Muguruma, Y. Hayashi, T. Shirasaka, R. Kawamata, H. Tanaka, Y. Yoshiwaka, A. Taguchi, K. Takayama, Y. Hayashi, *Bioorg. Med. Chem.* **2017**, *25*, 3623-3630.

[215] a) C. W. Pouton, *Eur. J. Pharm. Sci.* **2006**, *29*, 278-287; b) H. I. Chang, M. K. Yeh, *Int. J. Nanomed.* **2012**, *7*, 49-60.

[216] M. Xue, Y. Yang, X. Chi, X. Yan, F. Huang, *Chem. Rev.* **2015**, *115*, 7398-7501.

[217] a) R. Mitra, M. Thiele, F. Octa-Smolin, M. C. Letzel, J. Niemeyer, *Chem. Commun.* **2016**, *52*, 5977-5980; b) R. Mitra, H. Zhu, S. Grimme, J. Niemeyer, *Angew. Chem. Int. Ed.* **2017**, *56*, 11456-11459.

[218] J. E. Lewis, M. Galli, S. M. Goldup, *Chem. Commun.* **2016**, *53*, 298-312.

[219] F. B. Mallory, C. W. Mallory, **2005**, 1-456.

[220] M. L. Schulte, unpublished bachelor thesis, The Karlsruhe Institute of Technology (Karlsruhe), **2018**.

[221] A. Antoine John, Q. Lin, *J. Org. Chem.* **2017**, *82*, 9873-9876.

[222] L. Albert, A. Penalver, N. Djokovic, L. Werel, M. Hoffarth, D. Ruzic, J. Xu, L. O. Essen, K. Nikolic, Y. Dou, O. Vazquez, *ChemBioChem* **2019**.

[223] A. A. John, C. P. Ramil, Y. Tian, G. Cheng, Q. Lin, *Org. Lett.* **2015**, *17*, 6258-6261.

[224] a) Y. H. Tsai, S. Essig, J. R. James, K. Lang, J. W. Chin, *Nat. Chem.* **2015**, *7*, 554-561; b) S. J. Elsasser, R. J. Ernst, O. S. Walker, J. W. Chin, *Nat. Methods* **2016**, *13*, 158-164.

[225] C. Renner, L. Moroder, *ChemBioChem* **2006**, *7*, 868-878.

[226] a) R. Behrendt, C. Renner, M. Schenk, F. Wang, J. Wachtveitl, D. Oesterhelt, L. Moroder, *Angew. Chem. Int. Ed.* **1999**, *38*, 2771-2774; b) M. Schütt, S. S. Krupka, A. G. Milbradt, S. Deindl, E.-K. Sinner, D. Oesterhelt, C. Renner, L. Moroder, *Chem. Biol.* **2003**, *10*, 487-490; c) R. Behrendt, M. Schenk, H.-J. Musiol, L. Moroder, *J. Pept. Sci.* **1999**, *5*, 519-529; d) C. Renner, U. Kusebauch, M. Loweneck, A. G. Milbradt, L. Moroder, *J. Pept. Res.* **2005**, *65*, 4-14.

[227] A.-L. Leistner, unpublished bachelor thesis, The Karlsruhe Institute of Technology (Karlsruhe), **2017**.

[228] a) J. Kuil, L. T. van Wandelen, N. J. de Mol, R. M. Liskamp, *Bioorg. Med. Chem.* **2008**, *16*, 1393-1399; b) B. Priewisch, K. Ruck-Braun, *J. Org. Chem.* **2005**, *70*, 2350-2352; c) K. Rück-Braun, S. Kempa, B. Priewisch, A. Richter, S. Seedorff, L. Wallach, *Synthesis* **2009**; d) G. M. Murawska, C. Poloni, N. A. Simeth, W. Szymanski, B. L. Feringa, *Chem. Eur. J.* **2019**, *25*, 4965-4973.

[229] M. Hartmann, M. Berditsch, J. Hawecker, M. F. Ardakani, D. Gerthsen, A. S. Ulrich, *Antimicrob. Agents Chemother.* **2010**, *54*, 3132-3142.

[230] O. Babii, S. Afonin, L. V. Garmanchuk, V. V. Nikulina, T. V. Nikolaienko, O. V. Storozhuk, D. V. Shelest, O. I. Dasyukevich, L. I. Ostapchenko, V. Iurchenko, S. Zozulya, A. S. Ulrich, I. V. Komarov, *Angew Chem Int Edit* **2016**, *55*, 5493-5496.

[231] a) O. Babii, S. Afonin, L. V. Garmanchuk, V. V. Nikulina, T. V. Nikolaienko, O. V. Storozhuk, D. V. Shelest, O. I. Dasyukevich, L. I. Ostapchenko, V. Iurchenko, S. Zozulya, A. S. Ulrich, I. V. Komarov, *Angew. Chem. Int. Ed.* **2016**, *55*, 5493-5496; b) O. Babii, S. Afonin, A. Y. Ishchenko, T. Schober, A. O. Negelia, G. M. Tolstanova, L. V. Garmanchuk, L. I. Ostapchenko, I. V. Komarov, A. S. Ulrich, *J. Med. Chem.* **2018**, *61*, 10793-10813; c) I. V. Komarov, S. Afonin, O. Babii, T. Schober, A. S. Ulrich, *Chem. Eur. J.* **2018**, *24*, 11245-11254.

[232] a) R. W. Glaser, C. Sachse, U. H. Durr, P. Wadhwani, S. Afonin, E. Strandberg, A. S. Ulrich, *Biophys. J.* **2005**, *88*, 3392-3397; b) M. Irie, *Photochem. Photobiol. Sci.* **2010**, *9*, 1535-1542.
[233] G. B. Fields, *Methods Mol. Biol.* **1994**, *35*, 17-27.
[234] P. Nielsen, M. Egholm, R. Berg, O. Buchardt, *Science* **1991**, *254*, 1497-1500.
[235] M. Egholm, O. Buchardt, L. Christensen, C. Behrens, S. M. Freier, D. A. Driver, R. H. Berg, S. K. Kim, B. Norden, P. E. Nielsen, *Nature* **1993**, *365*, 566-568.
[236] P. Wittung, P. E. Nielsen, O. Buchardt, M. Egholm, B. Norden, *Nature* **1994**, *368*, 561-563.
[237] S. Shakeel, S. Karim, A. Ali, *J. Chem. Technol. Biotechnol.* **2006**, *81*, 892-899.
[238] H. Stender, *Expert Rev. Mol. Diagn.* **2003**, *3*, 649-655.
[239] K. Kim, H. Y. Shin, *J. Biochem. Mol. Biol.* **2000**, *33*, 321-325.
[240] a) B. Ketterer, F. Kadlubar, T. Flammang, T. Carne, G. Enderby, *Chem.-Biol. Interact.* **1979**, *25*, 7-21; b) C. Boulegue, M. Loweneck, C. Renner, L. Moroder, *ChemBioChem* **2007**, *8*, 591-594.
[241] S. H. Lee, E. Moroz, B. Castagner, J. C. Leroux, *J. Am. Chem. Soc.* **2014**, *136*, 12868-12871.
[242] R. P. Sinha, D.-P. Häder, *Photochem. Photobiol. Sci.* **2002**, *1*, 225-236.
[243] a) J. W. Karcher, unpublished master thesis, The Karlsruhe Institute of Technology (Karlsruhe), **2015**; b) T. Bantle, unpublished master thesis, The Karlsruhe Institute of Technology (Karlsruhe), **2017**.
[244] a) M. Mąkosza, A. Kwast, *J. Phys. Org. Chem.* **1998**, *11*, 341-349; b) S. D. Pintat, Stephen John; Moffat, David Festus Charles; Drummond, Alan Hastings; Hagemann, Thorsten (Chroma Therapeutics Ltd), WO2014170677A1, **2014**.
[245] a) B. F. Hoskins, R. Robson, *J. Am. Chem. Soc.* **1989**, *111*, 5962-5964; b) B. F. Hoskins, R. Robson, *J. Am. Chem. Soc.* **1990**, *112*, 1546-1554.
[246] B. F. Abrahams, B. F. Hoskins, D. M. Michail, R. Robson, *Nature* **1994**, *369*, 727-729.
[247] O. M. Yaghi, G. Li, H. Li, *Nature* **1995**, *378*, 703-706.
[248] L. E. Kreno, K. Leong, O. K. Farha, M. Allendorf, R. P. Van Duyne, J. T. Hupp, *Chem. Rev.* **2012**, *112*, 1105-1125.
[249] H. C. Zhou, J. R. Long, O. M. Yaghi, *Chem. Rev.* **2012**, *112*, 673-674.
[250] a) N. L. Rosi, J. Eckert, M. Eddaoudi, D. T. Vodak, J. Kim, M. O'Keeffe, O. M. Yaghi, *Science* **2003**, *300*, 1127-1129; b) L. J. Murray, M. Dinca, J. R. Long, *Chem. Soc. Rev.* **2009**, *38*, 1294-1314.
[251] S. Bai, X. Liu, K. Zhu, S. Wu, H. Zhou, *Nat. Energy* **2016**, *1*.
[252] a) F. Zhang, P. Zhao, M. Niu, J. Maddy, *Int. J. Hydrog. Energy* **2016**, *41*, 14535-14552; b) *Solid-state hydrogen storage : materials and chemistry*, 1. publ. ed., CRC Press, Boca Raton, Fla. [u.a.], **2008**.

[253] J. W. Brown, B. L. Henderson, M. D. Kiesz, A. C. Whalley, W. Morris, S. Grunder, H. Deng, H. Furukawa, J. I. Zink, J. F. Stoddart, O. M. Yaghi, *Chem. Sci.* **2013**, *4*.
[254] a) X. Meng, B. Gui, D. Yuan, M. Zeller, C. Wang, *Sci. Adv.* **2016**, *2*, e1600480; b) J. Park, L. B. Sun, Y. P. Chen, Z. Perry, H. C. Zhou, *Angew. Chem. Int. Ed.* **2014**, *53*, 5842-5846.
[255] a) C. Serre, *Angew. Chem. Int. Ed.* **2012**, *51*, 6048-6050; b) D. Hermann, H. A. Schwartz, M. Werker, D. Schaniel, U. Ruschewitz, *Chem. Eur. J.* **2019**, *25*, 3606-3616.
[256] a) H. A. B. Linke, D. Pramer, *Z. Naturforsch. B Chem. Sci.* **1969**, *24*, 997-998; b) D. R. Fahey, *Chem. Commun.* **1970**; c) D. R. Fahey, *J. Organomet. Chem.* **1971**, *27*, 283-292.
[257] S. Sun, J. Song, Z. Shan, R. Feng, *J. Electroanal. Chem.* **2012**, *676*, 1-5.
[258] Y. Jinno, M. Yamanaka, *Chem. Asian J.* **2012**, *7*, 1768-1771.
[259] D. Macaya, M. Spector, *Biomed. Mater.* **2012**, *7*, 012001.
[260] L. Zhang, Z. Cao, T. Bai, L. Carr, J. R. Ella-Menye, C. Irvin, B. D. Ratner, S. Jiang, *Nat. Biotechnol.* **2013**, *31*, 553-556.
[261] a) F. A. Plamper, W. Richtering, *Acc. Chem. Res.* **2017**, *50*, 131-140; b) Z. Hu, X. Xia, *Adv. Mater.* **2004**, *16*, 305-309.
[262] Y. Li, H. Meng, Y. Liu, A. Narkar, B. P. Lee, *ACS Appl. Mater. Interfaces* **2016**, *8*, 11980-11989.
[263] A. Schneeweis, C. C. Müller-Goymann, *Pharm. Res.* **1997**, *14*, 1726-1729.
[264] S. Hünig, H. Werner, *Liebigs Ann.* **1959**, *628*, 46-55.
[265] G. Lang, (L'Oréal), US4025301A, **1976**.
[266] A. Greaves, H. David, (L'Oréal), **2005**.
[267] a) M. A. Weaver, (Eastman Kodak Co), US4487719A, **1983**; b) R. B. Balsley, (Wyeth), US4081436A, **1978**.
[268] O. Thorn-Seshold, M. Borowiak, D. Trauner, J. Hasserodt, (C. N. d. l. R. S. Ludwig Maximilians Universität München, Universite Claude Bernard Lyon), US20170051149A1, **2014**.
[269] P. Gómez, E. Carrasco, P. Campos, P. Deleyto, M. Vega, J. J. Gómez-Reino, C. Conde, O. Gualillo, J. J. Pérez, Á. Messeguer, (Allinky Biopharma), WO2014094816A1, **2012**.
[270] W. Zhou, T. Kobayashi, H. Zhu, H. Yu, *Chem. Commun.* **2011**, *47*, 12768-12770.
[271] W. Zhou, H. Yu, *ACS Appl. Mater. Interfaces* **2012**, *4*, 2154-2159.
[272] K. i. Aoki, M. Nakagawa, K. Ichimura, *J. Am. Chem. Soc.* **2000**, *122*, 10997-11004.
[273] W. Lin, W. Lin, G. K. Wong, T. J. Marks, *J. Am. Chem. Soc.* **1996**, *118*, 8034-8042.
[274] M. Nakagawa, M. Rikukawa, K. Sanui, N. Ogata, *Supramolecular Science* **1998**, *5*, 83-87.
[275] N. Mas, A. Agostini, L. Mondragon, A. Bernardos, F. Sancenon, M. D. Marcos, R. Martinez-Manez, A. M. Costero, S. Gil, M. Merino-Sanjuan, P. Amoros, M. Orzaez, E. Perez-Paya, *Chem. Eur. J.* **2013**, *19*, 1346-1356.

[276] S. Mizukami, S. Watanabe, Y. Akimoto, K. Kikuchi, *J. Am. Chem. Soc.* **2012**, *134*, 1623-1629.
[277] J. Garcia-Amoros, W. A. Massad, S. Nonell, D. Velasco, *Org. Lett.* **2010**, *12*, 3514-3517.
[278] Z. H. Zhang, D. C. Burns, J. R. Kumita, O. S. Smart, G. A. Woolley, *Bioconjugate Chem.* **2003**, *14*, 824-829.
[279] A. Mourot, M. A. Kienzler, M. R. Banghart, T. Fehrentz, F. M. Huber, M. Stein, R. H. Kramer, D. Trauner, *ACS Chem. Neurosci.* **2011**, *2*, 536-543.
[280] J. Garcia-Amoros, S. Nonell, D. Velasco, *Chem. Commun.* **2012**, *48*, 3421-3423.
[281] J. Karcher, M. A. Bichelberger, M. Steinbiß, T. Bantle, J. Lackner, K. Nienhaus, G. U. Nienhaus, Z. L. Pianowski, KIT, Wiley-VCH, **2019**, p. 8.
[282] M. Steinbiß, unpublished master thesis, Karlsruhe Institute of Technology (Karlsruhe), **2014**.
[283] J. Garcia-Amoros, S. Nonell, D. Velasco, *Chem. Commun.* **2011**, *47*, 4022-4024.
[284] J. Epperlein, B. Blau, *Angew. Chem.* **1989**, *29*, 262-263.
[285] N. Nishimura, T. Sueyoshi, H. Yamanaka, E. Imai, S. Yamamoto, S. Hasegawa, *Bull. Chem. Soc. Jpn.* **1976**, *49*, 1381-1387.
[286] N. Nishimura, S. Kosako, Y. Sueishi, *Bull. Chem. Soc. Jpn.* **1984**, *57*, 1617-1625.
[287] V. D. Filimonov, M. Trusova, P. Postnikov, E. A. Krasnokutskaya, Y. M. Lee, H. Y. Hwang, H. Kim, K. W. Chi, *Org. Lett.* **2008**, *10*, 3961-3964.
[288] R. Patouret, T. M. Kamenecka, *Tetrahedron Lett.* **2016**, *57*, 1597-1599.
[289] J. Garcia-Amoros, M. Reig, A. Cuadrado, M. Ortega, S. Nonell, D. Velasco, *Chem. Commun.* **2014**, *50*, 11462-11464.
[290] C. W. Chang, Y. C. Lu, T. T. Wang, E. W. Diau, *J. Am. Chem. Soc.* **2004**, *126*, 10109-10118.
[291] a) K. Gille, H. Knoll, K. Quitzsch, *International Journal of Chemical Kinetics* **1999**, *31*, 337-350; b) N. Nishimura, T. Sueyoshi, H. Yamanaka, E. Imai, S. Yamamoto, S. Hasegawa, *Bull. Chem. Soc. Jpn.* **1976**, *49*, 1381-1387.
[292] M. Dong, A. Babalhavaeji, M. J. Hansen, L. Kalman, G. A. Woolley, *Chem. Commun.* **2015**, *51*, 12981-12984.
[293] M. Nakagawa, M. Rikukawa, M. Watanabe, K. Sanui, N. Ogata, *Bull. Chem. Soc. Jpn.* **1997**, *70*, 737-744.
[294] a) A. Aemissegger, V. Krautler, W. F. van Gunsteren, D. Hilvert, *J. Am. Chem. Soc.* **2005**, *127*, 2929-2936; b) V. Krautler, A. Aemissegger, P. H. Hunenberger, D. Hilvert, T. Hansson, W. F. van Gunsteren, *J. Am. Chem. Soc.* **2005**, *127*, 4935-4942.
[295] V. Tomar, G. Bhattacharjee, Kamaluddin, A. Kumar, *Bioorg. Med. Chem. Lett.* **2007**, *17*, 5321-5324.
[296] X.-Y. Zhang, Y.-F. Ma, Y.-G. Li, P.-P. Wang, Y.-L. Wang, Y.-F. Luo, *Front. Mater. Sci.* **2012**, *6*, 326-337.
[297] S. Bräse, in *2017* (Ed.: A. Bräse), KIT, IOC, **2019**, p. 70.

[298] a) Z. L. Pianowski, J. Karcher, K. Schneider, *Chem. Commun.* **2016**, *52*, 3143-3146; b) J. Karcher, Z. L. Pianowski, *Chem. Eur. J.* **2018**, *24*, 11605-11610.
[299] H. E. Gottlieb, V. Kotlyar, A. Nudelman, *J. Org. Chem.* **1997**, *62*, 7512-7515.
[300] a) T. Doura, Q. An, F. Sugihara, T. Matsuda, S. Sando, *Chem. Lett.* **2011**, *40*, 1357-1359; b) H. E. Gottlieb, V. Kotlyar, A. Nudelman, *The Journal of Organic Chemistry* **1997**, *62*, 7512-7515.
[301] R. Beckert, E. Fanghänel, K. Schwetlick, *Organikum: organisch-chemisches Grundpraktikum*, 23. ed., Wiley VCH, Weinheim, **2009**.
[302] a) J. E. Silver, C. A. Bailey, R. L. Lewis, S. R. Paeschke, *J. Am. Chem. Soc.* **2013**, *246*; b) W. C. Still, M. Kahn, A. Mitra, *J. Org. Chem.* **1978**, *43*, 2923-2925.
[303] U. Petersen, K. Grohe, E. Kuhle, H. J. Zeiler, K. G. Metzger, (Bayer AG), US4559341A, **1985**.
[304] W. A. Velema, J. P. van der Berg, W. Szymanski, A. J. Driessen, B. L. Feringa, *ACS Chem. Biol.* **2014**, *9*, 1969-1974.
[305] I. Wiegand, K. Hilpert, R. E. Hancock, *Nat. Protoc.* **2008**, *3*, 163-175.
[306] S. R. Liao, X. C. Qin, Z. Wang, D. Li, L. Xu, J. S. Li, Z. C. Tu, Y. Liu, *Eur. J. Med. Chem.* **2016**, *121*, 500-509.
[307] a) H. Tada, O. Shiho, K. Kuroshima, M. Koyama, K. Tsukamoto, *J. Immunol. Methods* **1986**, *93*, 157-165; b) T. Mosmann, *J. Immunol. Methods* **1983**, *65*, 55-63; c) R. Jover, X. Ponsoda, J. V. Castell, M. J. Gomezlechon, *Toxicol. In Vitro* **1994**, *8*, 47-54; d) J. Carmichael, W. G. DeGraff, A. F. Gazdar, J. D. Minna, J. B. Mitchell, *Cancer Res.* **1987**, *47*, 936-942; e) F. Denizot, R. Lang, *J. Immunol. Methods* **1986**, *89*, 271-277; f) B. G. Campling, J. Pym, P. R. Galbraith, S. P. Cole, *Leuk. Res.* **1988**, *12*, 823-831.
[308] C. P. Hald, unpublished bachelor thesis, The Karlsruhe Institute of Technology (Karlsruhe), **2017**.
[309] R. F. W. Jackson, M. Perez-Gonzalez, R. L. Danheiser, A. L. Crombie, *Org. Synth.* **2005**, *81*.
[310] N. Atmuri, *Org. Synth.* **2015**, *92*, 103-116.
[311] A.-L. Leistner, unpublished Vertiefer thesis, The Karlsruhe Institute of Technology (Karlsruhe), **2019**.
[312] Y. Yamazaki, K. Tanaka, B. Nicholson, G. Deyanat-Yazdi, B. Potts, T. Yoshida, A. Oda, T. Kitagawa, S. Orikasa, Y. Kiso, H. Yasui, M. Akamatsu, T. Chinen, T. Usui, Y. Shinozaki, F. Yakushiji, B. R. Miller, S. Neuteboom, M. Palladino, K. Kanoh, G. K. Lloyd, Y. Hayashi, *J. Med. Chem.* **2012**, *55*, 1056-1071.
[313] Z. Ding, H. Cheng, S. Wang, Y. Hou, J. Zhao, H. Guan, W. Li, *Bioorg. Med. Chem. Lett.* **2017**, *27*, 1416-1419.
[314] C. Pepper, H. J. Smith, K. J. Barrell, P. J. Nicholls, M. J. Hewlins, *Chirality* **1994**, *6*, 400-404.
[315] L. E. Webb, C.-F. Lin, *J. Am. Chem. Soc.* **1971**, *93*, 3818-3819.

[316] L. A. Carpino, A. El-Faham, F. Albericio, *Tetrahedron Lett.* **1994**, *35*, 2279-2282.
[317] U. Schöllkopf, U. Groth, C. Deng, *Angew. Chem. Int. Ed.* **1981**, *20*, 798-799.
[318] M. Cigl, A. Bubnov, M. Kašpar, F. Hampl, V. Hamplová, O. Pacherová, J. Svoboda, *J. Mater. Chem. C* **2016**, *4*, 5326-5333.
[319] P. Sabbatini, P. Wellendorph, S. Hog, M. H. Pedersen, H. Brauner-Osborne, L. Martiny, B. Frolund, R. P. Clausen, *J. Med. Chem.* **2010**, *53*, 6506-6510.
[320] S. A. Bateman, D. P. Kelly, J. M. White, R. F. Martin, *Aust. J. Chem.* **1999**, *52*.
[321] J. Sim, H. Yim, N. Ko, S. B. Choi, Y. Oh, H. J. Park, S. Park, J. Kim, *Dalton Trans* **2014**, *43*, 18017-18024.
[322] L. Peura, K. Malmioja, K. Laine, J. Leppanen, M. Gynther, A. Isotalo, J. Rautio, *Mol Pharm* **2011**, *8*, 1857-1866.
[323] P. Singh, C. Samorì, F. M. Toma, C. Bussy, A. Nunes, K. T. Al-Jamal, C. Ménard-Moyon, M. Prato, K. Kostarelos, A. Bianco, *J. Mater. Chem.* **2011**, *21*, 4850-4860.
[324] S. Gerritz, S. Shi, S. Zhu, (Bristol-Myers Squibb), WO2007002220A2, **2006**.
[325] R. Pascal, R. Sola, F. Labéguère, P. Jouin, *Eur. J. Org. Chem.* **2000**, *2000*, 3755-3761.
[326] P. S. Zhang, Joseph P.; Terefenko, Eugene Anthony; Trybulski, Eugene John, (Wyeth), WO2008073943A1, **2008**.
[327] G. Liang, B. Xu, C. Liu, Z. Huang, P. Cao, Z. Cai, Y. Liu, (Tianjin Institute of Pharmaceutical Research), US20090247470A1, **2009**.
[328] X. Y. Tian, J. W. Han, Q. Zhao, H. N. Wong, *Org. Biomol. Chem.* **2014**, *12*, 3686-3700.
[329] D. Mazzier, M. Maran, O. Polo Perucchin, M. Crisma, M. Zerbetto, V. Causin, C. Toniolo, A. Moretto, *Macromolecules* **2014**, *47*, 7272-7283.
[330] A. Szymańska, K. Wegner, L. Łankiewicz, *Helv. Chim. Acta* **2003**, *86*, 3326-3331.
[331] H. Endo, Y. Kanai, K. Saito, K. Tsujihara, (L. J-Pharma Co.), WO2003066574A1, **2004**.
[332] H. Garrido-Hernandez, K. D. Moon, R. L. Geahlen, R. F. Borch, *J. Med. Chem.* **2006**, *49*, 3368-3376.
[333] J. Dokic, M. Gothe, J. Wirth, M. V. Peters, J. Schwarz, S. Hecht, P. Saalfrank, *J. Phys. Chem. A* **2009**, *113*, 6763-6773.
[334] K. A. Kalesh, L. P. Tan, K. Lu, L. Gao, J. Wang, S. Q. Yao, *Chem. Commun.* **2010**, *46*, 589-591.
[335] Y. Ueno, J. Jose, A. Loudet, C. Perez-Bolivar, P. Anzenbacher, Jr., K. Burgess, *J. Am. Chem. Soc.* **2011**, *133*, 51-55.
[336] C. K. Lee, D. A. Davis, S. R. White, J. S. Moore, N. R. Sottos, P. V. Braun, *J. Am. Chem. Soc.* **2010**, *132*, 16107-16111.
[337] J. D. Reinheimer, J. D. Harley, W. W. Meyers, *J. Org. Chem.* **1963**, *28*, 1575-1579.

[338] H. David, A. Greaves, N. Daubresse, (L'Oréal), WO2006063867A3, **2005**.
[339] E. S. Hand, D. C. Baker, *Synthesis* **1989**, *1989*, 905-908.
[340] S. Ouarna, H. K'tir, S. Lakrout, H. Ghorab, A. Amira, Z. Aouf, M. Berredjem, N. Aouf, *Orient. J. Chem.* **2015**, *31*, 913-919.
[341] M. Barbero, M. Crisma, I. Degani, R. Fochi, P. Perracino, *Synthesis* **1998**, 1171-1175.
[342] T. Maisch, A. Späth, (TriOptoTec), US9185913B2, **2012**.
[343] B. Perly, S. Moutard, F. Pilard, (Commissariat A L'energie Atomique), WO2005042590A2, **2007**.
[344] S. P. Singh, A. Michaelides, A. R. Merrill, A. L. Schwan, *J. Org. Chem.* **2011**, *76*, 6825-6831.
[345] A. G. Ross, B. M. Benton, D. Chin, G. De Pascale, J. Fuller, J. A. Leeds, F. Reck, D. L. Richie, J. Vo, M. J. LaMarche, *Bioorg. Med. Chem. Lett.* **2015**, *25*, 3468-3475.
[346] B. Weber, E. Nickel, M. Horn, K. Nienhaus, G. U. Nienhaus, *J Phys Chem Lett* **2014**, *5*, 756-761.
[347] H. L. Jiang, D. Feng, T. F. Liu, J. R. Li, H. C. Zhou, *J. Am. Chem. Soc.* **2012**, *134*, 14690-14693.
[348] a) I. E. Gorman, R. L. Willer, L. K. Kemp, R. F. Storey, *Polymer* **2012**, *53*, 2548-2558; b) M. Tsotsalas, J. Liu, B. Tettmann, S. Grosjean, A. Shahnas, Z. Wang, C. Azucena, M. Addicoat, T. Heine, J. Lahann, J. Overhage, S. Brase, H. Gliemann, C. Woll, *J. Am. Chem. Soc.* **2014**, *136*, 8-11.
[349] L. Q. Xu, F. Yao, G. D. Fu, E. T. Kang, *Biomacromolecules* **2010**, *11*, 1810-1817.
[350] N. A. Al-Awadi, M. R. Ibrahim, M. H. Elnagdi, E. John, Y. A. Ibrahim, *Beilstein J. Org. Chem.* **2012**, *8*, 441-447.
[351] L. Rene, J. Poncet, G. Auzou, *Synthesis* **1986**, *1986*, 419-420.
[352] J. R. Pfister, *Synthesis* **1990**, *1990*, 689-690.

8 Appendix

8.1 Curriculum Vitae

Personal information:

Date of birth	22. March 1989
Place of birth	Achern, Germany
Nationality	German

WORK EXPERIENCE

Research Scientist Organic Chemistry
KIT, Institute of Organic Chemistry **January 2016 – October 2019, Karlsruhe**

- Establishing contact with 8 other scientific groups to evaluate bioactive small molecules and to get access to unique knowhow
- Tackling different synthetic problems daily with strong background in supramolecular chemistry, peptides and photopharmacology
- Scale-up of 5 synthetic pathways of heterocyclic chemistry on a multi-gram scale
- Managing 3 research projects simultaneously, coordinating up to 4 colleagues, organizing tasks and experiments
- Training and supervision of 4 international guest researchers, 5 master's & 4 bachelor's students
- Writing detailed semi-annual research reports, submitted 2 patents and publishing 2 articles with 4 more in the pipeline

Guest researcher Chemical Biology
Osaka University, Material & Life Science **February 2016 – March 2016, Osaka**

- Conducted research using biochemical and cell biological techniques e.g. transfection, chemical imaging and purification of proteins
- Examined gene expression in vivo using a fluorescent biorthogonal marker

Guest researcher Molecular Biology
KIT, Institute of Toxicology & Genetics **August 2017 – September 2017, Karlsruhe**

- Adopted molecular biological methods: MTT assays, PCR and gel electrophoresis
- Cultivated 5 laboratory cell cultures like HeLa and human cell lines
- Devised biochemical methods to quantify uptake, analysed data, experiments and started screening

Guest researcher Microbiology
KIT, Institute for Applied Biology **December 2017 – December 2017, Karlsruhe**

- Developed a biochemical assay for a light activated drug release system in E. coli cell cultures
- Conducted MIC assays to evaluate the antibacterial activity, chemical and physical properties of 7 compounds

TEAM PROJECTS

Photoswitchable diketopiperazines as an anti-cancer agent against non-small-cell lung carcinoma

- Coordinated team project with 6 members to develop anti-cancer agent activated by light
- Rational design of 2 potent heterocyclic compounds with good permeability and bioavailability properties
- Synthesis of drug candidates, identification of the lead structure and their strategic optimization

Stable anti-gene agent based on the photoswitchable nucleic acid PNA

- Synthesis of 2 photoswitches for the regulation of the growth of spinal cord in zebrafish
- Stability assays in blood serum to assess bioactive small molecules and in-depth analysis of results to achieve a half-life of 9 h

• EDUCATION

Doctor of Science in Chemistry and Biosciences

Organic Chemistry and Medicinal Chemistry • KIT • July 2019 • 1.0 (magna cum laude)

- Photomodulation of supramolecular systems containing bioactive small molecules and biopolymers

Master of Science in Chemical Biology

Organic Chemistry and Biochemistry • KIT • November 2015 • 1.1 (A)

- Master's thesis on organic synthesis of molecular photoswitches

Bachelor of Science in Chemical Biology

Biology and Physical Chemistry • KIT • November 2013 • 1.6 (B+)

- Bachelor's thesis on visualisation of glycocalyx on human cell lines and high throughput screening

Abitur (equivalent to A-levels)

Natural science • Heimschule Lender • Sasbach • June 2009 • 2.2 (B−)

COURSEWORK

Advanced Organic Chemistry

KIT • Teaching Advisor Preparative Chemistry January 2016 – October 2019, Karlsruhe

- Introducing four 4-step synthetic routes for students to spread my research and recruiting top talents for our group

INVOLVEMENT

Research Training Group GRK 2039

KIT • PhD student January 2016 – July 2019, Karlsruhe

- Data analysis, interpretation, visualization of results and presentation of 10 posters at international conferences, 2 talks about the latest research in Kyoto and Ischia
- Qualification programme, min 10 events each year e.g. seminar day Transfection & Microscopy

Family business

Agriculture • Lead worker **May 2010 – September 2010, Oberachern**

- Cultivated fruit and wine, distillation of fruit spirits, established a computer network and led 2 workers

Military police service

Battalion 452 • Financial Controller **July 2009 – March 2010, Stetten am kalten Markt**

- Scheduled meetings and prepared monthly financial reports with MS Office after the basic training

SKILLS

- Analytics • mass spectrometry • LC-MS • GC-MS • EA • SEC • HPLC • rheology
- Spectroscopy • NMR • IR • CD • UV-Vis • fluorescence spectroscopy
- Microscopy • TEM • SEM • fluorescence microscope • confocal microscope • bright field
- Adobe Creative Suite • OriginLab • ChemOffice • quick touch typing • programming Arduino
- Craftsmanship • constructing LED reactors • maintenance of HPLC • repairing of LC-MS
- Administration • project management • negotiating contracts • organizing events • prioritizing derivatives
- SAP (Supplier Relationship Management)
- Soft skills • motivating my colleagues • helping with personal issues • discussing scientific challenges
- Languages • fluency in English • German native speaker • qualification in Latin

PUBLICATIONS

Patents

- *Photoresponsive diketopiperazines for supramolecular systems and controlled release hydrogel systems.* Z. Pianowski, J. Karcher, K. Schneider, German Patent Office. **2015**, Germany. Patent: DE 10 2015 014 834 A1.
- *Diketopiperazine mit licht-aktivierter Zytotoxizität.* Z. Pianowski, A.-L. Leistner, S. Kirchner, S. Weber, A. Seliwjorstow, J. Karcher, German Patent Office. **2019**, Germany. Patent: DE 10 2019 005 005.3, *submitted.*

Articles

- *Photoresponsive self-healing supramolecular hydrogels for light-induced release of DNA and doxorubicin.* Z. Pianowski, J. Karcher, K. Schneider, *Chem. Commun.* **2016**, *52(15):3143-6.*
- *Photocontrol of drug release from supramolecular hydrogels with green light.* J. Karcher and Z. Pianowski, *Chem. Eur. J.* **2018**, *24(45):11605-10.*
- *Selective green-light-triggered release of an anticancer agent from supramolecular hydrogels.* J. Karcher, S. Kirchner, C. Hald, A.-L. Leistner, T. Bantle, Z. Pianowski, *submitted.*
- *Fluorinated azobenzenes switchable with red light.* A.-L. Leistner, S. Kirchner, J. Karcher, T. Bantle, M. Schulte and Z. Pianowski, *submitted.*
- *3-Arylazopyridinium derivatives switchable with green light.* J. Karcher, M. Bichelberger, M. Steinbiss, T. Bantle, K. Nienhaus, U. Nienhaus, Z. Pianowski, *in preparation.*
- *Supramolecular red-light-responsive supergelator*. J. Karcher, T. Bantle and Z. Pianowski, *in preparation.*

9 Acknowledgements

Thank you for the interest and all people who helped me to complete this thesis. First, I want to thank Dr. Zbigniew L. Pianowski for his support and clever ideas, beside your continuous mentoring and training. My colleague Susanne Kirchner and Tobias Bantle helped me a lot and we were both integrated in these projects.

In particular, I would like to mention and thank both for your help to correct this thesis and in general in the laboratory, for example with the reactions or HPLC. A part of this work including the experiment, synthesis, optimisation or as encounter of problems was performed by you and the students Anna-Lena Leistner, Mariam Schulte, Christian Hald, Michael Steinbiß and Saloni Dagal. Altogether, with the rest of the group Sven Weber, Angelika Seliwjorstow, Philipp Geng, Knut Schneider, Klaudia Bzdęga, Hulya Ulcar and Juliana Pfeifer, I want to thank you for admirable team spirit, the awesome atmosphere and productive work.

The same is true for the MZE group with their pleasant environment for writing and with the others the coffee breaks. Thank you for the funny and hilarious conversation at Chicco or AKK with Eduard Spuling, Janina Beck, Yannick Matt, Sven Weber, Rieke Schulte, Lisa Gramespacher, Angelika Seliwjorstow, Philipp Geng and Claudia Bizzarri. I would also like to thank Alexander Braun and Christina Retich for their help and advice, Stephan Münch for his help with the HPLC and the secretaries Christiane Lampert, Janine Bolz and Selin Samur for the bureaucratic support.

I want to thank Prof. Dr. Stefan Bräse (IOC KIT) for the infrastructural support and his constructive criticism and ideas, Dr. Gunnar Sturm and Prof. Dr. Johannes Gescher (IAB KIT) for the support in experiments on bacterial cells (MIC), Anna Meschkov and Prof. Dr. Ute Schepers (ITG KIT) for the support in experiments on mammalian cells (MTT tests), Mohammad F. Ardakani, Volker Zibat, Dr. Heike Störmer and PD Dr. Reinhard Schneider (LEM KIT Karlsruhe) for electron microscopy imaging, as well as Lukas Arens and Prof. Dr. Manfred Wilhelm for their support in rheological measurements.

In addition, I gratefully acknowledge the financial support from Deutsche Forschungsgemeinschaft (DFG) in form of an individual grant PI 1124/6-1 and participation in the Graduate Training School (Graduiertenkolleg) GRK 2039/1 and (Sonderforschungsbereich) SFB 1176. In the end I would like to thank my family for their support notably Stephan and Anna for building several LED-reactors, my mother for her culinary support, my father for his “lost” tools and Carolina, Anna and Larissa for correcting.